Advanced Research on Structure–Function Relationships of Membrane Proteins

Advanced Research on Structure–Function Relationships of Membrane Proteins

Guest Editors

Akira Naito

Izuru Kawamura

MDPI

Basel • Beijing • Wuhan • Barcelona • Belgrade • Novi Sad • Cluj • Manchester

Guest Editors

Akira Naito
Graduate School
of Engineering
Yokohama National
University
Yokohama
Japan

Izuru Kawamura
Graduate School
of Engineering
Yokohama National
University
Yokohama
Japan

Editorial Office
MDPI AG
Grosspeteranlage 5
4052 Basel, Switzerland

This is a reprint of the Special Issue, published open access by the journal *Membranes* (ISSN 2077-0375), freely accessible at: https://www.mdpi.com/journal/membranes/special_issues/Structure_Function_Membrane_Proteins.

For citation purposes, cite each article independently as indicated on the article page online and as indicated below:

Lastname, A.A.; Lastname, B.B. Article Title. *Journal Name* **Year**, *Volume Number*, Page Range.

ISBN 978-3-7258-8009-6 (Hbk)
ISBN 978-3-7258-8010-2 (PDF)
https://doi.org/10.3390/books978-3-7258-8010-2

Contents

About the Editors

Akira Naito

Akira Naito is a professor emeritus in the Graduate School of Engineering at Yokohama National University. He has been studied on structure–function relationships of biologically active molecules using mainly high-resolution solid-state NMR spectroscopy. First, he assessed the three-dimensional structure of [^{13}C, ^{15}N]-labeled Leu-enkephaline (Tyr-Gly-Gly-Phe-Leu) dehydrate on the basis of six sets of accurately determined ^{13}C-^{15}N interatomic distances using rotational echo double resonance (REDOR) and some additional constraints obtained from ^{13}C chemical shifts. Second, using the magnetically oriented vesicle systems, he investigated the dynamic structures of bee venom melittin bound to the vesicles by analyzing the anisotropic and isotropic ^{13}C chemical shifts of selectively ^{13}C-labeled carbonyl carbons of melittin under static and magic-angle spinning conditions. Third, he investigated myloid fibril formation mechanism of human calcitonin (hCT) via high-resolution solid-state ^{13}C NMR spectroscopy. Fourth, he detected photo-intermediates of photoreceptor membrane proteins using novel in situ photo-irradiation solid-state NMR spectroscopy. Fifth, he successfully developed and implemented a novel two-dimensional NMR experiment, using state–correlated two-dimensional NMR (SC-2D NMR) spectroscopy, between the nematic phase and the isotropic phase of 4'-methoxybenzylidene-4-acetoxyaniline (APAPA) via pulsed microwave heating.

Izuru Kawamura

Izuru Kawamura is a Professor of Yokohama National University. He received his Ph.D. in Engineering from Yokohama National University in 2007 (Supervisor: Prof. Akira Naito). After serving as a researcher at Yokohama National University and as a visiting researcher at the University of Guelph, Canada (Supervisors: Prof. V. Ladizhansky and Prof. L.S. Brown), he was appointed assistant professor at Yokohama National University in 2012, being promoted to associate professor in 2013. He has been a professor at the Faculty of Engineering, Yokohama National University, since 2024. His research interests include biophysical chemistry and solid-state nuclear magnetic resonance spectroscopy of membrane proteins and membrane-bound peptides. In particular, he has been actively advancing NMR research on microbial rhodopsins and antimicrobial peptides containing D-amino acids. He is also engaged in the development of bio-materials based on polymer science through structural and physicochemical analyses of cellulose nanofibers derived from food and agricultural waste. He received the Progress Award from the Nuclear Magnetic Resonance Society of Japan in 2023.

Preface

This Special Issue focus on advanced studies on structure–function relationships using advanced methods such as solid-state NMR, solution NMR, X-ray crystallography and cryo-electron microscopy to determine the high-resolution structure of membrane proteins. Within this Special Issue, it is shown that such advanced methods play important roles not only in determining the complete structure but also in making changes to the local structure and dynamic properties of membrane proteins. This information gives us insight into the structure–function relationship of membrane proteins.

Akira Naito and Izuru Kawamura
Guest Editors

membranes

Editorial

Advanced Research on Structure–Function Relationships of Membrane Proteins

Akira Naito * and Izuru Kawamura *

Graduate School of Engineering, Yokohama National University, 79-5 Tokiwadai, Hodogaya-ku, Yokohama 240-8501, Japan
* Correspondence: naito@ynu.ac.jp (A.N.); kawamura-izuru-wx@ynu.ac.jp (I.K.)

Membrane proteins embedded in biological membranes account for 30% of the proteins encoded in the human genome and play an essential role in maintaining the homeostasis of cells by functioning as transporters, for signal transaction and energy conversion, amongst other functions. Thus, knowledge of the atomic resolution structures of membrane proteins is extremely crucial to understanding their functions. However, it is difficult to determine the structure of membrane proteins at the atomic resolution, compared to soluble proteins, because of the degree of their crystallization.

Recent developments in the methodology of structure determination are providing high-resolution structures of membrane proteins, and several high-resolution structures of membrane proteins have already been reported (~4% of PDB). Solid-state NMR spectroscopy does not require crystallization and is not restricted by upper limits to the molecular weight and structure of the heptahelical membrane protein being determined [1]. Solution NMR spectroscopy has recently been used to determine the structure of membrane proteins using detergent micelles or nanodiscs as membrane mimic systems [2]. For X-ray crystallography, it is becoming possible to determine the structures of membrane proteins by using membrane or membrane mimic systems to form the crystals [3]. The recently developed time-resolved methods allow for a determination of the structures of the short-lived intermediates of membrane proteins [4]. Attention is now being focused on cryo-electron microscopy (cryo-EM) as a powerful method for determining the structure of membrane proteins using two-dimensional crystals [5]. The determination of membrane protein structure based on single-molecular cryo-EM observation has also recently become possible [6].

This Special Issue focuses on advanced studies of structure–function relationships using advanced methods to determine the high-resolution structures of membrane proteins, including developments in methodology. Within the scope of this Special Issue are not only determinations of complete structures but also of the functions related to the local structural changes and the dynamic properties of membrane proteins as summarized in Figure 1.

The article in this Special Issue by Florian Cymer et al. [7] reports that the single transmembrane helix of human carbonic anhydrases XII of membrane proteins contains a large number of amino acids with small side chains and suggests the critical roles of these small amino acids in the dimerization of the transmembrane domain. Using the GALLEX assay, the authors show that the transmembrane domain forms a strong transmembrane helix oligomer embedded in a biological membrane. The authors also showed that single or multiple mutations of small residues to isoleucine consistently increased the interaction propensities. A reduction of the helix flexibility and of the protein–lipid contacts may increase stability of helix–helix interactions within the membrane.

In the review by Koh Takeuchi et al. [8], a solution NMR has been highlighted to reveal the structural basis of the function of multi-spanning membrane proteins, such as ion channels, GPCRs, and transporters. Furthermore, membrane proteins are under conformational equilibrium between different conformational states that are associated with distinctive functional states. The relative population of the different conformational states can quantitatively reveal the functionality of the membrane protein under certain conditions. The exchange rates between the distinct conformational states described by solution NMR can ultimately define the time scale of the cell response as investigated for ion channels and GPCRs. Since solution NMR allows for the detecting and quantifying of proton dynamics at various timescales and at a semi-atomic resolution, the method is suitable for understanding the functions of the multi-spanning helical membrane proteins.

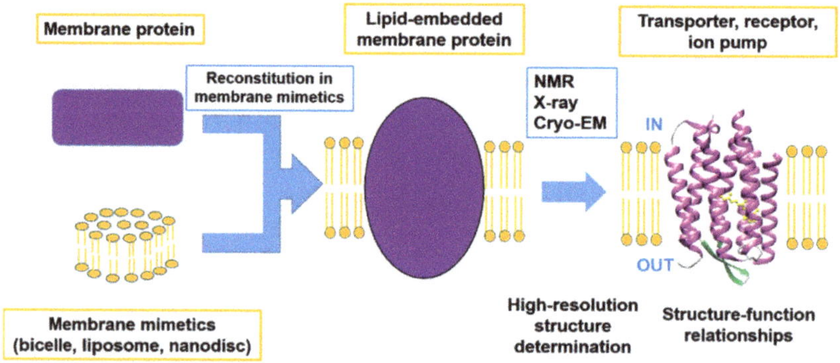

Figure 1. Schematic image of advanced research on the structure–function relationships of membrane proteins.

The article by Cuauhtemoc U. Gonzalez et al. [9] explores the research on the structural arrangement produced by concanavalin A (Con-A) binding to homomeric GluK2 receptors. Con-A stabilizes the active open-channel state of the kainate receptor and reduces the extent of desensitization. In this study, the authors used single-molecule fluorescence resonance energy transfer (smFRET) methods to investigate the conformational changes underlying kainate receptor modulation by Con-A. These studies showed that Con-A binding to Gluk2 receptor is located close to the subunit at the dimer–dimer interface at the amino-terminal domain and between the subunits at the agonist-binding domain. Based on these results, the authors conclude that Con-A modulation of kainate receptor function is mediated by a shift in the conformation of the kainate receptor toward a tightly packed extracellular domain.

The excellent review by Saman Majeed et al. [10] focuses on the most widely used membrane mimetics in structural and functional studies of integral membrane proteins (IMPs) at the molecular level. These membrane mimetics are detergents, liposomes, bicelles, nanodiscs/Lipodisqs, amphipols, and lipidic cubic phases. The authors also discuss the protocols for IMP reconstitution in membrane mimetics as well as the applicability of these membrane mimetic-IMP complexes in studies using a variety of biochemical, biophysical, structural, and structural-biological techniques. The diversity of these systems has grown significantly, and the widely ranging lipid membrane-mimetic platforms ca now provide high solubility, stability, more or less lipid-bilayer environments, and other specific properties that are utilized in studies using solution NMR, X-ray crystallography, cryo-EM, ESR, fluorescence spectroscopy assays, ligand binding, translocation assays, etc.

Chun-Hao Liu et al. [11] examine the effect of cholesterol on the interaction of viral protein R (Vpr) and lipid membranes. Using a calcein release assay, the authors found that the membrane permeability induced by the membrane binding of Vpr was significantly reduced in the presence of cholesterol in the membrane. Using solid-state NMR spectroscopy, Vpr was shown to experience inhomogeneous chemical environments as shown in the broad ^{13}CO NMR signal of Cys-76 residue ranging from 165–178 ppm, which can be attributed to the existence of multiple Vpr-membrane environments. The authors revealed that the presence of cholesterol in the membrane will alter the distribution of Vpr in the multiple membrane environments, which may explain the change of the Vpr induced membrane permeability in the presence of cholesterol.

In the review by Shinya Hanashima et al. [12], molecular level structural analyses of the interaction between membrane proteins and lipids or detergents that constitute biological or artificial model membranes are shown to be important for understanding the functions and physicochemical properties of membrane proteins and biomembranes. A determination of membrane protein structures at the molecular level is much more difficult when compared with that of soluble proteins, but the new technical development has accelerated the elucidation of the structure–function relationship of membrane proteins. The authors summarize the development of heavy atom derivative detergents and lipids that can be used for a molecular level structural analysis of membrane proteins, including their application for short lived intermediates with X-ray free-electron laser crystallography.

The article by Munehiro Kumashiro et al. [13] reports that the formation of β-strand oligomers of antimicrobial peptide magainin 2 (M2) contributed to the disruption of a phospholipid membrane. The authors measured the synchrotron-radiation circular dichroism and linear dichroism (LD) spectra of M2 in lipid membranes to characterize the conformation and orientation of M2 on the membrane. A singular value decomposition analysis supported the presence of the intermediate state, and a global fitting analysis revealed that M2 monomers with an α-helix structure assembled to form M2 oligomers with a β-strand-rich structure in the intermediate state, which was observed using LD spectra. Fluorescence spectroscopy showed that the formation of β-strand oligomers destabilized the membrane packing structure and induced the leakage of calcein. These results suggest that the formation of β-strand oligomers of M2 plays a crucial role in the disruption of the cell membrane.

The review of Nadezhda Barvitenko et al. [14] indicates that endothelial mechanosensors are the key upstream signaling proteins in shear stress (SS)-induced pro- and anti-inflammatory responses. As transmembrane proteins, mechanosensors not only experience fluid SS but also become regulated by the biomechanical properties of the lipid bilayer and the cytoskeleton. The authors highlight the apparent effects of pro-inflammatory factors (hypoxia, oxidative stress, hypercholesterolemia, and cytokines) on the biomechanics of the lipid bilayer and the cytoskeleton. An analysis of the data reveals that the formation of a vicious circle may occur, in which pro-inflammatory cytokines enhance and attenuate SS-induced pro-inflammatory and anti-inflammatory signaling, respectively.

Yee-Shan Ku et al. [15] investigate the phenomenon whereby multidrug and toxic compound extraction (MATE) transporters in eukaryotes have been characterized to be antiporters that mediate the transport of substrates in exchange for protons. By performing phylogenetic analyses and a protein sequence alignment of 14 representative plant species, the authors identified a distinctive N-terminal poly-glutamate motif among a cluster of MATE proteins in soybeans. Amongst them, GmMATE4 has the most consecutive glutamate residues at the N-terminus. A subcellular localization study showed that GmMATE4 was localized at the vacuolar membrane-like structure. Using yeast as the model, the authors showed that GmMATE4 mediated the transport of daidzein, genistein, glycitein, and glycitin. In summary, the authors demonstrated that GmMATE4 acts as an isoflavone transporter and revealed the functional significance of the N-terminal poly-glutamate motif of GmMATE4 in regulating the isoflavone transport activity.

In the article of Arisu Shigeta et al. [16], a novel technique for in situ photoirradiation of ^{13}C solid-state NMR was developed and used to observe a variety of photo-intermediates and photoreaction pathways in [20-^{13}C]retinal-WT-BR and its mutant [20-^{13}C, 14-^{13}C]retinal-D96N-BR. In WT-BR, the CS state converted to the CS* intermediate under green light irradiation at −20 °C and consequently converted to the AT state in the dark. The AT state converted to the N intermediate under green light irradiation. In D96N-BR, the CS state was converted to the CS* intermediate at −30 °C and consequently converted to the AT state. Simultaneously, the AT state converted to the M and L intermediates under green light irradiation at −30 °C and subsequently converted to the AT state in the dark.

The authors demonstrated that short-lived photo-intermediates could be observed in a stationary state using in situ photoirradiation solid-state NMR spectroscopy for WT-BR and D96N-BR, enabling insights into the light-driven proton pump activity of BR.

Matthew J. Laurence et al. [17] reviewed current nanodisc techniques and the fundamentals of fluorescence correlation spectroscopy (FCS). The development of nanodisc techniques, such as nanolipoprotein particles (NLPs) and styrene maleic acid polymers (SMALPs), allow membrane proteins to be expressed and isolated in solution as part of lipid bilayer rafts with defined, consistent nanometer sizes and compositions, thus enabling solution–based measurements. FCS is a relatively simple yet powerful optical microscopy-based technique that provides quantitative biophysical information, such as diffusion kinetics and concentrations, about individual or interacting membrane proteins in solution.

In conclusion, six research articles and five review articles have been included in this Special Issue. Structural, dynamics, and interaction studies on membrane proteins have been extensively explored in this Special Issue. The technical developments of solid-state NMR, solution NMR, X-ray crystallography, florescence resonance energy transfer (FRET), linear dichroism (LD), and fluorescence correlation spectroscopy (FCS), etc., provide structural and functional studies of membrane proteins that are treated by detergents, micelles, liposomes, and nanodiscs, etc., as membrane mimic systems.

Funding: This research received no external funding.

Acknowledgments: The editors would like to thank all the authors and reviewers for their contribution to make high-quality manuscripts in this Special Issue. The editors also would like to thank the editorial staff for all the assistance in the editorial processes of this Special Issue.

Conflicts of Interest: The authors declare no conflict of interest.

References

1. Wang, S.; Munro, R.A.; Shi, L.; Kawamura, I.; Okitsu, T.; Wada, A.; Kim, S.-Y.; Jung, K.-H.; Brown, L.; Ladizhansky, V. Solid-state NMR spectroscopy structure determination of a lipid-embedded heptahelical membrane protein. *Nat. Methods* **2013**, *10*, 1007–1012. [CrossRef] [PubMed]
2. Puthenveetil, R.; Vinogradova, O. Solution NMR: A powerful tool for structural and functional studies of membrane proteins in reconstituted environments. *J. Biol. Chem.* **2019**, *294*, 15914–15931. [CrossRef]
3. Sjodt, M.; Brock, K.; Dobihal, G.; Rohs, P.; Green, A.; Hopf, T.A.; Meeske, A.; Srisuknimit, V.; Kahne, D.; Walker, S.; et al. Structure of the peptidoglycan polymerase RodA resolved by evolutionary coupling analysis. *Nature* **2018**, *556*, 118–121. [CrossRef] [PubMed]
4. Nango, E.; Royant, A.; Kubo, M.; Nakane, T.; Wickstrand, C.; Kimura, T.; Tanaka, T.; Tono, K.; Song, C.; Tanaka, R.; et al. A three-dimensional movie of structural changes in bacteriorhodopsin. *Science* **2016**, *354*, 1552–1557. [CrossRef] [PubMed]
5. Murata, K.; Mitsuoka, K.; Hirai, T.; Walz, T.; Agre, P.; Heymann, J.B.; Engel, A.; Fujiyoshi, Y. Structural determinants of water permeation through aquaporin-1. *Nature* **2000**, *407*, 599–605. [CrossRef] [PubMed]
6. Liao, M.; Gao, E.; Julius, D.; Cheng, Y. Structure of the TRPV1 ion channel determined by electron cryo-microscopy. *Nature* **2013**, *504*, 107–112. [CrossRef] [PubMed]
7. Cymer, F.; Schneider, D. Small Residues Inhibit Homo-Dimerization of the Human Carbonic Anhydrase XII Transmembrane Domain. *Membranes* **2021**, *11*, 512. [CrossRef] [PubMed]
8. Takeuchi, K.; Kofuku, Y.; Imai, S.; Ueda, T.; Tokunaga, Y.; Toyama, Y.; Shiraishi, Y.; Shimada, I. Function-Related Dynamics in Multi-Spanning Helical Membrane Proteins Revealed by Solution NMR. *Membranes* **2021**, *11*, 604. [CrossRef] [PubMed]
9. Gonzalez, C.U.; Carrillo, E.; Berka, V.; Jayaraman, V. Structural Arrangement Produced by Concanavalin A Binding to Homomeric GluK2 Receptors. *Membranes* **2021**, *11*, 613. [CrossRef] [PubMed]
10. Majeed, S.; Ahmad, A.B.; Sehar, U.; Georgieva, E.R. Lipid Membrane Mimetics in Functional and Structural Studies of Integral Membrane Proteins. *Membranes* **2021**, *11*, 685. [CrossRef] [PubMed]
11. Liu, C.-H.; Huang, S.-J.; Yu, T.-Y. Cholesterol Modulates the Interaction between HIV-1 Viral Protein R and Membrane. *Membranes* **2021**, *11*, 784. [CrossRef] [PubMed]
12. Hanashima, S.; Nakane, T.; Mizohata, E. Heavy Atom Detergent/Lipid Combined X-ray Crystallography for Elucidating the Structure-Function Relationships of Membrane Proteins. *Membranes* **2021**, *11*, 823. [CrossRef] [PubMed]
13. Kumashiro, M.; Tsuji, R.; Suenaga, S.; Matsuo, K. Formation of β-Strand Oligomers of Antimicrobial Peptide Magainin 2 Contributes to Disruption of Phospholipid Membrane. *Membranes* **2022**, *12*, 131. [CrossRef] [PubMed]

14. Barvitenko, N.; Ashrafuzzaman, M.; Lawen, A.; Skverchinskaya, E.; Saldanha, C.; Manca, A.; Uras, G.; Aslam, M.; Pantaleo, A. Endothelial Cell Plasma Membrane Biomechanics Mediates Effects of Pro-Inflammatory Factors on Endothelial Mechanosensors: Vicious Circle Formation in Atherogenic Inflammation. *Membranes* **2022**, *12*, 205. [CrossRef] [PubMed]
15. Ku, Y.-S.; Cheng, S.-S.; Cheung, M.-Y.; Niu, Y.; Liu, A.; Chung, G.; Lam, H.-M. The Poly-Glutamate Motif of GmMATE4 Regulates Its Isoflavone Transport Activity. *Membranes* **2022**, *12*, 206. [CrossRef] [PubMed]
16. Shigeta, A.; Otani, Y.; Miyasa, R.; Makino, Y.; Kawamura, I.; Okitsu, T.; Wada, A.; Naito, A. Photoreaction Pathways of Bacteriorhodopsin and Its D96N Mutant as Revealed by in Situ Photoirradiation Solid-State NMR. *Membranes* **2022**, *12*, 279. [CrossRef] [PubMed]
17. Laurence, M.J.; Carpenter, T.S.; Laurence, T.A.; Coleman, M.A.; Shelby, M.; Liu, C. Biophysical Characterization of Membrane Proteins Embedded in Nanodiscs Using Fluorescence Correlation Spectroscopy. *Membranes* **2022**, *12*, 392. [CrossRef] [PubMed]

membranes

MDPI

Review

Biophysical Characterization of Membrane Proteins Embedded in Nanodiscs Using Fluorescence Correlation Spectroscopy

Matthew J. Laurence [1], Timothy S. Carpenter [1], Ted A. Laurence [2], Matthew A. Coleman [1,3], Megan Shelby [1,*] and Chao Liu [1,*]

1 Biosciences and Biotechnology Division, Lawrence Livermore National Laboratory, Livermore, CA 94550, USA; laurence3@llnl.gov (M.J.L.); carpenter36@llnl.gov (T.S.C.); coleman16@llnl.gov (M.A.C.)
2 Materials Science Division, Lawrence Livermore National Laboratory, Livermore, CA 94550, USA; laurence2@llnl.gov
3 Department of Radiation Oncology, University of California Davis, Sacramento, CA 95616, USA
* Correspondence: shelby4@llnl.gov (M.S.); liu82@llnl.gov (C.L.)

Abstract: Proteins embedded in biological membranes perform essential functions in all organisms, serving as receptors, transporters, channels, cell adhesion molecules, and other supporting cellular roles. These membrane proteins comprise ~30% of all human proteins and are the targets of ~60% of FDA-approved drugs, yet their extensive characterization using established biochemical and biophysical methods has continued to be elusive due to challenges associated with the purification of these insoluble proteins. In response, the development of nanodisc techniques, such as nanolipoprotein particles (NLPs) and styrene maleic acid polymers (SMALPs), allowed membrane proteins to be expressed and isolated in solution as part of lipid bilayer rafts with defined, consistent nanometer sizes and compositions, thus enabling solution-based measurements. Fluorescence correlation spectroscopy (FCS) is a relatively simple yet powerful optical microscopy-based technique that yields quantitative biophysical information, such as diffusion kinetics and concentrations, about individual or interacting species in solution. Here, we first summarize current nanodisc techniques and FCS fundamentals. We then provide a focused review of studies that employed FCS in combination with nanodisc technology to investigate a handful of membrane proteins, including bacteriorhodopsin, bacterial division protein ZipA, bacterial membrane insertases SecYEG and YidC, *Yersinia pestis* type III secretion protein YopB, yeast cell wall stress sensor Wsc1, epidermal growth factor receptor (EGFR), ABC transporters, and several G protein-coupled receptors (GPCRs).

Keywords: fluorescent correlation spectroscopy; membrane proteins; nanodiscs; cell-free expression

1. Isolation of Membrane Proteins

Membrane protein research is critical to understanding cell uptake (e.g., of nutrients, pharmaceuticals, invading pathogens and toxic materials), signaling across the membrane, and diverse other functions. Unfortunately, the hydrophobicity of membrane proteins contributes to their insolubility and instability outside their natural lipid bilayer environment, thus complicating their purification and subsequent structural and solution-based studies. Indeed, while they comprise almost 30% of all human proteins [1], membrane proteins make up only 3–4% of structures in the Protein Data Bank [2].

While the common approach of detergent micelle solubilization can enable solution studies of membrane proteins, these methods are fraught with difficulties in producing suitable quantities of pure, stable, soluble, monodisperse protein and often require truncation or modification of the full-length membrane protein to create a stable construct. Optimization of solubilization conditions while maintaining native fold can be difficult to impossible and may lead to the loss of functionally critical phospholipids, making aggregation challenging to avoid [3]. Newer amphiphilic reagents with advantages over conventional detergents continue to be developed, including the use of peptides [4–7], amphiphilic polymers [8], and maltose-neopentyl glycol (MNG) amphiphiles [9]. Alternatively, modifying the membrane protein of interest can also enable solution-based studies, for example by fusion with an N-terminal peptide that prevents membrane insertion and with a C-terminal amphipathic protein that keeps it soluble [10]. However, all of the above approaches remove the membrane protein of interest from its native lipid environment, which can significantly impact protein conformation and function.

As a result, a major focus of research has been the development of membrane mimetics that can be used to assess the in vitro behavior of membrane proteins while preserving their native structure and function, the most widespread of which are liposomes [11,12]. Unlike detergents, liposomes provide a more native environment in which membrane proteins are embedded in a lipid bilayer. They can be made in different sizes ranging from 20 nm to >1 μm depending on the preparation method [13–16]. However, liposomes present their own drawbacks and suffer from substantial issues with size heterogeneity and reproducible production of high-quality liposomes for different protein systems [13,17]. Liposomes are also often sparingly soluble and provide no control over how individual proteins inserted interact and oligomerize, thus confounding studies of protein association and binding that require isolated entities.

2. Nanodiscs as a Membrane Mimetic

An alternative method to simulate the membrane environment in solution is the use of membrane mimetics that form membrane protein supporting scaffolds in the form of a lipid "nanodisc". Nanodiscs consist of a section of lipid bilayer ~10 nm in diameter surrounded by a supporting belt or scaffold (Figure 1) [18]. Integral membrane proteins can be embedded in the bilayer to form soluble protein–nanodisc complexes. The composition of the bilayer can be varied to contain specific lipid mixtures, cholesterol, lipopolysaccharides, and synthetic lipids or membrane-incorporating species such as telodendrimers [19,20]. Thus, the composition can be specified to mimic cellular membranes, while species with specific functionality (e.g., biotinylation [21–24], Ni functionalization [25–28] or fluorescent labels) can also be incorporated. In some cases, lipids from native membranes can also be incorporated [29–35]. Compared to lipid vesicles, nanodiscs are easier to synthesize with monodisperse or discrete size distributions and do not introduce any unnatural curvature. Multiple types of nanodiscs have been developed in recent decades which principally differ based on the scaffold used to contain the lipid bilayer, including scaffolds consisting of apolipoproteins (to form nanolipoprotein particles, or NLPs), polymers, saposin, peptides, and DNA. Each platform has advantages and drawbacks for use in structural and biochemical studies, as discussed in a recent review [36].

Nanolipoprotein particles (NLPs), the most widely adopted version of nanodiscs, use membrane scaffold proteins (MSPs) to encircle and stabilize the lipid bilayer. The MSP usually consists of two copies of an apolipoprotein, an amphipathic α-helical protein with a physiological role in lipid-binding and transport. NLPs were first developed in 2002 by Bayburt, Grinkova, and Sligar while studying human membrane-binding proteins as a means to create realistic and structurally consistent artificial membranes [37]. Common MSPs use a truncated version of the human lipid-binding protein Apo A-1 (MSP1) and its various engineered derivatives, producing nanodiscs with diameters ranging from 10 to 15 nm [3,19,38]. More recently, Nasr et al. achieved even greater homogeneity in nanodisc size (11–12 nm) by covalently linking the C- and N-termini of MSPs. These "covalently circularized nanodiscs" have improved thermostability and proteolytic resistance, and the technology can be extended to produce a variety of defined sizes and geometries [39,40].

However, a limitation of MSP-based nanodiscs is that both membrane protein and apolipoprotein must be recombinantly expressed and solubilized by detergents before they are assembled into nanodiscs, which may cause denaturation and aggregation before the membrane protein of interest can be introduced into the membrane mimetic. Cell-free co-translation of the scaffold protein and target membrane protein in the presence of lipid, which leads to the embedding of the membrane protein into the bilayer and self-assembly of apolipoprotein scaffolded nanodiscs, enables the encapsulation of membrane proteins within homogenous populations of membrane mimetics particles while circumventing the need for detergent solubilization and purification before lipid bilayer insertion [41–43]. This avoids often lengthy optimization steps, both in detergent selection and in the engineering of stable constructs, to ensure the membrane protein is soluble and presents a native fold. Importantly, this both functionally expands the utility of nanodisc encapsulation to full-length membrane proteins previously not amenable to detergent solubilization and preserves native function for such systems [44,45].

A

B

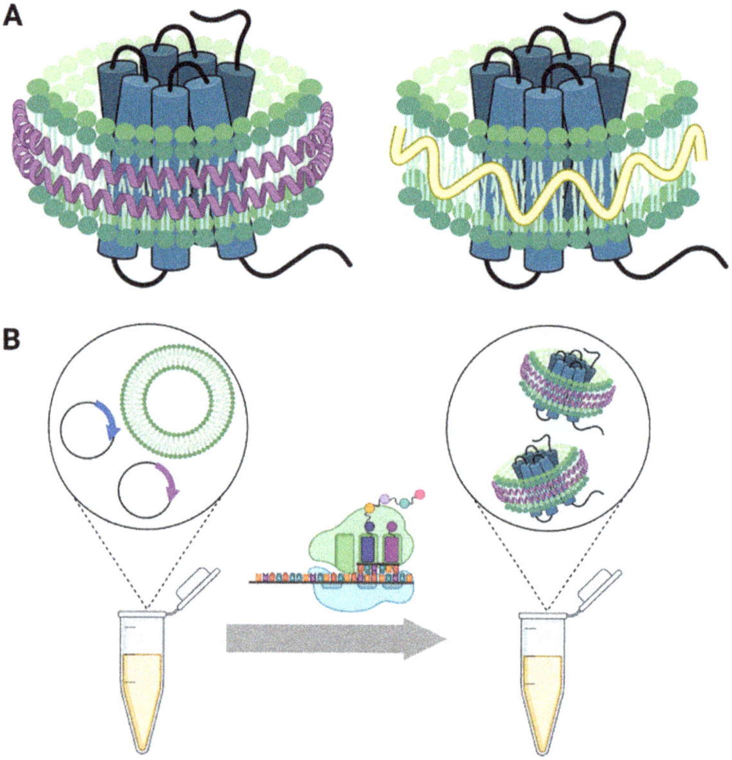

Figure 1. (**A**) Transmembrane protein embedded in a nanodisc lipid bilayer supported by (**left**) scaffold lipoproteins (NLPs) and (**right**) styrene maleic acid copolymers (SMALPs). (**B**) One-pot cell-free synthesis of membrane protein NLP complexes using a co-translational approach: transmembrane proteins and apolipoproteins are translated from cDNA simultaneously in the presence of liposomes using the translational machinery of an *E. coli* cell-free lysate, leading to the self-assembly of nanodisc loaded with the membrane protein of interest [41–43].

Another limitation of NLPs is that depending on the method of preparation, some heterogeneity in disc size and membrane protein insertion rate can remain. One study noted a broad size distribution [46], while several others reported the existence of the inserted membrane protein in a multimeric state [29,47]. Greater control over size and protein insertion rate can be achieved by various techniques and should be optimized for the particular protein under study. For example, with the ability to precisely prescribe the size of covalently circularized nanodiscs and lipid compositions, one can influence the number of proteins incorporated in the constrained bilayer [40]. It has also been demonstrated that with careful specification of plasmid levels and lipid compositions in cell-free expression systems, a monomeric incorporation of the membrane protein can be achieved in ~80% of NLPs [48]. Such techniques can be tuned to reduce heterogeneity and specify insertion rate as suitable for different types of studies.

Other varieties of nanodisc have been developed that use a co-polymer rather than lipid-binding proteins as the surrounding belt, as reviewed recently [34,49,50]. The most widely used polymer is styrene maleic acid (SMA), first shown to form nanodiscs on introduction to bacteriorhodopsin loaded liposomes by Knowles et al. in 2009 [30]. Newer co-polymers such as di-isobutylene maleic acid (DIBMA) are increasingly utilized to vary disc geometry and tune polymer/membrane interaction [51,52]. SMA removes a section of lipid bilayer to form styrene co-maleic acid lipid particles (SMALPs), a technique that has found greatest use in detergent-free solubilization of proteins of interest from native cell membranes for biochemical and structural studies [31–35,53–55]. Avoiding disassembly and reconstitution allows membrane proteins to exist in a disc of membrane as close as possible to their natural cellular conditions. SMALPs also display excellent thermal and temporal stability, keeping their protein monodisperse and functional for over a week at 4 °C [35,56]. While polymer nanodiscs do not have the benefit of the discretized size population of MSP-based nanodiscs, varying the co-polymer monomer ratios and new polymer designs are improving their size control, pH stability, and other spectroscopic and biochemical properties [57].

3. Fluorescence Correlation Spectroscopy

The interactions between membrane proteins and their ligands, agonists, and antagonists provide a basis for understanding their physiological functions and how they are affected by pathogens, cancer, and drugs. Techniques for studying these interactions include atomic force microscopy (AFM), surface plasmon resonance (SPR), circular dichroism spectroscopy, and fluorescence techniques such as fluorescence correlation spectroscopy (FCS), fluorescence recovery after photobleaching (FRAP), and Förster resonance energy transfer (FRET). Here, we first focus on FCS, then review the use of FCS combined with nanodisc technologies in studies of several membrane proteins.

In comparison to other techniques, the relative strength of fluorescence techniques are their specificity, sensitivity, and high-throughput potential. While the species of interest must be labeled with chromophores or bulky fluorescent proteins, this ensures that the signal observed is from the molecule of interest. In addition, multiple colors may be used, allowing multiple species to be traced simultaneously in the same sample. Furthermore, fluorescence techniques can be used in tandem with other forms of microspectroscopy [58].

Fluorescence correlation spectroscopy is one of the most common fluorescence techniques to study molecular dynamics (Figure 2). In its most common configuration, a confocal microscope is used to create a femtoliter-sized detection volume in a solution of fluorescent tagged molecules [59,60]. Individual photons emitted from the fluorescent molecules diffusing through the detection volume are detected using single photon counting detectors, which record the timestamps of the photon arrivals [61]. By performing the mathematical correlation operation, the data are converted into a correlation curve which can be fit to models for various diffusion processes.

FCS presents several advantages over related fluorescence systems. Fluorescence recovery after photobleaching (FRAP) measures protein diffusion kinetics by briefly photobleaching a region inside a solution, cell, or membrane containing fluorescently labeled proteins and measuring the time needed for the region to recover the fluorescence signal as non-bleached proteins diffuse in. Unlike FRAP, FCS can measure multicomponent diffusion and protein concentrations, requires lower protein concentrations, and is better suited for short time scale observations [62,63]. Förster resonance energy transfer (FRET) is a form of nonradiative energy transfer between two molecules used to determine if they are in proximity to each other. As the FRET efficiency is proportional to the sixth power of distance between the fluorescent donor and acceptor, FRET is a very sensitive method to probe for intermolecular binding and protein conformations [64–67]. In practice, however, binding measurements using FRET are much more challenging than using FCS because successful design of FRET pairs at specific positions on proteins requires precise knowledge of protein structures.

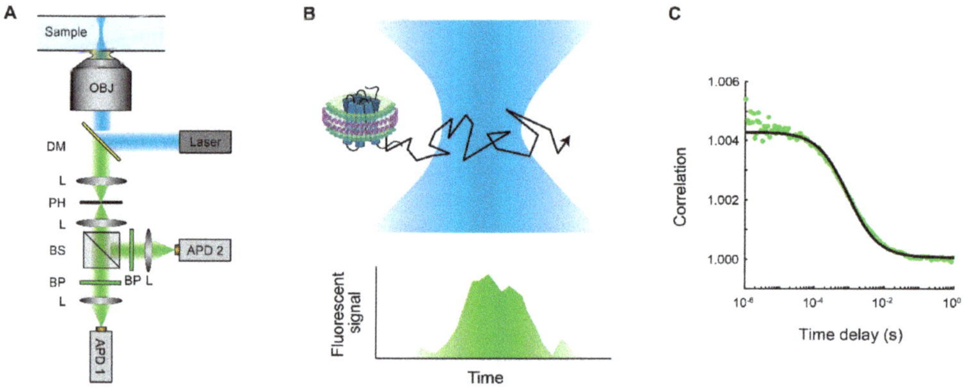

Figure 2. Fluorescence correlation spectroscopy (FCS) applied to membrane proteins embedded in nanodiscs. (**A**) FCS instrument schematic for a single-color system based on a confocal microscope. A tightly focused laser excitation combined with a confocal pinhole define a small, femtoliter-scale open detection volume. The emitted fluorescence signal is typically split and projected onto two detectors to eliminate detector dead time and afterpulsing effects. OBJ: objective; DM: dichroic mirror; L: lens; PH: pinhole; BS: beamsplitter; BP: bandpass filter; APD: avalanche photodiode. (**B**) As a molecule of the fluorescently labeled species, such as the membrane protein embedded in a nanodisc shown, diffuses through the open detection volume, the detected fluorescence signal fluctuates. Size of nanodisc in comparison to laser focus not drawn to scale. (**C**) The $g^{(2)}$ correlation is calculated from the fluorescence signal (green data points). A simple diffusion model (Equation (1)) (black line) is fitted to the data, giving the diffusion time and concentration of the diffusing species. Data shown are of the human β-2 adrenergic receptor conjugated with the green fluorescent protein and embedded inside an Apo-A1 encircled DMPC nanodisc. Fitted diffusion time $\tau_D = 0.92 \pm 0.02$ ms.

By attaching fluorescent tags to biological macromolecules, virtually any molecule in solution can be studied by FCS. Proteins can be either expressed as fusions with other fluorescent proteins or functionalized with chemical fluorescent dyes. The sample does not need to be removed from solution, allowing solution dynamics to remain unchanged. In addition to measurements in solution, FCS may be measured inside cells [68–71] allowing findings from simpler solution-based assays to be validated in cell-based experiments.

Analysis of the correlation curve from FCS measurements gives insight into the physical properties of a system. The shape of the correlation curve depends on an effective detection profile produced by the laser excitation volume combined with the detection pinhole. The simplest model for FCS is that of a single species diffusing in a 2D Gaussian detection profile, with correlation C given by

$$C(\tau) = 1 + \frac{1}{N\left(1 + \frac{\tau}{\tau_D}\right)} \tag{1}$$

where τ is the time delay, τ_D is the diffusion time, and N is the average number of fluorescent molecules in the detection volume. τ_D represents the average time for a molecule to diffuse out of the open detection volume in the solution. The diffusion constant D, which is dependent on factors including molecular size and structure, can be calculated by $D = \frac{r^2}{4\tau_D}$, where r is the width of the 2D Gaussian profile as determined through a calibration. The concentration of the species can be calculated from N and the calibrated volume. In this model, diffusion in the axial dimension (along the optical path) is ignored since the timescale of diffusion out of the open detection volume in that axis is much longer.

Deviations from the simple model above can indicate the presence of additional species with different diffusion times [72], an anomalously diffusing species, or other photophysical behaviors such as triplet-state excitation [73]. The diffusion of two distinct species with average numbers N_1 and N_2 and diffusion times τ_{D1} and τ_{D2} can be modeled with correlation function

$$C(\tau) = 1 + \frac{N_1}{(N_1 + N_2)^2\left(1 + \frac{\tau}{\tau_{D1}}\right)} + \frac{N_2}{(N_1 + N_2)^2\left(1 + \frac{\tau}{\tau_{D2}}\right)} \tag{2}$$

The presence of multiple species may be due to different entities (e.g., different proteins) or different states of the same entity (e.g., oligomers or aggregates). The presence of two species is expected when a biomolecule is present that binds to the fluorescently tagged species, for example in cases of two interacting proteins or drug binding to protein. Under sub-saturating concentrations of the non-fluorescent molecule, one species would be the bound complex (with a longer diffusion time) while the second would be the free fluorescently labeled molecule. A shift in the correlation curve towards longer diffusion times is a simple readout of binding when probing for molecular interactions. Note, however, that the diffusion times of the two species must differ by at least a factor of 1.6 to be distinguishable under typical conditions [72]. Since diffusion time is inversely proportional to the cubed root of mass, this criterion stipulates that the masses of the two species must differ minimally by a factor of 4. Thus, a critical strategy when investigating potential interactions (e.g., during drug screening) is the placement of the fluorescent label on the small ligand so that a large change in mass occurs upon binding of the larger protein.

FCS can also be expanded to use multiple colors to observe multiple molecular species simultaneously. In fluorescence cross-correlation spectroscopy (FCCS), two or more molecular species that may interact are labeled with spectrally dissimilar fluorescent tags and excited with separate lasers [74]. The distinct fluorescence signals can be autocorrelated individually as in single-color FCS or cross correlated. Fitting the cross-correlation curve gives the concentration of bound pairs, which can be used with the concentrations of each unbound species to calculate the binding coefficient [61]. Alternating laser excitation and pulsed-interleaved excitation allow for the elimination of many sources of spurious cross-correlations [75–77].

Conventional FCS only measures diffusion at a single spot of hundreds of nanometers to 1 μm in size, much smaller than many target areas of interest in biological systems, thus missing potential spatial heterogeneity in the sample. Various methods have been developed that extend conventional FCS to provide greater spatial information. Scanning FCS performs measurements as the observation volume is continuously moved through the sample, providing spatial information while reducing photobleaching [78]. Imaging correlation spectroscopy makes use of EMCCD cameras that achieve single-photon sensitivity over a much wider area. Instead of tracking photon counts from a single point, the camera records an image containing FCS data for hundreds or thousands of pixels simultaneously [79,80]. Acquiring many correlations in parallel across space not only provides location-dependent information but also increases the statistics while reducing acquisition time, thus allowing high-throughput measurements of diffusing species and their interactions.

FCS has several limitations as a technique in general and when applied to the study of membrane proteins. As with all fluorescence techniques, one must first ensure that the fluorescent label does not interfere with the structure or function of the protein under study. The requirement that two species have masses that differ by at least a factor of 4 to be distinguishable in single-color FCS has been discussed above [72]. Accurate concentration determination requires an accurate calibration of the detection volume as well as sufficient signal-to-noise ratio. While the use of higher laser powers improves the signal-to-noise ratio, it can also lead to triplet-state excitation and enlargement of the detection volume [73,81,82]. High laser powers can also lead to photobleaching of fluorophores, a common issue with many fluorescence measurements. Photobleaching is especially detrimental in experiments involving slowly diffusing molecules that require long acquisition times, in which case lower laser powers should be used. In addition, slowly diffusing aggregations of fluorescently labeled proteins, a problem common with membrane proteins, produce strong signals that can dramatically skew correlation curves even with a comparatively low concentration of aggregates. This again underscores the importance of methods such as nanodiscs to producing soluble non-aggregated membrane proteins. Finally, proper interpretation of FCS data requires fitting to a suitable model. Photobleaching, triplet-state excitation, aggregations, and complexities arising from measurements taken inside cells or on membranes (e.g., constrained diffusion, immobilization, and active transport) all result in correlation curves requiring models beyond simple diffusion [83,84].

4. FCS Measurements of Membrane Proteins Embedded in Nanodiscs

Nanodiscs permit the application of solution-based techniques such as FCS to the study of individual membrane proteins embedded in well-defined bilayers. Starting in the past decade, FCS has been applied to the study of various membrane proteins embedded in nanodiscs, as reviewed below and summarized in Table 1.

Table 1. Membrane proteins embedded in nanodiscs and studied by FCS.

Membrane Protein	Species	Expression and Assembly System	Nanodisc Type	Fluorescent Label	Diffusion Characteristic (FCS)	Reference
Bacteriorhodopsin (bR)	*H. salinarum*	Cell-free co-translation	DMPC + ApoΔ49A1	BODIPY-lysine during translation	τ_D = 0.35 ms	Gao et al., 2011 [47]
ZipA	*E. coli*	Proteins from *E. coli*, then mixed with lipids	*E. coli* lipid extract + MSP1D1	Lissamine rhodamine B on lipids	$D = 32 \pm 4\ \mu m^2/s$	Hernandez-Rocamora et al., 2012 [85]
SecYEG	*E. coli*	Proteins from *E. coli*, then mixed with lipids	DOPC, DOPE, DOPG, CL + MSP1D1	Alexa Fluor 488 on protein	$D = 27 \pm 3\ \mu m^2/s$	Wu et al., 2012 [86]
YidC	*E. coli*	Proteins from *E. coli*, then mixed with lipids	DOPC, DOPE, DOPG, CL + MSP1D1	Alexa Fluor 488 or Atto 647N on protein	$D = 31 \pm 2\ \mu m^2/s$	Kedrov et al., 2013 [87]
YidC	*E. coli*	Proteins from *E. coli*, then mixed with lipids	DOPC, DOPE, DOPG + MSP1D1	Alexa Fluor 488 on protein	$D = 39 \pm 2\ \mu m^2/s$	Geng et al., 2015 [88]
YopB	*Y. pestis*	Cell-free co-translation	DMPC + ApoΔ49A1	Binding to LcrV-GFP	$\tau_D = 0.63 \pm 0.06$ ms	Ly et al., 2014 [89]
Epidermal growth factor receptor (EGFR)	*H. sapiens*	Cell-free co-translation	DMPC + ApoΔ49A1	SNAP fusion construct labeled with SNAP surface 594	$\tau_D = 0.167 \pm 0.002$ ms	Quinn et al., 2019 [48]
ABCB1/P-glycoprotein	*M. musculus*	Membrane protein from *P. pastoris*, MSP from *E. coli*, then mixed with lipids	DMPC + MSP1D1	Binding to small BODIPY-linked ligands	$\tau_D = 3.0 \pm 0.2$ ms	Li et al., 2017 [90]
ABCG2/BCRP (breast cancer resistance protein)	*H. sapiens*	HEK293T expression, then native membrane extraction by SMA	SMALP	GFP fusion construct	$D = 31 \pm 4\ \mu m^2/s$ (likely dimer)	Horsey et al., 2020 [29]
Neurokinin-1 receptor (NK1R)	*H. sapiens*	Cell-free co-translation	DMPC + ApoΔ49A1	GFP fusion construct	$\tau_D = 0.51 \pm 0.37$ ms	Gao et al., 2012 [45]

Table 1. *Cont.*

Membrane Protein	Species	Expression and Assembly System	Nanodisc Type	Fluorescent Label	Diffusion Characteristic (FCS)	Reference
Adenosine A_{2A} receptor (A2AR)	*H. sapiens*	*P. pastoris* expression, then native membrane extraction by SMA	SMALP	Binding to small red BODIPY-linked ligand	τ_D = 0.63 ± 0.02 ms D = 30 ± 4 $\mu m^2/s$	Grime et al., 2020 [91]
Neurotensin receptor 1 (NTSR1)	*H. sapiens*	HEK293T expression, then native membrane extraction by SMA	SMALP	mRuby fusion construct	τ_D = 5.88 ms D = 16 $\mu m^2/s$ (broad size distribution and large residuals noted)	Dathe et al., 2019 [46]
Wsc1	*S. cerevisiae*	*S. cerevisiae* expression, then native membrane extraction by SMA	SMALP	GFP fusion construct	D = 50 ± 4.6 $\mu m^2/s$	Voskoboynikova et al., 2021 [92]
Cytochrome *c* oxidase (CytcO)	*R. sphaeroides*	Membrane protein from *R. sphaeroides*, MSP from *E. coli*, then mixed with lipids	DOPG + MSP1D1 or MSP1E3D1	Fluorescein on lipids or protein		Xu et al., 2016 [93]

Cell-free co-translation: proteins expressed in the presence of lipids for self-assembly of nanodiscs during expression. SMALP: styrene maleic acid lipid particles. ApoΔ49A1: truncated apolipoprotein A1. DMPC: 1,2-dimyristoyl-sn-glycero-3-phosphocholine. DOPC: dioleoylphosphatidylcholine. DOPE: dioleoylphosphatidylethanolamine. DOPG: dioleoylphosphatidylglycerol. CL: cardiolipin. τ_D: diffusion time. D: diffusion coefficient.

In the first FCS characterization of a nanodisc-embedded membrane protein, Gao et al. measured the diffusion times of bacteriorhodopsin (bR) embedded in nanodiscs made of a DMPC-lipid bilayer surrounded by truncated Apo-A1 [47]. Empty and bR-loaded nanodiscs had diffusion times ~0.5 ms. The authors also measured the diffusion times of several particles (e.g., large vesicles, beads, and fluorescent dyes) of known sizes as determined by dynamic light scattering. From this calibration of diffusion time vs. particle size, the ~0.5 ms diffusion times of empty and bR-nanodiscs correspond to diameters of ~10 nm, agreeing with the value expected of Apo-A1 nanodiscs.

Since then, FCS has been used to characterize interactions between several other nanodisc-embedded bacterial membrane proteins and their soluble binding partners. Hernandez-Rocamora et al. measured the binding of nanodisc-embedded *E. coli* ZipA, which provides essential membrane tethering during cell division, to FtsZ in different polymerization and nucleotide states [85]. Empty nanodiscs made of *E. coli* polar lipid extract surrounded by MSP1D1 and discs containing one ZipA protein had diffusion times of ~0.5 ms, corresponding to diameters of 10 and 13 nm, respectively, as expected. The addition of FtsZ resulted in a rightward shift of the correlation curve from which the fraction of bound species was determined.

Wu et al. [86], Kedrov et al. [87], and Geng et al. [88] elucidated the specific nature of binding between the ribosome and *E. coli*'s two membrane insertases, SecYEG and YidC and, in the process, revealed the importance of using a native-like lipid environment when attempting to uncover physiological binding behavior. Membrane insertases associate with translating ribosomes and integrate nascent polypeptides into the lipid bilayer. By FCS measurements of SecYEG embedded in nanodiscs, Wu et al. found that SecYEG binds ribosome strongly only in the presence of a nascent chain (ribosome-bound nascent chain complex, RNC) as seen by a rightward shift in the diffusion curve [86]. This requirement of nascent chain was not fully apparent if not for the use of nanodiscs, as ribosome without nascent chain was still able to bind the majority of SecYEG solubilized in detergent. The authors further found that the addition of the SecA protein, which also binds SecYEG, competitively inhibits binding of RNC to SecYEG-nanodiscs, as seen by decreases in the fraction of RNC-SecYEG-nanodisc species in FCS diffusion curves. In contrast, this competitive binding behavior of SecA and RNC to SecYEG was not observed when SecYEG was solubilized in detergent rather than nanodiscs; in detergent, almost all SecYEG was bound to RNC despite the presence of excess SecA. In a parallel study to that of SecYEG, Kedrov et al. found that when embedded in nanodiscs, YidC binding to ribosome, as measured by a rightward shift in the FCS correlation curve, also requires the presence of a translating nascent chain [87]. In contrast, YidC in detergent exhibited several artifactual interaction characteristics, such as binding of ribosome without nascent chain at non-physiological acidic pH. Geng et al. further explored the structural basis of the binding between YidC-nanodiscs and RNC through FCS measurements of YidC mutant constructs and found that YidC's cytosolic loop C2 and C terminus are both crucial in this interaction [88]. All the above results underscore the importance of nanodiscs in providing a native-like environment to study physiological membrane protein interactions.

Ly et al. used FCS to measure the interactions between *Yersinia pestis* membrane-bound YopB and soluble LcrV, two proteins of the bacteria's type III secretion system critical for host cell invasion [89]. Binding of YopB-nanodisc to eGFP-LcrV resulted in a rightward shift in the correlation curve from which the fraction of bound species was determined (Figure 3). A full binding curve obtained by varying concentrations of YopB-nanodisc produced a dissociation constant ~20 nM. In addition to enabling FCS measurements, YopB produced concurrently with nanodiscs in the cell-free expression system had the added advantage of greater protein solubility [44].

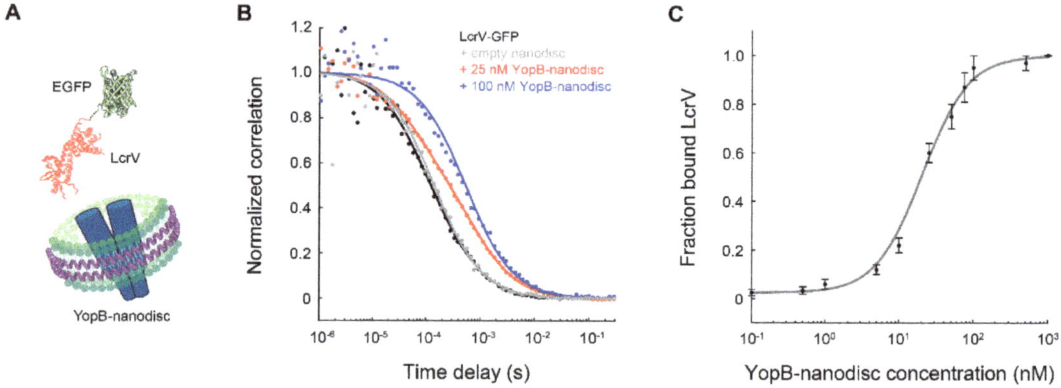

Figure 3. Binding of nanodisc-embedded membrane proteins to smaller, fluorescently labeled ligands are easily detectable by FCS due to a large change in diffusion time. (**A**) Cartoon of the *Y. pestis* membrane protein YopB (a multimer) inserted into a 10 nm nanodisc and its cognate protein LcrV (PDB: 1R6F) labeled with EGFP. (**B**) FCS correlation curves of LcrV-EGFP in the presence of different concentrations of YopB-nanodisc (data selected from Ly et al. [89]). Addition of YopB-nanodisc shifts the correlation curve of LcrV-EGFP to the right. Intermediate concentrations of YopB-nanodisc result in a diffusion curve with two species present, one of free LcrV-EGFP and one bound to YopB-nanodisc. Curves are fitted to Equation (2) to obtain the diffusion times and fractions of the two species. (**C**) Fraction of LcrV-EGFP bound to YopB-nanodisc at different YopB-nanodisc concentrations. A fit to the Hill equation gives dissociation constant $K_D = 20 \pm 2$ nM and Hill coefficient $n = 1.4$ [89].

In addition to studying the above bacterial proteins, FCS in combination with nanodisc technology has been used to characterize various eukaryotic membrane proteins. Voskoboynikova et al. used FCS to demonstrate successful production of monodisperse Wsc1, a yeast transmembrane cell wall stress sensor, embedded in SMALPs by SMA extraction from native *S. cerevisiae* cell membrane [92]. Other studies further characterized by FCS the interactions between eukaryotic membrane proteins and their substrates. Quinn et al. measured FCS on nanodisc-embedded full-length epidermal growth factor receptor (EGFR) produced by cell-free expression [48]. Remarkably, the authors reported the ability to detect microsecond shifts in diffusion times upon ATP binding to EGFR-nanodisc and upon phosphorylated EGFR-nanodisc binding to a specific antibody. Li et al. [90] and Horsey et al. [29] used FCS to study ligand binding of ATP binding cassette (ABC) transporters, which couple ATP hydrolysis to the binding and transport of substrates across membranes and which play major roles in drug interactions and resistance. Li et al. found that the binding of several fluorescent-labeled substrates to nanodisc-embedded ABCB1/P-glycoprotein is dependent on different nucleotide states of the protein [90]. Horsey et al. measured the binding of fluorescent-prazosin to ABCG2/BCRP (breast cancer resistance protein) embedded in SMALP by SMA extraction from the native HEK293 cell membrane [29].

Finally, FCS combined with nanodisc technology has been used to study G protein-coupled receptors (GPCRs), the largest family of membrane proteins in humans and major drug targets. Dathe et al. purified SMALPs containing neurotensin receptor 1 (NTSR1) and found its diffusion time by FCS to be much larger than expected along with large fitting residuals, suggesting a broad rather than monodisperse distribution of sizes [46]. Other studies investigated ligand binding to GPCRs. Gao et al. produced several nanodisc-bound human GPCRs by cell-free expression, including neurokinin-1 receptor (NK1R), adrenergic receptor ADRB2, and dopamine receptor DRD1 [45]. The authors proceeded to measure by FCS a full binding curve between NK1R-nanodisc and its fluorescently labeled ligand Substance P, from which a dissociation constant ~83 nM was obtained. More recently, Grime et al. measured the binding of human adenosine A_{2A} receptor embedded in SMALP to its fluorescently labeled antagonist xanthine amine congener [91]. The use of FCS to measure interactions between a small fluorescently-labeled ligand and a large nanodisc-embedded GPCR invites many future studies for potential drug development.

Beyond the determination of the binding and diffusion biophysics of membrane proteins, the combination of FCS and nanodisc technology has been used to study the protonation effect of membranes. Lipid membrane can enrich [H^+] near its surface in a phenomenon known as the local membrane proton-collecting antenna effect, serving to accelerate proton uptake by membrane-bound proton transporters such as bacteriorhodopsin and cytochrome *c* oxidase (Cyt*c*O). Xu et al. measured FCS curves of fluorescein-labeled nanodiscs of different sizes, in the absence or the presence of membrane-embedded Cyt*c*O, and under various buffer conditions [93]. The curves were fitted to a multistate model with components including normal diffusion and fluorescence decay, the latter serving as the readout of protonation levels as fluorescein fluorescence is highly sensitive to pH. While previous studies had measured the proton antenna effect for large 30 nm vesicles, the use of nanodiscs allowed the authors to detect the effect for membranes as small as 9 nm in diameter. The authors also found that the presence of Cyt*c*O in the membrane decreased the effect, requiring a larger membrane (12 nm in diameter) to achieve the same protonation levels.

A direct comparison of diffusion times is not possible across all the above studies since different FCS setups have slightly different focal volumes. Rather, conclusions are drawn within a study based on changes in correlation curves under different biological conditions. Nevertheless, it is informative to visualize FCS curves of nanodiscs-membrane proteins in relation to other species across a wide range of sizes. As expected, correlation curves and diffusion times of membrane proteins embedded in ~10 nm-nanodiscs lie between those of individual fluorescent proteins (GFP and mCherry) and polystyrene microspheres with diameters ~40–100 nm (Figure 4).

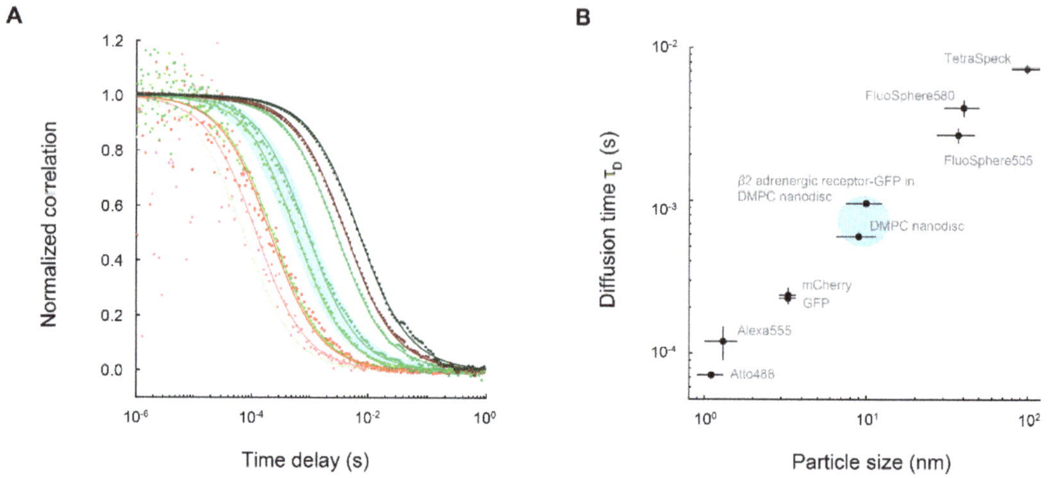

Figure 4. FCS correlation curves and diffusion times of nanodisc-embedded membrane proteins among a large range of particle sizes. (**A**) A representative correlation curve, from data collected for 1 min, for each species given in (**B**). Blue shaded area represents the range in which curves for membrane proteins embedded in nanodiscs are expected to reside. Green shades represent data collected with 488 nm laser excitation, and red shades represent data collected with 561 nm laser excitation. Data were collected on a homemade FCS system with a calibrated detection volume of ~3 μm^3. (**B**) Diffusion times of different species plotted against particle size (diameter). Error bars on diffusion times are standard deviations based on multiple measurements. Error bars on size are estimated from diameters of the species. Blue shaded area represents the range in which nanodisc-embedded membrane proteins are expected to reside. Atto488 (Sigma-Aldrich, St. Louis, MS, USA, product #41051); Alexa555: Alexa Fluor 555 NHS ester (Thermo Fisher Scientific, Carlsbad, CA, USA, product #A20009); FluoSphere505 and FluoSphere580: FluoSpheres fluorescent carboxylate-modified microspheres 0.04 μm (Thermo Fisher Scientific, product #F10720); TetraSpeck: TetraSpeck fluorescent microspheres 0.1 μm (Thermo Fisher Scientific, product #T7279).

5. Summary and Future Directions

Despite its age, FCS remains an effective and widely used biophysical technique to study protein biophysics and binding in solution as well as within cells. Nanodiscs enable FCS measurements of in-solution, purified, individual membrane proteins residing in a native-like lipid environment. While only a handful of studies have used this combination as reviewed above, the simplicity of FCS combined with the versatility of nanodiscs presents great potential for future studies of the plethora of membrane proteins. FCS may be applied to any of the ~100 membrane proteins that have been successfully reconstituted in nanodiscs (as listed in several recent reviews [3,34]) as well as important nanodisc-embedded membrane proteins to be successfully isolated in the future. Candidates for the latter include other bacterial type III secretion proteins which play critical roles in host cell invasion, in FCS binding studies analogous to those performed for *Y. pestis* YopB [89].

Future studies may leverage the ability to precisely control lipid composition in nanodiscs to investigate the effects of lipids on membrane protein function and structure. Membrane lipids have immense diversity, with the human body containing thousands of different lipids with significant implications for cellular shape and function [94,95]. Studies have broadly established that membrane lipid composition, such as cholesterol content and phospholipid type, affects membrane protein behavior through both direct lipid-protein interactions and indirect mechanisms such as membrane fluidity and curvature [96–102]. In recent years, computational research in lipid bilayers has also had a focus on increasing the complexity of the system to more closely replicate various biological membranes [103–107].

This coordinated drive towards complex realism in experimental and computational work opens the opportunities for both disciplines to inform the other. Of particular interest are studies of the major drug targets GPCRs embedded in nanodiscs of various prescribed lipid compositions. The effects of the surrounding membrane composition on interactions with G proteins, affinities for ligands, and affinities for therapeutic drugs of many GPCRs remain to be investigated by computational tools, FCS, and other experimental techniques.

The FCS-nanodisc assay shows great promise for future improvements. Advancements in detector technology will continue to increase the range of concentrations over which FCS can be performed by increasing photon count limits and time resolution [108,109]. New types of spatial extent FCS such as line scanning FCS or imaging FCS have not yet been applied to nanodisc measurements, but future work may find them useful in enhancing the throughput power of the assay over a short time interval. Continued improvements to nanodisc technology, such as further development and use of polymers and membrane scaffold proteins, will facilitate the ease of protein isolation and increase the fidelity of the mimetic to the native membrane. Future studies employing such nanodiscs will give FCS access to a broader selection of context-relevant membrane proteins and the complexes they form.

Author Contributions: All authors contributed to writing and editing of the review. All authors have read and agreed to the published version of the manuscript.

Funding: This work was performed, in part, under the auspices of the U.S. Department of Energy by Lawrence Livermore National Laboratory under Contract DE-AC52-07NA27344. Partial support was provided by NIH grants 1R01GM117342 and U19AI144184 and by the LLNL Laboratory Directed Research and Development programs (21-ERD-039 and 21-ERD-047).

Institutional Review Board Statement: Not applicable.

Informed Consent Statement: Not applicable.

Data Availability Statement: Data presented is contained within this paper.

Acknowledgments: The authors would like to thank Steven Hoang-Phou and Bonnee Rubinfeld for providing reagents used in experiments and Edmond Lau for help in constructing the YopB structure model.

Conflicts of Interest: The authors declare no conflict of interest.

References

1. Almén, M.S.; Nordström, K.J.V.; Fredriksson, R.; Schiöth, H.B. Mapping the human membrane proteome: A majority of the human membrane proteins can be classified according to function and evolutionary origin. *BMC Biol.* **2009**, *7*, 50. [CrossRef] [PubMed]
2. RCSB Protein Data Bank. Available online: https://www.rcsb.org/stats/growth/growth-protein (accessed on 19 January 2022).
3. Rouck, J.E.; Krapf, J.E.; Roy, J.; Huff, H.C.; Das, A. Recent advances in nanodisc technology for membrane protein studies (2012–2017). *FEBS Lett.* **2017**, *591*, 2057–2088. [CrossRef] [PubMed]
4. McGregor, C.-L.; Chen, L.; Pomroy, N.C.; Hwang, P.; Go, S.; Chakrabartty, A.; Privé, G.G. Lipopeptide detergents designed for the structural study of membrane proteins. *Nat. Biotechnol.* **2003**, *21*, 171–176. [CrossRef] [PubMed]
5. Schafmeister, C.E.; Miercke, L.J.W.; Stroud, R.M. Structure at 2.5 Å of a Designed Peptide that Maintains Solubility of Membrane Proteins. *Science* **1993**, *262*, 734–738. [CrossRef] [PubMed]
6. Tao, H.; Lee, S.C.; Moeller, A.; Roy, R.S.; Siu, F.Y.; Zimmermann, J.; Stevens, R.C.; Potter, C.S.; Carragher, B.; Zhang, Q. Engineered nanostructured β-sheet peptides protect membrane proteins. *Nat. Methods* **2013**, *10*, 759–761. [CrossRef]
7. Zhao, X.; Nagai, Y.; Reeves, P.J.; Kiley, P.; Khorana, H.G.; Zhang, S. Designer short peptide surfactants stabilize G protein-coupled receptor bovine rhodopsin. *Proc. Natl. Acad. Sci. USA* **2006**, *103*, 17707–17712. [CrossRef]
8. Tribet, C.; Audebert, R.; Popot, J.-L. Amphipols: Polymers that keep membrane proteins soluble in aqueous solutions. *Proc. Natl. Acad. Sci. USA* **1996**, *93*, 15047–15050. [CrossRef]
9. Chae, P.S.; Rasmussen, S.G.F.; Rana, R.R.; Gotfryd, K.; Chandra, R.; Goren, M.A.; Kruse, A.C.; Nurva, S.; Loland, C.J.; Pierre, Y.; et al. Maltose–neopentyl glycol (MNG) amphiphiles for solubilization, stabilization and crystallization of membrane proteins. *Nat. Methods* **2010**, *7*, 1003–1008. [CrossRef]
10. Mizrachi, D.; Chen, Y.; Liu, J.; Peng, H.-M.; Ke, A.; Pollack, L.; Turner, R.J.; Auchus, R.J.; Delisa, M.P. Making water-soluble integral membrane proteins in vivo using an amphipathic protein fusion strategy. *Nat. Commun.* **2015**, *6*, 6826. [CrossRef]

11. Klöpfer, K.; Hagn, F. Beyond detergent micelles: The advantages and applications of non-micellar and lipid-based membrane mimetics for solution-state NMR. *Prog. Nucl. Magn. Reson. Spectrosc.* **2019**, *114–115*, 271–283. [CrossRef]
12. Warschawski, D.E.; Arnold, A.; Beaugrand, M.; Gravel, A.; Chartrand, É.; Marcotte, I. Choosing membrane mimetics for NMR structural studies of transmembrane proteins. *Biochim. Biophys. Acta (BBA)-Biomembr.* **2011**, *1808*, 1957–1974. [CrossRef] [PubMed]
13. Girard, P.; Pécréaux, J.; Lenoir, G.; Falson, P.; Rigaud, J.-L.; Bassereau, P. A New Method for the Reconstitution of Membrane Proteins into Giant Unilamellar Vesicles. *Biophys. J.* **2004**, *87*, 419–429. [CrossRef] [PubMed]
14. Jesorka, A.; Orwar, O. Liposomes: Technologies and Analytical Applications. *Annu. Rev. Anal. Chem.* **2008**, *1*, 801–832. [CrossRef] [PubMed]
15. Venetië, R.; Leunissen-Bijvelt, J.; Verkleij, A.J.; Ververgaert, P.H.J.T. Size determination of sonicated vesicles by freeze-fracture electron microscopy, using the spray-freezing method. *J. Microsc.* **1980**, *118*, 401–408. [CrossRef]
16. Rideau, E.; Dimova, R.; Schwille, P.; Wurm, F.R.; Landfester, K. Liposomes and polymersomes: A comparative review towards cell mimicking. *Chem. Soc. Rev.* **2018**, *47*, 8572–8610. [CrossRef]
17. Litschel, T.; Schwille, P. Protein Reconstitution Inside Giant Unilamellar Vesicles. *Annu. Rev. Biophys.* **2021**, *50*, 525–548. [CrossRef]
18. Denisov, I.G.; Sligar, S.G. Nanodiscs in Membrane Biochemistry and Biophysics. *Chem. Rev.* **2017**, *117*, 4669–4713. [CrossRef]
19. Ritchie, T.K.; Grinkova, Y.V.; Bayburt, T.H.; Denisov, I.G.; Zolnerciks, J.K.; Atkins, W.M.; Sligar, S.G. *Methods in Enzymology*; Düzgünes, N., Ed.; Academic Press: Cambridge, MA, USA, 2009; Volume 464, pp. 211–231.
20. He, W.; Luo, J.; Bourguet, F.; Xing, L.; Yi, S.K.; Gao, T.; Blanchette, C.; Henderson, P.T.; Kuhn, E.; Malfatti, M.; et al. Controlling the diameter, monodispersity, and solubility of ApoA1 nanolipoprotein particles using telodendrimer chemistry. *Protein Sci.* **2013**, *22*, 1078–1086. [CrossRef]
21. Bayburt, T.H.; Sligar, S.G. Self-assembly of single integral membrane proteins into soluble nanoscale phospholipid bilayers. *Protein Sci.* **2003**, *12*, 2476–2481. [CrossRef]
22. Dominik, P.K.; Borowska, M.T.; Dalmas, O.; Kim, S.S.; Perozo, E.; Keenan, R.; Kossiakoff, A.A. Conformational Chaperones for Structural Studies of Membrane Proteins Using Antibody Phage Display with Nanodiscs. *Structure* **2016**, *24*, 300–309. [CrossRef]
23. Goluch, E.D.; Shaw, A.W.; Sligar, S.G.; Liu, C. Microfluidic patterning of nanodisc lipid bilayers and multiplexed analysis of protein interaction. *Lab Chip* **2008**, *8*, 1723–1728. [CrossRef]
24. Haruyama, T.; Sugano, Y.; Kodera, N.; Uchihashi, T.; Ando, T.; Tanaka, Y.; Konno, H.; Tsukazaki, T. Single-Unit Imaging of Membrane Protein-Embedded Nanodiscs from Two Oriented Sides by High-Speed Atomic Force Microscopy. *Structure* **2019**, *27*, 152–160.e3. [CrossRef]
25. Fischer, N.O.; Rasley, A.; Corzett, M.; Hwang, M.H.; Hoeprich, P.D.; Blanchette, C.D. Colocalized Delivery of Adjuvant and Antigen Using Nanolipoprotein Particles Enhances the Immune Response to Recombinant Antigens. *J. Am. Chem. Soc.* **2013**, *135*, 2044–2047. [CrossRef] [PubMed]
26. Blanchette, C.D.; Fischer, N.O.; Corzett, M.; Bench, G.; Hoeprich, P.D. Kinetic Analysis of His-Tagged Protein Binding to Nickel-Chelating Nanolipoprotein Particles. *Bioconjugate Chem.* **2010**, *21*, 1321–1330. [CrossRef] [PubMed]
27. Fischer, N.; Infante, E.; Ishikawa, T.; Blanchette, C.D.; Bourne, N.; Hoeprich, P.D.; Mason, P.W. Conjugation to Nickel-Chelating Nanolipoprotein Particles Increases the Potency and Efficacy of Subunit Vaccines to Prevent West Nile Encephalitis. *Bioconjugate Chem.* **2010**, *21*, 1018–1022. [CrossRef] [PubMed]
28. Fischer, N.O.; Blanchette, C.D.; Chromy, B.A.; Kuhn, E.A.; Segelke, B.W.; Corzett, M.; Bench, G.; Mason, P.W.; Hoeprich, P.D. Immobilization of His-Tagged Proteins on Nickel-Chelating Nanolipoprotein Particles. *Bioconjugate Chem.* **2009**, *20*, 460–465. [CrossRef]
29. Horsey, A.J.; Briggs, D.A.; Holliday, N.D.; Briddon, S.J.; Kerr, I.D. Application of fluorescence correlation spectroscopy to study substrate binding in styrene maleic acid lipid copolymer encapsulated ABCG2. *Biochim. Biophys. Acta (BBA)-Biomembr.* **2020**, *1862*, 183218. [CrossRef]
30. Knowles, T.J.; Finka, R.; Smith, C.; Lin, Y.-P.; Dafforn, T.; Overduin, M. Membrane Proteins Solubilized Intact in Lipid Containing Nanoparticles Bounded by Styrene Maleic Acid Copolymer. *J. Am. Chem. Soc.* **2009**, *131*, 7484–7485. [CrossRef]
31. Dörr, J.M.; Koorengevel, M.C.; Schäfer, M.; Prokofyev, A.V.; Scheidelaar, S.; van der Cruijsen, E.A.W.; Dafforn, T.R.; Baldus, M.; Killian, J.A. Detergent-free isolation, characterization, and functional reconstitution of a tetrameric K^+ channel: The power of native nanodiscs. *Proc. Natl. Acad. Sci. USA* **2014**, *111*, 18607–18612. [CrossRef]
32. Pollock, N.L.; Lee, S.C.; Patel, J.; Gulamhussein, A.A.; Rothnie, A.J. Structure and function of membrane proteins encapsulated in a polymer-bound lipid bilayer. *Biochim. Biophys. Acta (BBA)-Biomembr.* **2018**, *1860*, 809–817. [CrossRef]
33. Long, A.R.; O'Brien, C.C.; Malhotra, K.; Schwall, C.T.; Albert, A.D.; Watts, A.; Alder, N.N. A detergent-free strategy for the reconstitution of active enzyme complexes from native biological membranes into nanoscale discs. *BMC Biotechnol.* **2013**, *13*, 41. [CrossRef] [PubMed]
34. Brown, C.J.; Trieber, C.; Overduin, M. Structural biology of endogenous membrane protein assemblies in native nanodiscs. *Curr. Opin. Struct. Biol.* **2021**, *69*, 70–77. [CrossRef]
35. Logez, C.; Damian, M.; Legros, C.; Dupré, C.; Guéry, M.; Mary, S.; Wagner, R.; M'Kadmi, C.; Nosjean, O.; Fould, B.; et al. Detergent-free Isolation of Functional G Protein-Coupled Receptors into Nanometric Lipid Particles. *Biochemistry* **2016**, *55*, 38–48. [CrossRef]
36. Günsel, U.; Hagn, F. Lipid Nanodiscs for High-Resolution NMR Studies of Membrane Proteins. *Chem. Rev. accepted.* **2021**. [CrossRef] [PubMed]

37. Bayburt, T.H.; Grinkova, Y.V.; Sligar, S.G. Self-Assembly of Discoidal Phospholipid Bilayer Nanoparticles with Membrane Scaffold Proteins. *Nano Lett.* **2002**, *2*, 853–856. [CrossRef]
38. Bayburt, T.H.; Sligar, S.G. Membrane protein assembly into Nanodiscs. *FEBS Lett.* **2010**, *584*, 1721–1727. [CrossRef] [PubMed]
39. Nasr, M.L.; Baptista, D.; Strauss, M.; Sun, Z.-Y.J.; Grigoriu, S.; Huser, S.; Plückthun, A.; Hagn, F.; Walz, S.H.T.; Hogle, J.M.; et al. Covalently circularized nanodiscs for studying membrane proteins and viral entry. *Nat. Methods* **2017**, *14*, 49–52. [CrossRef] [PubMed]
40. Nasr, M.L.; Wagner, G. Covalently circularized nanodiscs; challenges and applications. *Curr. Opin. Struct. Biol.* **2018**, *51*, 129–134. [CrossRef]
41. Cappuccio, J.A.; Hinz, A.K.; Kuhn, E.A.; Fletcher, J.E.; Arroyo, E.S.; Henderson, P.T.; Blanchette, C.D.; Walsworth, V.L.; Corzett, M.H.; Law, R.J.; et al. Cell-Free Expression for Nanolipoprotein Particles: Building a High-Throughput Membrane Protein Solubility Platform. *Methods Mol. Biol.* **2009**, *498*, 273–295. [CrossRef]
42. Cappuccio, J.A.; Blanchette, C.D.; Sulchek, T.A.; Arroyo, E.S.; Kralj, J.M.; Hinz, A.K.; Kuhn, E.A.; Chromy, B.A.; Segelke, B.W.; Rothschild, K.; et al. Cell-free Co-expression of Functional Membrane Proteins and Apolipoprotein, Forming Soluble Nanolipoprotein Particles. *Mol. Cell. Proteom.* **2008**, *7*, 2246–2253. [CrossRef]
43. Roos, C.; Kai, L.; Proverbio, D.; Ghoshdastider, U.; Filipek, S.; Dötsch, V.; Bernhard, F. Co-translational association of cell-free expressed membrane proteins with supplied lipid bilayers. *Mol. Membr. Biol.* **2013**, *30*, 75–89. [CrossRef] [PubMed]
44. Coleman, M.A.; Cappuccio, J.A.; Blanchette, C.D.; Gao, T.; Arroyo, E.S.; Hinz, A.K.; Bourguet, F.A.; Segelke, B.; Hoeprich, P.D.; Huser, T.; et al. Expression and Association of the Yersinia pestis Translocon Proteins, YopB and YopD, Are Facilitated by Nanolipoprotein Particles. *PLoS ONE* **2016**, *11*, e0150166. [CrossRef] [PubMed]
45. Gao, T.; Petrlova, J.; He, W.; Huser, T.; Kudlick, W.; Voss, J.; Coleman, M.A. Characterization of De Novo Synthesized GPCRs Supported in Nanolipoprotein Discs. *PLoS ONE* **2012**, *7*, e44911. [CrossRef] [PubMed]
46. Dathe, A.; Sielaff, H.; Heitkamp, T.; Börsch, M.; Pérez, I.; Westphal, A.; Reuter, S.; Mrowka, R. Observing monomer: Dimer transitions of neurotensin receptors 1 in single SMALPs by homoFRET and in an ABELtrap. In *Single Molecule Spectroscopy and Superresolution Imaging XII*; SPIE: Bellingham, WA, USA, 2019; Volume 10884, pp. 47–57. [CrossRef]
47. Gao, T.; Blanchette, C.D.; He, W.; Bourguet, F.; Ly, S.; Katzen, F.; Kudlicki, W.A.; Henderson, P.T.; Laurence, T.A.; Huser, T.; et al. Characterizing diffusion dynamics of a membrane protein associated with nanolipoproteins using fluorescence correlation spectroscopy. *Protein Sci.* **2011**, *20*, 437–447. [CrossRef] [PubMed]
48. Quinn, S.D.; Srinivasan, S.; Gordon, J.; He, W.; Carraway, K.L.; Coleman, M.A.; Schlau-Cohen, G.S. Single-Molecule Fluorescence Detection of the Epidermal Growth Factor Receptor in Membrane Discs. *Biochemistry* **2019**, *58*, 286–294. [CrossRef]
49. Unger, L.; Ronco-Campaña, A.; Kitchen, P.; Bill, R.M.; Rothnie, A.J. Biological insights from SMA-extracted proteins. *Biochem. Soc. Trans.* **2021**, *49*, 1349–1359. [CrossRef]
50. Simon, K.S.; Pollock, N.L.; Lee, S.C. Membrane protein nanoparticles: The shape of things to come. *Biochem. Soc. Trans.* **2018**, *46*, 1495–1504. [CrossRef]
51. Oluwole, A.O.; Danielczak, B.; Meister, P.-D.D.A.; Babalola, J.O.; Vargas, C.; Keller, S. Solubilization of Membrane Proteins into Functional Lipid-Bilayer Nanodiscs Using a Diisobutylene/Maleic Acid Copolymer. *Angew. Chem. Int. Ed.* **2017**, *56*, 1919–1924. [CrossRef]
52. Gulamhussein, A.A.; Uddin, R.; Tighe, B.J.; Poyner, D.R.; Rothnie, A.J. A comparison of SMA (styrene maleic acid) and DIBMA (di-isobutylene maleic acid) for membrane protein purification. *Biochim. Biophys. Acta (BBA)-Biomembr.* **2020**, *1862*, 183281. [CrossRef]
53. Lee, S.C.; Knowles, T.; Postis, V.; Jamshad, M.; Parslow, R.A.; Lin, Y.-P.; Goldman, A.; Sridhar, T.J.K.P.; Overduin, M.; Muench, S.P.; et al. A method for detergent-free isolation of membrane proteins in their local lipid environment. *Nat. Protoc.* **2016**, *11*, 1149–1162. [CrossRef]
54. Esmaili, M.; Overduin, M. Membrane biology visualized in nanometer-sized discs formed by styrene maleic acid polymers. *Biochim. Biophys. Acta (BBA)-Biomembr.* **2018**, *1860*, 257–263. [CrossRef] [PubMed]
55. Juarez, J.F.B.; Harper, A.J.; Judge, P.J.; Tonge, S.R.; Watts, A. From polymer chemistry to structural biology: The development of SMA and related amphipathic polymers for membrane protein extraction and solubilisation. *Chem. Phys. Lipids* **2019**, *221*, 167–175. [CrossRef] [PubMed]
56. Jamshad, M.; Charlton, J.; Lin, Y.-P.; Routledge, S.; Bawa, Z.; Knowles, T.; Overduin, M.; Dekker, N.; Dafforn, T.; Bill, R.; et al. G-protein coupled receptor solubilization and purification for biophysical analysis and functional studies, in the total absence of detergent. *Biosci. Rep.* **2015**, *35*, e00188. [CrossRef] [PubMed]
57. Ravula, T.; Hardin, N.Z.; Ramamoorthy, A. Polymer nanodiscs: Advantages and limitations. *Chem. Phys. Lipids* **2019**, *219*, 45–49. [CrossRef]
58. Flores, S.M.; Toca-Herrera, J.L. The new future of scanning probe microscopy: Combining atomic force microscopy with other surface-sensitive techniques, optical microscopy and fluorescence techniques. *Nanoscale* **2009**, *1*, 40–49. [CrossRef]
59. Magde, D.; Elson, E.; Webb, W.W. Thermodynamic Fluctuations in a Reacting System—Measurement by Fluorescence Correlation Spectroscopy. *Phys. Rev. Lett.* **1972**, *29*, 705–708. [CrossRef]
60. Eigen, M.; Rigler, R. Sorting single molecules: Application to diagnostics and evolutionary biotechnology. *Proc. Natl. Acad. Sci. USA* **1994**, *91*, 5740–5747. [CrossRef]

61. Schwille, P.; Haustein, E. *Fluorescence Correlation Spectroscopy: An Introduction to its Concepts and Applications*; Biophysical Society: Rockville, MD, USA, 2002.
62. Chenette, E.J. FCS and FRAP: Illuminating cellular processes. *Nat. Cell Biol.* **2009**, *11*, S13–S14. [CrossRef]
63. Koppel, D.; Axelrod, D.; Schlessinger, J.; Elson, E.; Webb, W. Dynamics of fluorescence marker concentration as a probe of mobility. *Biophys. J.* **1976**, *16*, 1315–1329. [CrossRef]
64. Roy, R.; Hohng, S.; Ha, T. A practical guide to single-molecule FRET. *Nat. Methods* **2008**, *5*, 507–516. [CrossRef]
65. Sasmal, D.K.; Pulido, L.E.; Kasal, S.; Huang, J. Single-molecule fluorescence resonance energy transfer in molecular biology. *Nanoscale* **2016**, *8*, 19928–19944. [CrossRef]
66. Mazal, H.; Haran, G. Single-molecule FRET methods to study the dynamics of proteins at work. *Curr. Opin. Biomed. Eng.* **2019**, *12*, 8–17. [CrossRef] [PubMed]
67. Krainer, G.; Keller, S.; Schlierf, M. Structural dynamics of membrane-protein folding from single-molecule FRET. *Curr. Opin. Struct. Biol.* **2019**, *58*, 124–137. [CrossRef] [PubMed]
68. Briddon, S.; Kilpatrick, L.E.; Hill, S.J. Studying GPCR Pharmacology in Membrane Microdomains: Fluorescence Correlation Spectroscopy Comes of Age. *Trends Pharmacol. Sci.* **2018**, *39*, 158–174. [CrossRef] [PubMed]
69. Briddon, S.; Hill, S. Pharmacology under the microscope: The use of fluorescence correlation spectroscopy to determine the properties of ligand–receptor complexes. *Trends Pharmacol. Sci.* **2007**, *28*, 637–645. [CrossRef] [PubMed]
70. Bacia, K.; Kim, S.A.; Schwille, P. Fluorescence cross-correlation spectroscopy in living cells. *Nat. Methods* **2006**, *3*, 83–89. [CrossRef]
71. Kim, S.A.; Heinze, K.G.; Schwille, P. Fluorescence correlation spectroscopy in living cells. *Nat. Methods* **2007**, *4*, 963–973. [CrossRef]
72. Meseth, U.; Wohland, T.; Rigler, R.; Vogel, H. Resolution of Fluorescence Correlation Measurements. *Biophys. J.* **1999**, *76*, 1619–1631. [CrossRef]
73. Widengren, J.; Mets, U.; Rigler, R. Fluorescence correlation spectroscopy of triplet states in solution: A theoretical and experimental study. *J. Phys. Chem.* **1995**, *99*, 13368–13379. [CrossRef]
74. Schwille, P.; Meyer-Almes, F.; Rigler, R. Dual-color fluorescence cross-correlation spectroscopy for multicomponent diffusional analysis in solution. *Biophys. J.* **1997**, *72*, 1878–1886. [CrossRef]
75. Kapanidis, A.N.; Laurence, T.A.; Lee, N.K.; Margeat, E.; Kong, A.X.; Weiss, S. Alternating-Laser Excitation of Single Molecules. *Accounts Chem. Res.* **2005**, *38*, 523–533. [CrossRef] [PubMed]
76. Müller, B.K.; Zaychikov, E.; Bräuchle, C.; Lamb, D.C. Pulsed Interleaved Excitation. *Biophys. J.* **2005**, *89*, 3508–3522. [CrossRef] [PubMed]
77. Miller, A.E.; Hollars, C.W.; Lane, S.M.; Laurence, T.A. Fluorescence Cross-Correlation Spectroscopy as a Universal Method for Protein Detection with Low False Positives. *Anal. Chem.* **2009**, *81*, 5614–5622. [CrossRef] [PubMed]
78. Ruan, Q.; Cheng, M.A.; Levi, M.; Gratton, E.; Mantulin, W.W. Spatial-Temporal Studies of Membrane Dynamics: Scanning Fluorescence Correlation Spectroscopy (SFCS). *Biophys. J.* **2004**, *87*, 1260–1267. [CrossRef] [PubMed]
79. Krieger, J.W.; Singh, A.P.; Bag, N.; Garbe, C.S.; Saunders, T.; Langowski, J.; Wohland, T. Imaging fluorescence (cross-) correlation spectroscopy in live cells and organisms. *Nat. Protoc.* **2015**, *10*, 1948–1974. [CrossRef]
80. Brown, C.M.; Dalal, R.B.; Hebert, B.; Digman, M.A.; Horwitz, A.R.; Gratton, E. Raster image correlation spectroscopy (RICS) for measuring fast protein dynamics and concentrations with a commercial laser scanning confocal microscope. *J. Microsc.* **2008**, *229*, 78–91. [CrossRef]
81. Ries, J.; Schwille, P. Fluorescence correlation spectroscopy. *BioEssays* **2012**, *34*, 361–368. [CrossRef]
82. Enderlein, J.; Gregor, I.; Patra, D.; Dertinger, T.; Kaupp, U.B. Performance of Fluorescence Correlation Spectroscopy for Measuring Diffusion and Concentration. *ChemPhysChem* **2005**, *6*, 2324–2336. [CrossRef]
83. Kahya, N.; Schwille, P. Fluorescence correlation studies of lipid domains in model membranes (Review). *Mol. Membr. Biol.* **2006**, *23*, 29–39. [CrossRef]
84. Chiantia, S.; Ries, J.; Schwille, P. Fluorescence correlation spectroscopy in membrane structure elucidation. *Biochim. Biophys. Acta (BBA)-Biomembr.* **2009**, *1788*, 225–233. [CrossRef]
85. Hernández-Rocamora, V.M.; Reija, B.; García, C.; Natale, P.; Alfonso, C.; Minton, A.P.; Zorrilla, S.; Rivas, G.; Vicente, M. Dynamic Interaction of the Escherichia coli Cell Division ZipA and FtsZ Proteins Evidenced in Nanodiscs. *J. Biol. Chem.* **2012**, *287*, 30097–30104. [CrossRef] [PubMed]
86. Wu, Z.C.; de Keyzer, J.; Kedrov, A.; Driessen, A.J. Competitive Binding of the SecA ATPase and Ribosomes to the SecYEG Translocon. *J. Biol. Chem.* **2012**, *287*, 7885–7895. [CrossRef] [PubMed]
87. Kedrov, A.; Sustarsic, M.; de Keyzer, J.; Caumanns, J.J.; Wu, Z.C.; Driessen, A.J. Elucidating the Native Architecture of the YidC: Ribosome Complex. *J. Mol. Biol.* **2013**, *425*, 4112–4124. [CrossRef]
88. Geng, Y.; Kedrov, A.; Caumanns, J.J.; Crevenna, A.; Lamb, D.; Beckmann, R.; Driessen, A.J.M. Role of the Cytosolic Loop C2 and the C Terminus of YidC in Ribosome Binding and Insertion Activity. *J. Biol. Chem.* **2015**, *290*, 17250–17261. [CrossRef]
89. Ly, S.; Bourguet, F.; Fischer, N.O.; Lau, E.Y.; Coleman, M.A.; Laurence, T.A. Quantifying Interactions of a Membrane Protein Embedded in a Lipid Nanodisc using Fluorescence Correlation Spectroscopy. *Biophys. J.* **2014**, *106*, L05–L08. [CrossRef] [PubMed]
90. Li, M.J.; Nath, A.; Atkins, W. Differential Coupling of Binding, ATP Hydrolysis, and Transport of Fluorescent Probes with P-Glycoprotein in Lipid Nanodiscs. *Biochemistry* **2017**, *56*, 2506–2517. [CrossRef]

91. Grime, R.L.; Goulding, J.; Uddin, R.; Stoddart, L.A.; Hill, S.J.; Poyner, D.R.; Briddon, S.J.; Wheatley, M. Single molecule binding of a ligand to a G-protein-coupled receptor in real time using fluorescence correlation spectroscopy, rendered possible by nano-encapsulation in styrene maleic acid lipid particles. *Nanoscale* **2020**, *12*, 11518–11525. [CrossRef] [PubMed]
92. Voskoboynikova, N.; Karlova, M.; Kurre, R.; Mulkidjanian, A.; Shaitan, K.; Sokolova, O.; Steinhoff, H.-J.; Heinisch, J. A Three-Dimensional Model of the Yeast Transmembrane Sensor Wsc1 Obtained by SMA-Based Detergent-Free Purification and Transmission Electron Microscopy. *J. Fungi* **2021**, *7*, 118. [CrossRef]
93. Xu, L.; Öjemyr, L.N.; Bergstrand, J.; Brzezinski, P.; Widengren, J. Protonation Dynamics on Lipid Nanodiscs: Influence of the Membrane Surface Area and External Buffers. *Biophys. J.* **2016**, *110*, 1993–2003. [CrossRef]
94. Harayama, T.; Riezman, H. Understanding the diversity of membrane lipid composition. *Nat. Rev. Mol. Cell Biol.* **2018**, *19*, 281–296. [CrossRef]
95. Stillwell, W. *An Introduction to Biological Membranes: Composition, Structure and Function*; Elsevier: Amsterdam, The Netherlands, 2016.
96. Dawaliby, R.; Trubbia, C.; Delporte, C.; Masureel, M.; Van Antwerpen, P.; Kobilka, B.B.K.; Govaerts, C. Allosteric regulation of G protein–coupled receptor activity by phospholipids. *Nat. Chem. Biol.* **2016**, *12*, 35–39. [CrossRef]
97. Inagaki, S.; Ghirlando, R.; White, J.F.; Gvozdenovic-Jeremic, J.; Northup, J.K.; Grisshammer, R. Modulation of the Interaction between Neurotensin Receptor NTS1 and Gq Protein by Lipid. *J. Mol. Biol.* **2012**, *417*, 95–111. [CrossRef] [PubMed]
98. Zocher, M.; Zhang, C.; Rasmussen, S.G.F.; Kobilka, B.K.; Müller, D.J. Cholesterol increases kinetic, energetic, and mechanical stability of the human β_2-adrenergic receptor. *Proc. Natl. Acad. Sci. USA* **2012**, *109*, E3463–E3472. [CrossRef] [PubMed]
99. Gutierrez, M.G.; Mansfield, K.S.; Malmstadt, N. The Functional Activity of the Human Serotonin 5-HT 1A Receptor Is Controlled by Lipid Bilayer Composition. *Biophys. J.* **2016**, *110*, 2486–2495. [CrossRef] [PubMed]
100. Yoshida, K.; Nagatoishi, S.; Kuroda, D.; Suzuki, N.; Murata, T.; Tsumoto, K. Phospholipid Membrane Fluidity Alters Ligand Binding Activity of a G Protein-Coupled Receptor by Shifting the Conformational Equilibrium. *Biochemistry* **2019**, *58*, 504–508. [CrossRef] [PubMed]
101. Rosholm, K.R.; Leijnse, N.; Mantsiou, A.; Tkach, V.; Pedersen, S.L.; Wirth, V.F.; Oddershede, L.B.; Jensen, K.; Martinez, K.; Hatzakis, N.S.; et al. Membrane curvature regulates ligand-specific membrane sorting of GPCRs in living cells. *Nat. Chem. Biol.* **2017**, *13*, 724–729. [CrossRef]
102. Wootten, D.; Christopoulos, A.; Marti-Solano, M.; Babu, M.M.; Sexton, P.M. Mechanisms of signalling and biased agonism in G protein-coupled receptors. *Nat. Rev. Mol. Cell Biol.* **2018**, *19*, 638–653. [CrossRef]
103. Marrink, S.J.; Corradi, V.; Souza, P.C.; Ingólfsson, H.I.; Tieleman, D.P.; Sansom, M.S. Computational Modeling of Realistic Cell Membranes. *Chem. Rev.* **2019**, *119*, 6184–6226. [CrossRef]
104. Shahane, G.; Ding, W.; Palaiokostas, M.; Orsi, M. Physical properties of model biological lipid bilayers: Insights from all-atom molecular dynamics simulations. *J. Mol. Model.* **2019**, *25*, 76. [CrossRef]
105. Ingólfsson, H.I.; Bhatia, H.; Zeppelin, T.; Bennett, W.F.D.; Carpenter, K.A.; Hsu, P.-C.; Dharuman, G.; Bremer, P.-T.; Schiøtt, B.; Lightstone, F.C.; et al. Capturing biologically complex tissue-specific membranes at different levels of compositional complexity. *J. Phys. Chem. B* **2020**, *124*, 7819–7829. [CrossRef]
106. Ingólfsson, H.I.; Carpenter, T.S.; Bhatia, H.; Bremer, P.-T.; Marrink, S.J.; Lightstone, F.C. Computational Lipidomics of the Neuronal Plasma Membrane. *Biophys. J.* **2017**, *113*, 2271–2280. [CrossRef] [PubMed]
107. Ingólfsson, H.I.; Melo, M.N.; van Eerden, F.J.; Arnarez, C.; Lopez, C.A.; Wassenaar, T.A.; Periole, X.; de Vries, A.H.; Tieleman, D.P.; Marrink, S.J. Lipid Organization of the Plasma Membrane. *J. Am. Chem. Soc.* **2014**, *136*, 14554–14559. [CrossRef] [PubMed]
108. Laurence, T.A.; Ly, S.; Bourguet, F.; Fischer, N.O.; Coleman, M.A. Fluorescence Correlation Spectroscopy at Micromolar Concentrations without Optical Nanoconfinement. *J. Phys. Chem. B* **2014**, *118*, 9662–9667. [CrossRef] [PubMed]
109. Schneider, F.; Hernandez-Varas, P.; Lagerholm, B.C.; Shrestha, D.; Sezgin, E.; Roberti, M.J.; Ossato, G.; Hecht, F.; Eggeling, C.; Urbančič, I. High photon count rates improve the quality of super-resolution fluorescence fluctuation spectroscopy. *J. Phys. D: Appl. Phys.* **2020**, *53*, 164003. [CrossRef] [PubMed]

membranes

MDPI

Article

Photoreaction Pathways of Bacteriorhodopsin and Its D96N Mutant as Revealed by in Situ Photoirradiation Solid-State NMR

Arisu Shigeta [1], Yuto Otani [1], Ryota Miyasa [1], Yoshiteru Makino [1], Izuru Kawamura [1,*], Takashi Okitsu [2], Akimori Wada [2] and Akira Naito [1,*]

[1] Graduate School of Engineering, Yokohama National University, 79-5 Tokiwadai, Hodogaya-ku, Yokohama 240-8501, Japan; arisushigeta@gmail.com (A.S.); otani.yuto.dv@gmail.com (Y.O.); r.miyasa@gmail.com (R.M.); ymakino@med.kobe-u.ac.jp (Y.M.)

[2] Laboratory of Organic Chemistry for Life Science, Kobe Pharmaceutical University, 4-19-1 Motoyamakitamachi, Higashinada-ku, Kobe 658-8558, Japan; okitsu@kobepharma-u.ac.jp (T.O.); w.akimori5139@gmail.com (A.W.)

* Correspondence: kawamura-izuru-wx@ynu.ac.jp (I.K.); naito@ynu.ac.jp (A.N.)

Abstract: Bacteriorhodopsin (BR) functions as a light-driven proton pump that transitions between different states during the photocycle, such as *all-trans* (AT; BR568) and 13-*cis*, 15-*syn* (CS; BR548) state and K, L, M_1, M_2, N, and O intermediates. In this study, we used in situ photoirradiation ^{13}C solid-state NMR to observe a variety of photo-intermediates and photoreaction pathways in [20-^{13}C]retinal-WT-BR and its mutant [20-^{13}C, 14-^{13}C]retinal-D96N-BR. In WT-BR, the CS state converted to the CS* intermediate under photoirradiation with green light at −20 °C and consequently converted to the AT state in the dark. The AT state converted to the N intermediate under irradiation with green light. In D96N-BR, the CS state was converted to the CS* intermediate at −30 °C and consequently converted to the AT state. Simultaneously, the AT state converted to the M and L intermediates under green light illumination at −30 °C and subsequently converted to the AT state in the dark. The M intermediate was directly excited to the AT state by UV light illumination. We demonstrated that short-lived photo-intermediates could be observed in a stationary state using in situ photoirradiation solid-state NMR spectroscopy for WT-BR and D96N-BR, enabling insight into the light-driven proton pump activity of BR.

Keywords: bacteriorhodopsin; light driven proton pump; photo-intermediate; photoirradiation solid-state NMR

1. Introduction

Bacteriorhodopsin (BR) is a retinal-containing transmembrane H^+ pump found in the purple membrane (PM) of the archaeon *Halobacterium salinarum*, where it generates a proton gradient across the cytoplasmic membrane for ATP synthesis [1–3]. The 26 kDa BR protein contains seven transmembrane α-helices and a retinal group covalently bonded to the ε-amino group of Lys216 in helix G. In the extracellular region, an extensive three-dimensional hydrogen-bonded network of protein residues and seven water molecules connects the buried retinal Schiff base (SB) and the proton acceptor Asp85 to the membrane surface [4]. Energy from the light absorbed by the protein causes isomerization of the retinal from all-*trans* (AT; BR568) to 13-*cis*, 15-*anti* (CS; BR548) configurations to induce conformational changes in the protein, including a change in the tilt angle of the E and F helices. These structural changes suggest that protonation of Asp85 initiates a cascade of atomic displacements in the extracellular region, causing the release of a proton to the surface (see Figure 1A) [5,6].

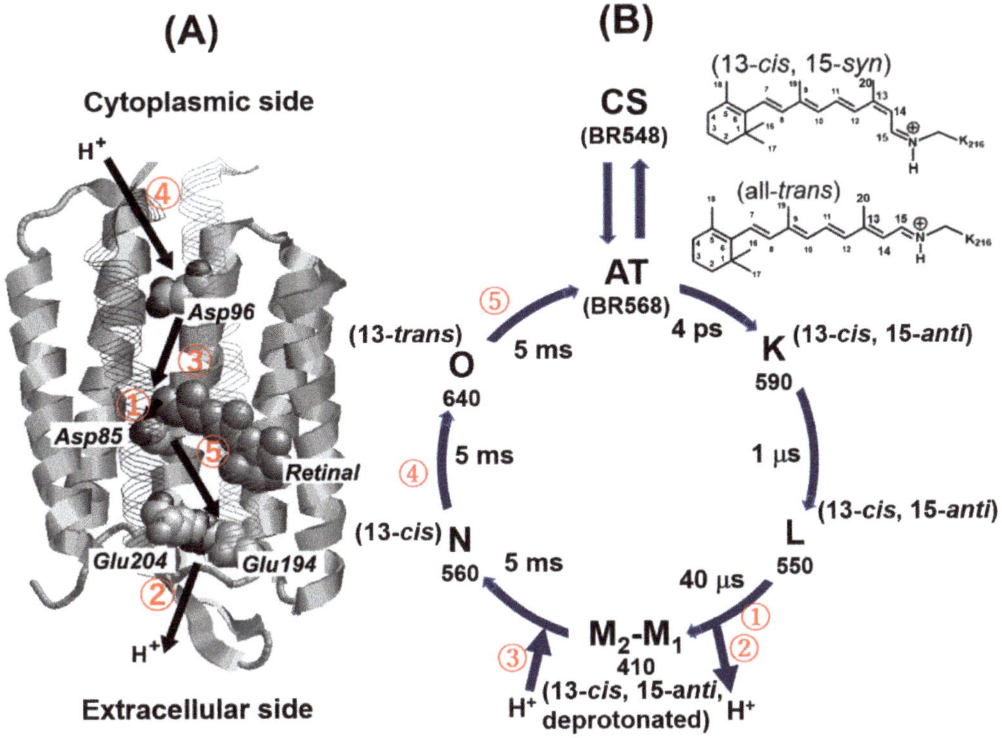

Figure 1. (**A**) Structure and proton transport pathways of bacteriorhodopsin (BR) (PDB: 1C3W). (**B**) Schematic of the BR photocycle. Approximate half-life for each transition and the absorption maximum of each intermediate are shown for the K, L, M_1 (early M), M_2 (late M), N, and O intermediates [7].

Recently, a time-resolved serial femtosecond crystallographic structural study of BR using an X-ray-free electron laser revealed structural events that occur nanoseconds and milliseconds after photoactivation [7–9]. This time-resolved structural analysis revealed the sub-picosecond to millisecond retinal isomerization, proton release, proton uptake steps, and how the structural changes in BR achieve uni-directional membrane transport of H^+ against a transmembrane concentration gradient.

The ordered sequence of conformational changes that occurs in BR as a result of light absorption is known as the photocycle (see Figure 1B) [7]. Upon illumination with green light (560 nm) at room temperature, the population of the AT state increases, a condition known as the light-adapted (LA) state. During the AT photocycle, the AT state is excited to the K-intermediate under green light illumination and is consequently relaxed through the L, M_1 (early M) and M_2 (late M), N, and O intermediates and ultimately returns to the AT (BR_{568}) state (see Figure 1B).

During transition from the L to M intermediates, a H^+ is transported from a SB to Asp85 via W402 (step ①), and a H^+ is simultaneously released from a proton release group (PRG) such as Glu194 or Glu204 on the extracellular side (step ②). During transition from the M to N intermediate, a H^+ is transported from Asp96 to a protonated Schiff base (SBH), which causes re-protonation of the SB (step ③). In the transition from the N to O intermediate, a H^+ is transported from the cytoplasmic side to Asp96, which causes uptake of H^+ from the cytoplasmic side to BR (step ④). In the last step of the photocycle, Asp85 re-protonates a PRG such as Glu194 or Glu204 in the transition from the O intermediate to the AT state (step ⑤). These stepwise changes in the sidechains induce a change in the pKa value (of up to 3 pH units) that results in H^+ flux to the extracellular side of BR (see Figure 1A) [6–11].

In contrast to the AT photo-cycle that is related to proton pump activity, the photoreaction of the CS state (CS photocycle) has not been extensively studied. In previous studies [12–15], the CS state was irradiated with an Nd laser (530 nm) and an intermediate with a maximum absorbance of 610 nm was detected in the CS photocycle. As the same intermediate is not detected in the AT photocycle, it was thus determined to be an intermediate transformed from the CS state and assigned as the batho-13-*cis*-bR [13] or "610"-intermediate [12]. The intermediates are subsequently converted to the AT state through a leaking process from the CS photocycle. Therefore, one of the objectives in this study is to observe the photo-intermediates using in situ photoirradiation solid-state NMR of the CS photocycle with atomic resolution.

When the O intermediate transforms to the AT state in the photocycle of BR, a H^+ is transported from Asp85 to a PRG through a hydrogen bond network consisting of Asp212 and Tyr185. As breakdown of the hydrogen bond network with the substitution of Tyr185 to Phe185 prevents H^+ transport, it is thus predicted that the half-life of the O intermediate is significantly increased [16]. In a spectroscopic study of the Y185F-BR mutant, the dark-adapted (DA) state consisted of a CS:AT molar ratio of 3:1 [17,18]. Light-adaptation produces the AT state and the O intermediate [19]. Photoexcitation at −103 and −23 °C of the O intermediate produces an N intermediate, while the M intermediate is not found [20,21].

In the photoreaction cycle shown in Figure 1B, the photoreaction L ↔ M refers to proton exchange between SB and Asp85, while M ↔ N refers to proton exchange between Asp96 and SB. Therefore, replacement of Asp96 with Asn96 (D96N-BR) prevents the accumulation of any intermediate subsequent to the M intermediates, such as the N and O intermediates, while retaining proton pump activity [22–25]. Thus, the half-lives of the M and L intermediates are significantly increased in the D96N-BR mutant.

It is important to characterize the photo-intermediates to clarify the light-driven proton pump activity of BR. Photo-intermediates are normally detected based on the wavelength of the absorbance maximum using flash photolysis with pulsed laser light. For molecular characterization, solid-state NMR spectroscopy provides a powerful alternative for detecting intermediates during the photocycle. The L, M, and N intermediates have been detected by measuring ^{13}C NMR signals of [^{13}C]retinal-BR as well as ^{15}N NMR signals of [ε-^{15}N]Lys^{216}-BR at a temperature range of 0 to −120 °C under illumination with light (λ > 540 for N and M, λ > 610 nm for L) [26,27]. The 13-*cis* form of [20-^{13}C]retinal-BR can be distinguished from the 13-*trans* form, and the 15-*syn* form of [14-^{13}C]retinal-BR can be distinguished from the 15-*anti* configurations based on the γ-position effects [28–32]. In another study, two protein structures corresponding to the AT and CS state retinal configurations were detected by observing rotational echo double resonance (REDOR) filtered ^{13}C NMR signals of [1-^{13}C]Tyr185-BR [33].

An in situ photoirradiation apparatus for the solid-state NMR spectrometer has been developed in which light passes through an optical fiber in the probe and the sample is irradiated from outside the rotor tube [31,34]. This system has been used to characterize an early M intermediate in the BR photocycle. The intermediate of the BR was trapped by irradiation of the LA state with light at >610 nm in a temperature range of −95 to −130 °C for at least 2 h [26,27]. In combination with the dynamic nuclear polarization (DNP) method, the heterogeneity of the dark-adapted BR and distortion in the K intermediates were revealed, and four distinctive L intermediates were detected [32,35]. In addition, photoirradiation DNP-enhanced solid-state NMR has been used to detect the photoreaction site of channelrhodopsin-2 combined with ^{13}C double quantum filtering experiments [36].

Another type of in situ photoirradiation solid-state NMR apparatus was developed by our group, in which the sample is irradiated from inside the rotor tube through a glass rod inserted into the rotor [33,37–45]. This allows irradiation of the sample with extremely high efficiency and enables observation of the photo-intermediates as a stationary trapped state and the photoreaction processes of photoreceptor membrane proteins. In situ photoirradiation is particularly useful for the study of photo-cycles of retinal proteins such as sensory rhodopsin I (SR I) [38] and sensory rhodopsin II (SR II) [37,39], in which the late intermediates have long half-lives.

In this study, we focused on real-time detection of the photo-intermediate of WT-BR and D96N-BR using in situ photoirradiation solid-state NMR spectroscopy. Photoreaction pathways under photo-illumination with green and UV lights can be detected at low temperatures in such experiments. Notably, the photocycles of the CS and AT states were simultaneously investigated using in situ photoirradiation solid-state NMR experiments. Our aim is to gain insight into the structures and functions of the photo-intermediates using this methodology.

2. Materials and Methods

2.1. Materials

The *Halobacterium salinarum* retinal-deficient strain E1001 was grown in a temporary medium with the exogenous addition of [20-^{13}C]retinal at a final concentration of 5 μM to yield selectively labeled [20-^{13}C]retinal-WT-BR in the purple membrane (PM). Consequently, the PM with WT-BR was isolated according to a previously reported method [46].

The *H. salinarum* transformant for the D96N mutant, which was constructed using site-directed mutagenesis of bacterio-opsin, was used in this study. (The transformants were a gift by Dr. Satoru Tuzi). This *H. salinarum* (D96N-BR) strain was grown in a temporary medium containing [ε-^{15}N]-Lys to yield selectively labeled [ε-^{15}N]-Lys-D96N-BR in the PM. The PM was then isolated according to a previously reported method [46].

D96N bacterio-opsins (D96N-BOs) were prepared by photo-bleaching of the corresponding D96N-BR in hydroxylamine solution (500 mM, pH 7) at 4 °C under white light illumination for one day. Isotopically labeled regenerated [20-^{13}C, 14-^{13}C]retinal-D96N-BRs were obtained by the addition of [20-^{13}C, 14-^{13}C]retinal to the D96N-BOs obtained from bleached D96N-BR in a process termed in vitro retinal reconstitution [47]. Retinal bleaching and reconstitution were confirmed by measuring the absorbance of D96N-BR.

All BR samples were then suspended in 5 mM *N*-(2-hydroxyethyl)piperazine-*N*′-2-ethan sulfonic acid (HEPES) buffer and 10 mM NaCl at pH 7. For NMR measurements, 5 mM Tris buffer with 300 mM NaCl at pH 9.0 was used.

2.2. Methods

2.2.1. ^{13}C and ^{15}N Solid-State NMR Experiments

High-resolution solid-state ^{13}C and ^{15}N NMR spectra were recorded by cross polarization-magic angle spinning (CP-MAS) on a Chemagnetics CMX-400 Infinity FT-NMR spectrometer operated at 100.1 MHz for ^{13}C, 40.3 MHz for ^{15}N and 398.1 MHz for ^{1}H. The spinning frequency was set at 4 kHz, and the probe temperature was kept at 20, −20, −30 and −60 °C using a temperature controller. The ^{13}C chemical shifts were externally referenced to 176.03 ppm of glycine carbonyl carbon with respect to the tetramethylsilane (TMS) signal at 0 ppm. The ^{15}N chemical shift value was referenced to 11.59 ppm of the amino nitrogen of glycine as an external reference that corresponds to $^{15}NH_4NO_3$ at 0 ppm.

2.2.2. In Situ Photoirradiation Solid-State NMR Experiments

In situ photoirradiation was carried out using an optical fiber from outside the magnet through a tightly sealed cap made of a glass rod glued to a zirconia rotor (Figure S1). Photoirradiation was performed from the top of the spinner without touching the optical fiber to the glass cap of the rotor [37,38]. In this system, MAS operates under photo-illumination to generate high-resolution solid-state NMR signals. Thus, the MAS frequency was set at 4 kHz. The tip of the glass cap was inserted deeply and was ground to cast the light in random directions, including perpendicular to the glass rod. This allows the light to penetrate into samples that form a film on the inner wall from top to bottom of the rotor in which the absorbance is quite uniform and high (Figure S1). Using a CMX-400 Infinity NMR spectrometer equipped with this photoirradiation system, we were able to irradiate samples in the rotor with 100 mW green LED light (520 nm) and 50 mW UV LED light (365 nm) sources [37,38].

In the NMR measurements, a CP pulse sequence was used with a contact time of 1 ms, followed by acquisition with a 50 kHz TPPM proton decoupling pulse [48]. Typically, 20,000 transients were accumulated for the dark condition and light illumination with 520 or 365 nm LED light.

2.2.3. Observation of Photo-Intermediates and Photoreaction Pathways of BR

In the first step, the photoreaction pathways of the dark-adapted (DA) state which consists of AT and CS states of BR with a molar ration of ~ 1:1, were observed under dark conditions (D1) and subsequently observed under photo-illumination conditions (L1). In this experiment, the photoreaction pathways starting from the AT and CS states can be observed simultaneously. In the second step, the photo-reaction pathways of the light-adapted (LA) state, which consists of only the AT state, were observed, so the photoreaction pathway starting from the AT state alone can be clearly observed.

When BR is irradiated with green LED light, retinal absorbs the light energy and is excited from the AT state to the K intermediate and then subsequently relaxes to the L, M, N and O intermediates and finally returns to the AT state; this pathway is referred to as the photocycle (Figure 1B). It has been reported that the N intermediate exhibits a longer half-life in BR at −20 °C [27]. In this case, the N intermediate can be trapped in a stationary phase under continuous light illumination, as the light-activated production rate of the N intermediate is possibly faster than the relaxation decay rate. Thus, in a photocycle system, the photo-activation pathway of BR can be observed by accumulating the ^{13}C CP-MAS NMR spectra initially under a dark condition (D1) and subsequently under irradiation with green LED light (L1) using the same acquisition parameters. The differences in spectra (L1-D1) between the dark (D1) and green light (L1) conditions provide a background-free spectrum, and photo-reacted 20-^{13}C and 14-^{13}C retinal signals distinctively appeared without protein and lipid signals, in which the reactant species exhibit negative peaks and the product species exhibit positive peaks. Thus, the signal assignments of photo intermediates and photoreaction pathways can be clearly characterized by inspecting the differences in spectra.

The retinal configuration of photo-intermediates can be determined by observing [20-^{13}C]retinal-WT-BR and [20-^{13}C, 14-^{13}C]retinal-D96N-BR. Chemical shift values of [20-^{13}C]retinal distinguish between 13-*trans* and 13-*cis*, and [14-^{13}C]retinal distinguishes between 15-*anti* and 15-*syn* configurations.

3. Results and Discussion

3.1. Photoreaction Pathway of [20-^{13}C]retinal-BR under Illumination with Green Light at 20 °C

The DA state of BR was prepared by keeping the sample in the dark at 20 °C for one day. Afterward, ^{13}C CP MAS NMR spectra of [20-^{13}C]retinal-BR in the DA state were observed in the dark at 20 °C as shown in Figure 2A(D1). A CS signal appeared at 22.1 ppm and an AT signal appeared at 13.1 ppm (Table 1) with an intensity ratio of 1:1. When illuminating with green light, the CS signal decreased and the AT signal increased (Figure 2A(L1)). The difference in spectra between L1 and D1 is shown in Figure 2A(L1-D1). As mentioned in the Methods section, a negative peak represents reactant species (CS state) and a positive peak represents product species (AT state). Thus, the CS state converted completely to the AT state upon illumination with green light at 20 °C to produce an LA state (L1). After the sample was subsequently kept in the dark for one day, NMR signals were acquired to observe the DA state, as shown in Figure 2B(D2). Consequently, the difference in spectra between D2 and L1 was obtained, as shown in Figure 2B(D2-L1) which shows that one-half of the AT state changed to the CS state. In summary, only the CS state in the DA state changed to the AT state to produce the LA state under illumination with green light, and one-half of the AT state in the LA state changed to the CS state to produce the DA state under dark conditions.

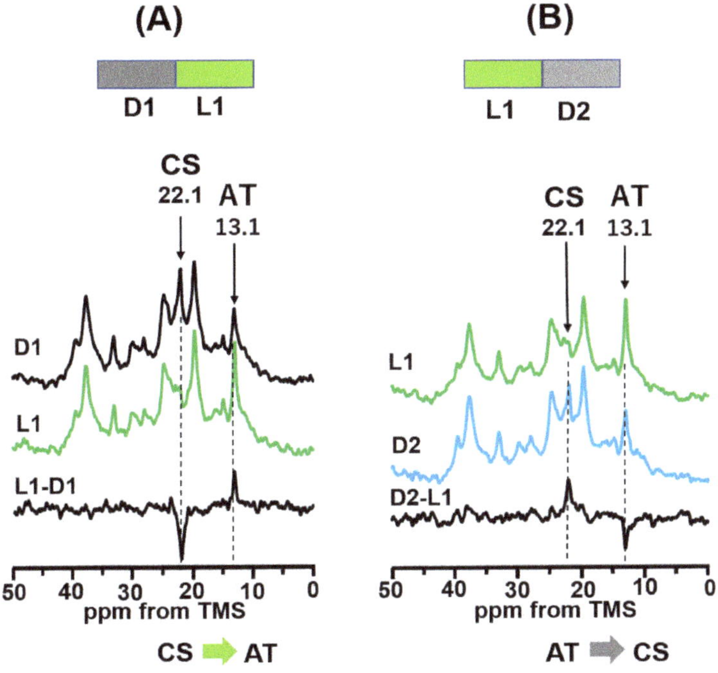

Figure 2. (**A**) Photo reaction pathway of [20-^{13}C]retinal-WT-BR under green light (520 nm) illumination at 20 °C. ^{13}C CP-MAS NMR spectra were recorded in the dark (D1; 20-^{13}C) and under green light (L1; 20-^{13}C), and the difference in spectra between L1 and D1 (L1-D1; 20-^{13}C) indicated that the CS state changed to the AT state. (**B**) Photoreaction pathway in the dark after green light illumination at 20 °C. ^{13}C CP-MAS spectra were recorded under green light (L1; 20-^{13}C) and in the dark one day after L1 (D2; 20-^{13}C), and the difference in spectra between D2 and L1 (D2-L1; 20-^{13}C) indicated that the AT state changed to the CS state.

Table 1. ^{13}C and ^{15}N chemical shift values (ppm) of individual states of [20-^{13}C, 14-^{13}C]-retinal and [ε-^{15}N]Lys-bacteriorhodopsin(WT-BR) and D96N-BR.

Protein	State	Temp.	20-^{13}C	14-^{13}C	[ε-^{15}N]Lys^{216}	Configuration
WT-BR	AT(BR568)	20 °C	13.1			all-*trans*
		−20 °C	13.3			all-*trans*
	CS(BR548)	20 °C	22.1			13-*cis*
		−20 °C	22.3			13-*cis*
	CS*	−20 °C	19.6			13-*cis*
	N	−20 °C	19.8			13-*cis*
D96N	AT(BR568)	−30 °C	13.1	121.8		all-*trans*
		−60 °C			148	all-*trans*
	CS(BR548)	−30 °C	22.6	110.1		13-*cis*, 15-*syn*
		−60 °C			153	13-cis, 15-syn
	CS*	−30 °C	20.0	115.8		13-*cis*
	M	−30 °C	20.0	124.1		13-*cis*, 15-*anti*
		−60 °C			300	13-cis, 15-*anti*
	L	−30 °C	18.6	115.8		13-*cis*,

CS*: State of intermediate that is transformed from CS state under green light illumination.

3.2. Photoreaction Pathways of [20-^{13}C]retinal-BR under Illumination with Green Light at −20 °C

After the WT-BR was kept in the dark at 20 °C for one day to produce the DA state, the ^{13}C CP-MAS NMR spectrum of [20-^{13}C]retinal-WT-BR was observed in the dark at −20 °C as shown in Figure 3A(D1). CS and AT signals appeared at 22.3 and 13.3 ppm, respectively, in the dark-adapted (DA) state (Table 1). A signal (*) appearing at 20.0 ppm was assigned to the alkyl carbons of lipids in the PM. Subsequently, Figure 3A(L1) was recorded under illumination with green light at −20 °C to observe the photoreaction of WT-BR from the DA state. The CS state was significantly reduced, and * signals increased significantly.

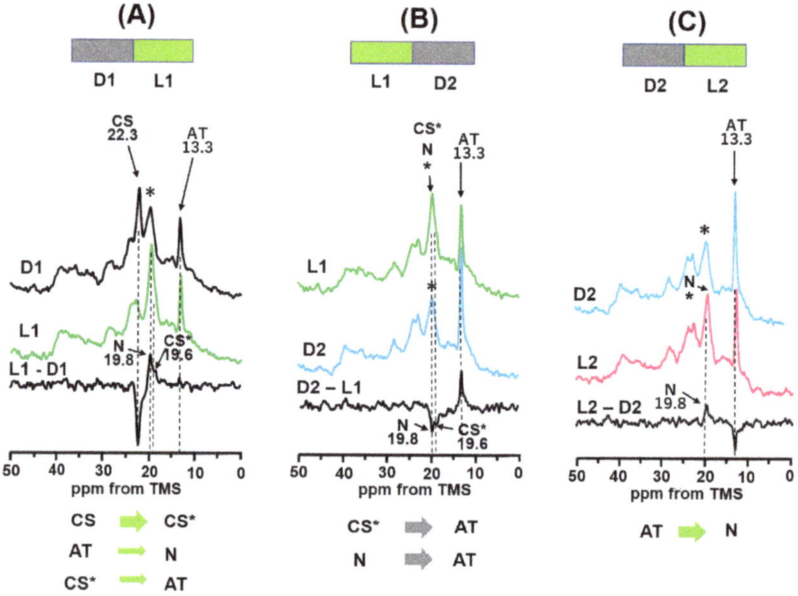

Figure 3. (**A**) Photo reaction pathway of [20-^{13}C]retinal-WT-BR under green light (520 nm) illumination at −20 °C. ^{13}C CP-MAS NMR spectra were recorded in the dark (D1; 20-^{13}C) and subsequently under green light (L1; 20-^{13}C), and the difference in spectra between L1 and D1 (L1-D1; 20-^{13}C) indicated that the CS state changed to the CS* intermediate. (**B**) Photoreaction pathway in the dark after green light illumination at −20 °C. ^{13}C CP-MAS NMR spectra were recorded under green light (L1; 20-^{13}C) and subsequently in the dark after L1 (D2; 20-^{13}C), and the difference in spectra between the D2 and L1 spectra (D2-L1; 20-^{13}C) indicated that the CS* intermediate changed to the AT state. (**C**) Photoreaction pathway under green light illumination at −20 °C after the D2 state. ^{13}C CP-MAS NMR spectra were recorded in the dark (D2; 20-^{13}C) after L1 and subsequently under green light (L2; 20-^{13}C), and the difference in spectra between L2 and D2 (L2-D2; 20-^{13}C) indicated that the AT state changed to the N intermediate.

Figure 3A(L1-D1) shows the difference in spectra between L1 and D1 and indicates that the positive major signals of the CS* intermediate appeared at 19.6 ppm, and a positive minor signal of the N intermediate appeared at 19.8 ppm at −20 °C; these were not present at 20 °C. Specifically, the CS state is transformed to the CS* intermediate under green light illumination. The CS* intermediate forms a 13-*cis* configuration in reference to the chemical shift value of 20-^{13}C as shown in Table 1. Thus, we demonstrated that the ^{13}C NMR signal of the short-lived CS* intermediate could be observed in a stationary state at −20 °C using in situ photoirradiation solid-state NMR. Simultaneously, the N intermediate was transformed from the AT state during the AT photocycle as described later.

After illumination with green light, signals were acquired in dark conditions (Figure 3B(D2)). We found that the CS* and N intermediates were reduced, and the AT state was increased.

Thus, the difference in spectra between D2 and L1 (Figure 2B(D2-L1)) clearly shows that the CS* and N intermediates were converted to the AT state to produce an LA state (D2).

This LA state was again illuminated with green light to observe the AT photocycle independently, as shown in Figure 3C(L2). The difference in spectra between L2 and D2 (Figure 3C(L2-D2)) indicates that a small fraction of the AT state converted to the N intermediate at 19.8 ppm. This result indicates that the AT photocycle exists, although only the N intermediate was trapped at −20 °C. However, the other intermediates (K, L, M and O) could not be detected at −20 °C, because their half-lives are too short to be detected using in situ photoirradiation solid-state NMR. It is thus noted that transition from the AT state to the N intermediate occurred in the process from D1 to L1. This result makes it clear that a positive peak at ~19.6 ppm (CS*) in Figure 3A(L1-D1) overlapped with the peak of the N intermediate at 19.8 ppm. In addition, the leakage process from the CS* to AT state coexisted under green light illumination. This resulted in an NMR peak for the AT state that apparently did not change, as shown in Figure 3A(L1-D1).

In the CS photocycle, the CS state changed to the CS* intermediate under illumination with green light, and simultaneously the CS* intermediate partly changed to the AT state in the leakage process from the CS photocycle (Figure 4B). Subsequently, the CS* intermediate changed to the AT state in the dark to produce the LA state. In the AT photocycle, although the AT state changed to the K, L, M, N, and O intermediates and finally changed to the AT state at −20 °C, only the N intermediate was detected by in situ photoirradiation solid state NMR.

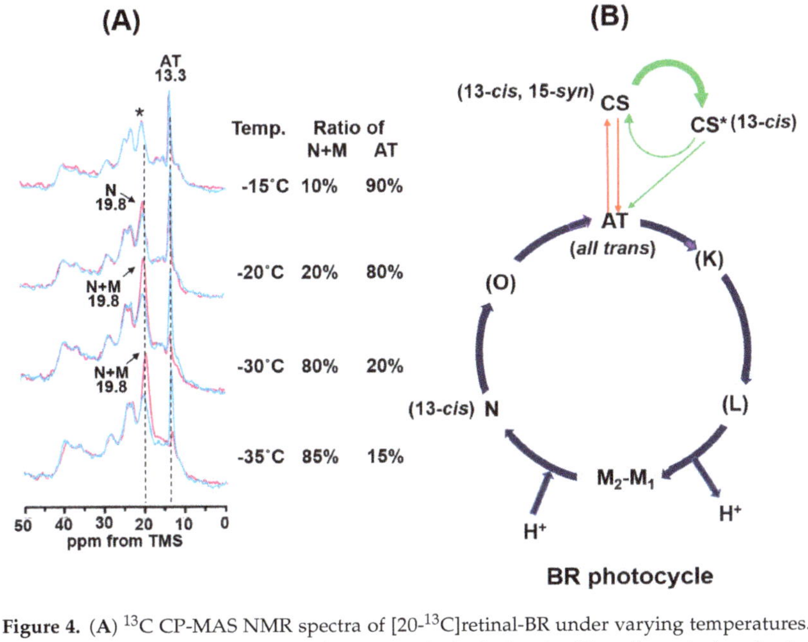

Figure 4. (**A**) ^{13}C CP-MAS NMR spectra of [20-^{13}C]retinal-BR under varying temperatures. Blue spectra were recorded in the dark for the LA state (retinal takes the AT configuration), and red spectra were recorded under green light illumination. The AT state changed to the N intermediate at −20 °C and to the N and M intermediates at −30 and −35 °C. (**B**) Photoreaction cycle of BR. In the AT photocycle, only the N-intermediate was detected. In the CS photocycle, the CS state was transferred to the CS* intermediate under green light illumination and consequently changed to the AT state upon dark exposure through a leakage process from the CS* intermediate to the AT state. (K), (L) and (O) intermediates were not detected using in situ photo-irradiation solid-state NMR.

As shown in Figure 4A, in the AT photocycle, the trapped fraction of the N intermediate is significantly increased at a temperature lower than −20 °C. It has been reported that the M-intermediate significantly appears at a temperature lower than −20 °C and pH 10 and that 60% of the AT state converted to the M intermediate at −30 °C [27]. Considering this previous report, we revealed that the M intermediate significantly increased at 19.8 ppm in addition to the N intermediate at −30 °C. Furthermore, 85% of the AT state converted to the M and N intermediates at −35 °C. This may cause a slightly higher field shift of the (N+M+*) signal at −35 °C.

3.3. *Photo-Reaction Cycle of [20-^{13}C]retinal-BR*

In situ photoirradiation solid-state NMR experiments demonstrated that unstable intermediates can be observed in a stationary state in real-time under photo-illumination. In the CS photocycle, the CS state is not directly converted to the AT state, but is converted through the CS* intermediate to the AT state under green light illumination, as shown in Figure 4B. The NMR signal of the CS* intermediate was observed for the first time in the CS photocycle of WT-BR using in situ photoirradiation solid-state NMR. At −20 °C, the CS* intermediate became a major component under green light illumination. As summarized in Table 1, the chemical shift value of the CS* intermediate is 19.6 ppm and that of CS is 22.3 ppm for (20-^{13}C), thus the CS* intermediate has the same (13-*cis*) configuration as the CS state, although it differs slightly. In the active photocycle starting from the AT state (AT photocycle), photo-intermediates (K, L, M, N, O) could not be detected at 20 °C using in situ photoirradiation solid-state NMR because of their short half-lives. However, the N intermediate could be trapped at −20 °C, which is the only intermediate previously reported to be trapped at that temperature [26,27]. The fraction of the N+M intermediate significantly increased at a temperature lower than −20 °C, because the M intermediate starts appearing at below −20 °C and dominated at −30 °C.

A summary of the photocycle starting from the DA state is given in Figure 4B. Upon photoirradiation of the DA state, the CS state transforms to the CS* intermediate in the CS photocycle under green light illumination, and the CS* intermediate partly transformed through a leakage process from the CS photocycle. Simultaneously, the AT state undergoes a photocycle through the K, L, M, N, and O intermediates; however, only the N and M intermediates could be observed using in situ photoirradiation NMR at a temperature lower than −20 °C. Next, we attempted to observe the M intermediate using D96N-BR, which has a longer half-life of the M intermediate than WT-BR.

3.4. *Photo-Reaction Pathways as Revealed by [20-^{13}C, 14-^{13}C]retinal-D96N-BR under Green Light Illumination from the DA to LA State at −30 °C*

The DA state of D96N-BR was prepared by keeping the sample in the dark at 20 °C for one day. Afterward, the ^{13}C CP MAS NMR spectra of [20-^{13}C, 14-^{13}C]retinal-D96N-BR in the DA state were observed in the dark at −30 °C as shown in Figure 5A(D1). In the D1 spectrum, AT and CS signals of [20-^{13}C and 14-^{13}C] retinal appeared at (13.1 and 121.8 ppm) and (22.6 and 110.1 ppm), respectively. Figure 5A(L1) shows the ^{13}C CP-MAS NMR spectrum of [20-^{13}C, 14-^{13}C]retinal-D96N-BR under illumination with green light (L1). In the L1 spectrum, the AT and CS signals were significantly reduced, and the signals around 20 ppm were increased.

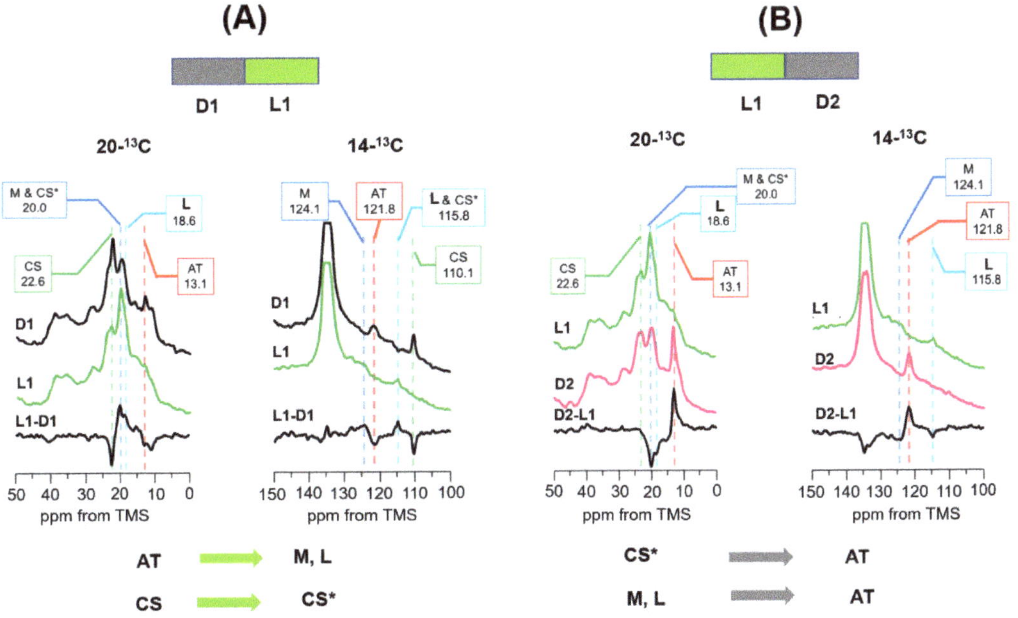

Figure 5. (**A**) Photoreaction pathways of [20-^{13}C, 14-^{13}C]retinal-D96N-BR under green light (520 nm) illumination at −30 °C. ^{13}C CP-MAS NMR spectra were recorded in the dark (D1; 20-^{13}C and 14-^{13}C) and then under green light (L1; 20-^{13}C and 14-^{13}C), and the difference in spectra between L1 and D1 (L1-D1; 20-^{13}C and 14-^{13}C) indicated that the AT state changed to the M and L intermediates and the CS state changed to the CS* intermediate. (**B**) Photo reaction pathway of [20-^{13}C, 14-^{13}C]retinal-D96N-BR in the dark after green light illumination at −30 °C. ^{13}C CP-MAS NMR spectra were recorded under green light (L1; 20-^{13}C and 14-^{13}C) and then in the dark (D2; 20-^{13}C and 14-^{13}C), and the difference in spectra between D2 and L1 (D2-L1; 20-^{13}C and 14-^{13}C) indicated that the CS* intermediate changed to the AT state and the M and L intermediates changed to the AT state.

The difference in spectra between L1 and D1 (Figure 5A(L1-D1)) indicated that the CS and AT peaks are negative, and the CS* (20.0 ppm; 20-^{13}C, 115.8 ppm; 14-^{13}C), M (20.0 ppm, 124.1 ppm) and L (18.6, 115.8 ppm) intermediate peaks are positive. The M intermediate was assigned by comparing the ^{13}C chemical shift value of 20-^{13}C and 14-^{13}C with those previously reported for the M-intermediate [31,37] (Table S1). It has been reported that an N intermediate is not observed in D96N-BR, because the proton donor D96 is replaced by N96 [22–25]. Therefore, another peak was assigned to the L intermediate and the chemical shifts agree with previously reported values [31] (Table S1). These results clearly indicate that the CS state is mostly converted to the CS* intermediate under green light illumination at −30 °C. At the same time, the AT state is converted to the L and M intermediates. Notably, the L, M, and CS* intermediates with short half-lives were observed in real time as stationary states using in situ photoirradiation solid-state NMR. It is thus revealed that the CS state is not directly converted to the AT state, but is converted through the CS* intermediate similar to the case of WT-BR at −20 °C (as described in Section 3.2).

After green light illumination, the NMR signals were observed in subsequent dark condition (Figure 5B(D2)) which may produce a LA state. The difference in spectra between D2 and L1 is shown in Figure 5B(D2-L1). The spectra clearly indicate that all CS* intermediate were converted to the AT state, and, L and M intermediates were converted to the AT state to form the LA condition, which consists entirely of the AT state.

3.5. Photoreaction Pathways as Revealed by [14-^{13}C, 20-^{13}C]retinal-D96N-BR under Green Light Illumination from LA (D2) to DA (D3) States

Figure 6A (D2) shows the ^{13}C CP-MAS NMR spectrum of the LA state, in which all retinal assumes the AT state. After green light illumination, the AT state was reduced and the L and M intermediates were increased (L2). Figure 6A(L2-D2) shows the difference in spectra between L2 and D2 for [20-^{13}C, 14-^{13}C] retinal-D96N-BR at −30 °C. It can be clearly seen that the AT state is transformed to the L and M intermediates during the AT photocycle.

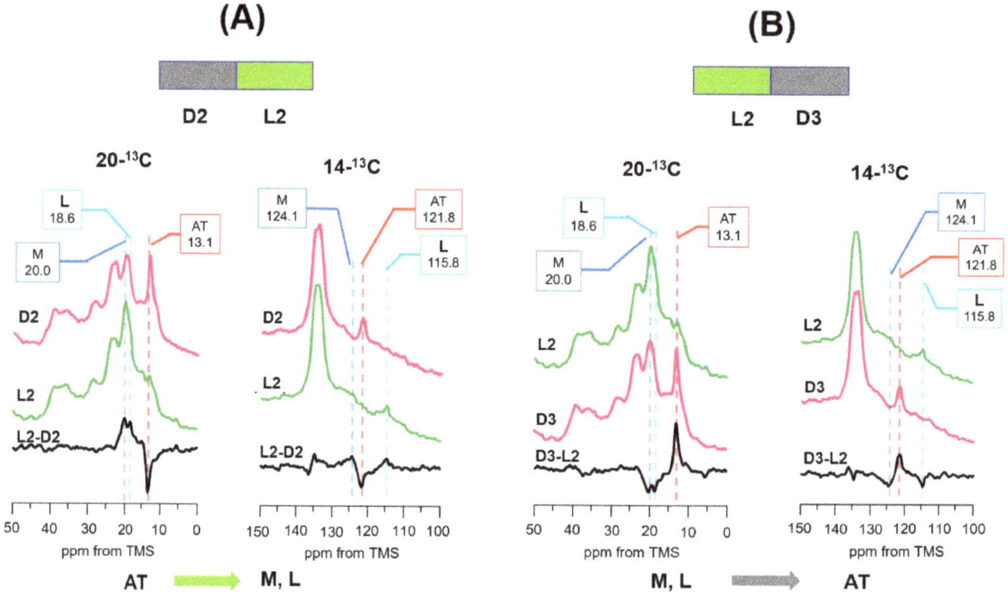

Figure 6. (**A**) Photo reaction pathways of [20-^{13}C, 14-^{13}C]retinal-D96N-BR under green light illumination from the light-adapted D2 condition at −30 °C. ^{13}C CP-MAS NMR spectra were recorded in the dark (D2; 20-^{13}C and 14-^{13}C) and then green light (L2; 20-^{13}C and 14-^{13}C), and the difference in spectra between L2 and D2 (L2-D2; 20-^{13}C and 14-^{13}C) indicated that the AT state changed to the M and L intermediates. (**B**) Photoreaction pathways of [20-^{13}C, 14-^{13}C]retinal-D96N-BR in the dark from the L2 condition at −30 °C. ^{13}C CP NMR spectra were recorded under green light (L2; 20-^{13}C and 14-^{13}C) and then in the dark (D3; 20-^{13}C and 14-^{13}C), and the difference in spectra between D3 and L2 (D3-L2; 20-^{13}C and 14-^{13}C) indicated that the M and L intermediates changed to the AT state.

Figure 6B(L2) shows the ^{13}C CP-MAS NMR signal of [20-^{13}C, 14-^{13}C]retinal-D96N-BR at −30 °C under green light illumination as the LA state. Following green light illumination, NMR signals were recorded in the dark (D3). Figure 6B(D3-L2) shows the difference in spectra between D3 and L2. The results revealed that the L and M intermediates were reduced, and the AT state was increased. This shows that the L and M intermediates are transformed to the AT state. As the CS state does not exist in the LA state, the CS* intermediate did not appear in this process.

3.6. Photoreaction Pathways as Revealed by [20-^{13}C, 14-^{13}C]retinal-D96N-BR under UV Light Illumination following Green Light Illumination

After green light illumination, the L2 spectrum was recorded and found to mainly contain the L and M intermediates, as shown in Figure 7A(L2). Subsequently, the L2 state under green light illumination was changed to illumination with UV light, as shown in Figure 7A(UL). The difference in spectra between UL and L2 is shown in Figure 7A(UL-L2).

The L and M intermediates decreased and the AT state increased. The spectral difference between UL and L2 for [20-^{13}C, 14-^{13}C]retinal-D96N-BR indicated that the M intermediate was reduced more significantly than the L intermediate; this is more clearly seen in the spectrum for [14-^{13}C]retinal-D96N-BR (Figure 7A(UL-L2)), where the L intermediate did not completely disappear. Consequently, following UV light illumination, the spectrum in the dark (D4) was acquired, as shown in Figure 7B(D4). Figure 7B(D4-UL) shows the difference in spectra between D4 and UL, which indicated that the remaining L intermediate transformed to the AT state and the M intermediate did not remain. This result indicates that there is a direct pathway from the M intermediate to the AT state by UV light illumination.

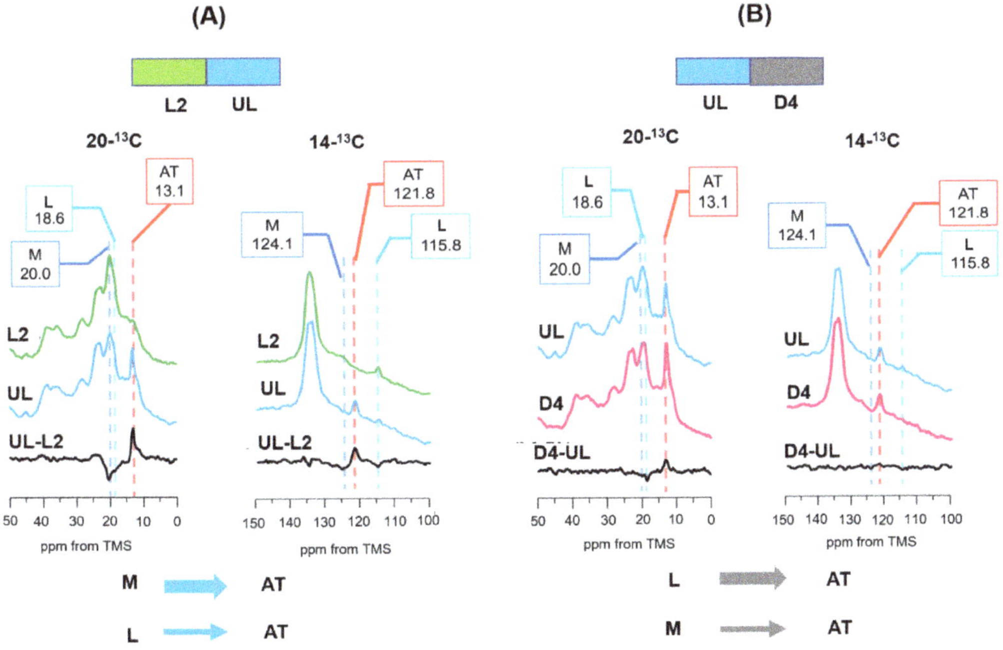

Figure 7. (**A**) Photoreaction pathway of [20-^{13}C, 14-^{13}C]retinal-D96N-BR under UV light (365 nm) illumination after green light illumination at −30 °C. ^{13}C CP-MAS NMR spectra were recorded under green light (L2; 20-^{13}C and 14-^{13}C) and then under UV light (UL; 20-^{13}C and 14-^{13}C), and the difference in spectra between UL and L2 (UL-L2) indicated that the M-intermediate changed more significantly to the AT state compared to the L intermediate. (**B**) Photoreaction pathway of [20-^{13}C, 14-^{13}C]retinal-D96N-BR in the dark after UV light illumination at −30 °C. ^{13}C CP-MAS NMR spectra were recorded under UV light (UL; 20-^{13}C and 14-^{13}C) and then in the dark (D4; 20-^{13}C and 14-^{13}C), and the difference in spectra between D4 and UL (D4-UL; 20-^{13}C and 14-^{13}C) indicated that the L intermediates changed to the AT state, and a small amount of the M-intermediate also changed to the AT state.

3.7. Photo-Reaction Pathways as Revealed by [ε-^{15}N]Lys^{216}-D96N-BR

Figure 8A shows the ^{15}N CP-MAS NMR spectra of [ε-^{15}N]Lys-D96N-BR in the dark state (black line) and under illumination with green light (green line) at −60 °C. In the black line spectrum, the AT state appeared at 148 ppm and the CS state appeared at 153 ppm, which were assigned according to the previously reported ^{15}N chemical shift values of [ε-^{15}N]-Lys^{216}-WT-BR[30] (Figure 8A black line and Table S1). After illumination with green light, the protonated [ε-^{15}N]-Lys^{216} NMR signal of the AT and CS states decreased and the signal of the deprotonated [ε-^{15}N]-Lys^{216} at 300 ppm increased, which was assigned to the M intermediate. The difference in the spectra between the green and black lines

(Figure 8A(bottom)) shows that the AT state (negative peak) converted to the M intermediate (positive peak). As the signal of the M intermediate appeared at 150 ppm lower field than those of the CS and AT states which form an protonated SB with [ε-^{15}N]-Lys^{216}, the M intermediate forms the deprotonated SB bond with [ε-^{15}N]Lys^{216}, which is in agreement with the previously reported chemical shift values of [ε-^{15}N]Lys^{216} in a photo-excited M intermediate under green light illumination [30] (Table S1).

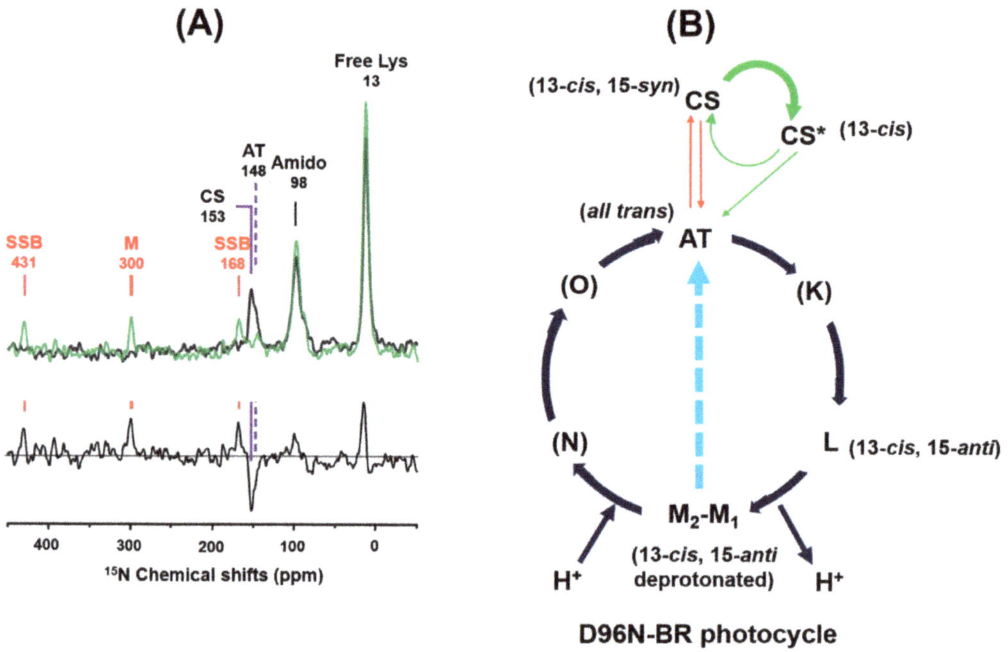

Figure 8. (**A**) ^{15}N CP-MAS NMR spectra of [ε-^{15}N]Lys-D96N-BR were recorded at −60 °C in the dark (black line) and under green light illumination (green line), and the difference in spectra between the black line and green lines was obtained as shown in the bottom line. The difference spectrum of [ε-^{15}N]Lys^{216}-D96N-BR indicated that the AT state changed to the M intermediate. (**B**) Photoreaction cycle of D96N-BR. In the AT photocycle, the L and M intermediates were detected, and the M intermediate directly changed to the AT state. In the CS photocycle, the CS state was changed to the CS* intermediate and then to the AT intermediate. (K), (N) and (O) intermediates were not detected using in situ photoirradiation solid-state NMR.

3.8. Photo Reaction Pathways in the D96N-BR Mutant

When DA (D1) D96N-BR was illuminated with green light, the CS state transformed to the CS* intermediate, which was stable at −30 °C and gradually transformed to the AT state through a leakage pathway from the CS* intermediate to the AT state. Therefore, the CS state is not directly transformed to the AT state, but passes through the CS* intermediate and subsequently changes to the AT state. Simultaneously, the AT state changed to the L intermediate then subsequently changed to the M intermediate during the AT photocycle. The M intermediate did not change to the N intermediate, instead changing to the AT state. The M intermediate is further directly excited to the AT state by UV light illumination (Figure 8B). A previous report showed that the back reaction from the M intermediate to the AT state under UV light illumination is significant for WT-BR, though not for D96N-BR [49]. Further work is therefore needed to clarify the existence of a photo reaction pathway from the M intermediate to the AT state under UV light illumination for D96N-BR.

Chemical shift values of the CS* intermediate appeared at 20.0 and 115.8 ppm for 20-^{13}C and 14-^{13}C, respectively. By comparing the chemical shift values of CS (22.9 ppm, 110.1 ppm), the CS* retinal configuration can be determined to be the 13-*cis* retinal configuration as shown in Table 1. Chemical shift values of the M intermediate were determined to be 20.0 ppm for 20-^{13}C and 124.1 ppm for 14-^{13}C, and it was assigned to the 13-*cis*, 15-*anti* retinal configuration. Chemical shift values of the L intermediate were determined to be 18.6 ppm for 20-^{13}C and 115.6 ppm for 14-^{13}C and it was assigned to the 13-*cis* retinal configuration, which agrees well with the previously reported chemical shift values for the L intermediate (Table S1) [31].

4. Conclusions

The ^{13}C CP/MAS NMR signals of [20-^{13}C]retinal-BR and [20-^{13}C, 14-^{13}C]retinal-D96N-BR were measured under green light illumination. The photochemical reaction pathway for [20-^{13}C]retinal-BR was investigated by illuminating with green light at 20 °C. We clearly observed that the CS state was completely converted to the AT state to produce the LA state. Following exposure to a dark state for one day, half of the AT state converted to the CS state to produce the DA state. In the next step, D1(DA) was illuminated with green light at −20 °C. We observed that the CS state was converted to the CS* intermediate. This result indicates that the CS state is not directly converted to the AT state by green light illumination at −20 °C, but is converted to the AT state via the CS* intermediate. After green light exposure, the ^{13}C CP-MAS NMR spectra were acquired in the dark. In this photoreaction pathway, the CS* intermediate was converted to the AT state, which produced an LA state. This LA state was illuminated with green light at −20 °C, and the AT state partly converted to the N intermediate through the AT photo cycle; the M intermediate significantly increased at temperatures lower than –20 °C Figure 4B summarizes the WT-BR photocycle obtained in this work.

Photoreaction pathways were observed for [20-^{13}C, 14-^{13}C]retinal-D96N-BR by green light illumination at −30 °C. We observed that the CS state was converted to the CS* intermediate and then changed to the AT state in the dark, and the AT state converted to the M and L intermediates, because the half-lives of the M and L intermediates are significantly increased in D96N-BR. The M and L intermediates were converted to the AT state in the dark to produce the LA state. This LA state (full AT state) was converted to the M and L intermediates, but not the CS* intermediate under green light exposure because the CS state does not exist in the LA state. We demonstrated that the unstable CS*, L and M intermediates could be observed in a steady state under illumination with green light at −30 °C. Also, the M intermediate was directly excited to the AT state by UV light. Thus, a photo-reaction pathway exists from the M intermediate to the AT state under illumination with UV light. The photocycle of D96N-BR obtained in this work is summarized in Figure 8B.

The present study demonstrates that in situ photoirradiation solid-state NMR is a powerful means to detect NMR signals in short-lived photo-intermediates in the DA and LA states of WT-BR and D96N-BR. In the CS photocycle, NMR signals of the CS* intermediate were stationarily detected in WT-BR and D96N-BR for the first time. In the AT photocycle, N and M intermediates were stationarily detected in WT-BR at −30 °C, and the L and M intermediates were stationarily detected in D96N-BR at −30 °C. Thus, not only the retinal structure but also the protein structures of photo-intermediates can be determined using in situ photoirradiation solid-state NMR spectroscopy.

Supplementary Materials: The following is available online at https://www.mdpi.com/article/10.3390/membranes12030279/s1. Table S1. ^{13}C and ^{15}N chemical shift values (ppm) of individual states of [20-^{13}C, 14-^{13}C]-retinal and [ε-^{15}N]Lys-bacteriorhodopsin(BR) and Y185F-BR reported in the literature. Figure S1. Schematic diagram of the in situ photoirradiation solid state NMR spectrometer. Figure S2. Full scale ^{13}C CP-MAS NMR spectra of [Tyr^{13}CO] and [20-^{13}C, 14-^{13}C]retinal-D96N-BR were recorded in the dark (D1; dark-adapted state) and then under green light (L1) illumination at −30 °C. Figure S3. Full scale ^{13}C CP-MAS NMR spectra of [Tyr^{13}CO] and [20-^{13}C, 14-^{13}C]retinal-

D96N-BR were recorded in the dark (D2; light-adapted state) and then under green light (L2) illumination at −30 °C.

Author Contributions: A.S., sample preparation and NMR measurement; Y.O., sample preparation and NMR measurement; R.M., sample preparation and NMR measurement; Y.M., NMR measurement; I.K., conception and writing of the manuscript; T.O., sample preparation, A.W., sample preparation, A.N. supervision, conception and writing of the manuscript. All authors have read and approved the final version of the manuscript.

Funding: This research was funded by Grants-in-Aid for Scientific Research in an Innovative Area (26102514 and 16H00828), and by a Grant-in-Aid for Scientific Research (C) (15K06963) and Research (B) (18H02387) from the Ministry of Education, Culture, Sports, Science and Technology of Japan, and by the Yokohama Academic Foundation.

Institutional Review Board Statement: Not applicable.

Informed Consent Statement: Not applicable.

Data Availability Statement: All data generated in this study are available within this manuscript and Supplementary Materials.

Acknowledgments: The authors thank N. Yamaguchi for providing financial support and S. Tuzi for gifting the *H. salinarum* transformant for the D96N mutant.

Conflicts of Interest: The authors declare no conflict of interest.

References

1. Lanyi, J.K. Mechanism of Ion Transport across Membranes. *J. Biol. Chem.* **1997**, *272*, 31209–31212. [CrossRef]
2. Facciotti, M.T.; Rouhani-Manshadi, S.; Glaeser, R.M. Energy transduction in transmembrane ion pumps. *Trends Biochem. Sci.* **2004**, *29*, 445–451. [CrossRef] [PubMed]
3. Ernst, O.P.; Lodowski, D.T.; Elstner, M.; Hegemann, P.; Brown, L.S.; Kandori, H. Microbial and Animal Rhodopsins: Structures, Functions, and Molecular Mechanisms. *Chem. Rev.* **2014**, *114*, 126–163. [CrossRef] [PubMed]
4. Luecke, H.; Schobert, B.; Richter, H.-T.; Cartailler, J.-P.; Lanyi, J.K. Structure of bacteriorhodopsin at 1.55 Å resolution. *J. Mol. Biol.* **1999**, *291*, 899–911. [CrossRef]
5. Luecke, H.; Schobert, B.; Cartailler, J.-P.; Richter, H.-T.; Rosengarth, A.; Needleman, R.; Lanyi, J.K. Coupling photoisomerization of retinal to directional transport in bacteriorhodopsin. *J. Mol. Biol.* **2000**, *300*, 1237–1255. [CrossRef]
6. Lanyi, J.K. Proton transfers in the bacteriorhodopsin photocycle. *Biochim. Biophys. Acta* **2006**, *1757*, 1012–1018. [CrossRef] [PubMed]
7. Nango, E.; Royant, A.; Kubo, M.; Nakane, T.; Wickstrand, C.; Kimura, T.; Tanaka, T.; Tono, K.; Song, C.; Tanaka, R.; et al. A three-dimensional movie of structural changes in bacteriorhodopsin. *Science* **2016**, *354*, 1552–1557. [CrossRef] [PubMed]
8. Nogly, P.; Weinert, T.; James, D.; Carbajo, S.; Ozerov, D.; Furrer, A.; Gashi, D.; Borin, V.; Skopintsev, P.; Jaeger, K.; et al. Retinal isomerization in bacteriorhodopsin captured by a femtosecond x-ray laser. *Science* **2018**, *361*, eaat0094. [CrossRef]
9. Weinert, T.; Skopintsev, P.; James, D.; Dworkowski, F.; Panepucci, E.; Kekilli, D.; Furrer, A.; Brünle, S.; Mous, S.; Ozerov, D.; et al. Proton uptake mechanism in bacteriorhodopsin captured by serial synchrotron crystallography. *Science* **2019**, *365*, 61–65. [CrossRef]
10. Lanyi, J.K. Molecular Mechanism of Ion Transport in Bacteriorhodopsin: Insights from Crystallographic, Spectroscopic, Kinetic, and Mutational Studies. *J. Phys. Chem. B* **2000**, *104*, 11441–11448. [CrossRef]
11. Morgan, J.E.; Vakkasoglu, A.S.; Lanyi, J.K.; Lugtenburg, J.; Gennis, R.B.; Maeda, A. Structure Changes upon Deprotonation of the Proton Release Group in the Bacteriorhodopsin Photocycle. *Biophys. J.* **2012**, *103*, 444–452. [CrossRef] [PubMed]
12. Kalisky, O.; Goldschmidt, C.; Ottolenghi, M. On the photocycle and light adaptation of dark-adapted bacteriorhodopsin. *Biophys. J.* **1977**, *19*, 185–189. [CrossRef]
13. Iwasa, T.; Tokunaga, F.; Yoshizawa, T. Photochemical reaction of 13-cis-bacteriorhodopsin studied by low temperature spectrophotometry. *Photochem. Photobiol.* **1981**, *33*, 539–545. [CrossRef]
14. Roepe, P.D.; Ahl, P.L.; Herzfeld, J.; Lugtenburg, J.; Rothschild, K. Tyrosine protonation changes in bacteriorhodopsin. A Fourier transform infrared study of BR548 and its primary photoproduct. *J. Biol. Chem.* **1988**, *263*, 5110–5117. [CrossRef]
15. Logunov, I.; Humphrey, W.; Schulten, K.; Sheves, M. Molecular dynamics study of the 13-cis form (bR548) of bacteriorhodopsin and its photocycle. *Biophys. J.* **1995**, *68*, 1270–1282. [CrossRef]
16. Richter, H.-T.; Needleman, R.; Lanyi, J.K. Perturbed interaction between residues 85 and 204 in Tyr-185→Phe and Asp-85→Glu bacteriorhodopsins. *Biophys. J.* **1996**, *71*, 3392–3398. [CrossRef]
17. Dunach, M.; Marti, T.; Khorana, H.G.; Rothschild, K. Uv-visible spectroscopy of bacteriorhodopsin mutants: Substitution of Arg-82, Asp-85, Tyr-185, and Asp-212 results in abnormal light-dark adaptation. *Proc. Natl. Acad. Sci. USA* **1990**, *87*, 9873–9877. [CrossRef]

18. Song, L.; Yang, D.; El-Sayed, M.A.; Lanyi, J.K. Retinal Isomer Composition in Some Bacteriorhodopsin Mutants under Light and Dark Adaptation Conditions. *J. Phys. Chem.* **1995**, *99*, 10052–10055. [CrossRef]
19. Sonar, S.; Krebs, M.P.; Khorana, H.G.; Rothschild, K. Static and time-resolved absorption spectroscopy of the bacteriorhodopsin mutant Tyr-185.fwdarw. Phe: Evidence for an equilibrium between bR570 and an O-like species. *Biochemistry* **1993**, *32*, 2263–2271. [CrossRef]
20. Rath, P.; Krebs, M.P.; He, Y.; Khorana, H.G.; Rothschild, K.J. Fourier transform Raman spectroscopy of the bacteriorhodopsin mutant Tyr-185→Phe: Formation of a stable O-like species during light adaptation and detection of its transient N-like photoproduct. *Biochemistry* **1993**, *32*, 2272–2281. [CrossRef]
21. He, Y.; Krebs, M.P.; Fischer, W.B.; Khorana, H.G.; Rothschild, K.J. FTIR difference spectroscopy of the bacteriorhodopsin mutant Tyr-185→fwdarw. Phe: Detection of a stable O-like species and characterization of its photocycle at low temperature. *Biochemistry* **1993**, *32*, 2282–2290. [CrossRef] [PubMed]
22. Zimányi, L.; Varo, G.; Chang, M.; Ni, B.; Needleman, R.; Lanyi, J.K. Pathways of proton release in the bacteriorhodopsin photo-cycle. *Biochemistry* **1992**, *31*, 8535–8543. [CrossRef] [PubMed]
23. Zimányi, L.; Kulcsár, Á.; Lanyi, J.K.; Sears, D.F.; Saltiel, J. Intermediate spectra and photocycle kinetics of the Asp96 → Asn mutant bacteriorhodopsin determined by singular value decomposition with self-modeling. *Proc. Natl. Acad. Sci. USA* **1999**, *96*, 4414–4419. [CrossRef] [PubMed]
24. Kikukawa, T.; Araiso, T.; Mukasa, K.; Shimozawa, T.; Kamo, N. The molecular motion of bacteriorhodopsin mutant D96N in the purple membrane. *FEBS Lett.* **1995**, *377*, 502–504. [CrossRef]
25. Yao, B.; Zheng, Y.; Wang, Y.; Lei, M.; Chen, G.; Hampp, N. Kinetic spectra of light-adaptation dark-adaptation and M-intermediate of BR-D96N. *Opt. Commun.* **2003**, *218*, 125–130. [CrossRef]
26. Lakshmi, K.V.; Farrar, M.R.; Raap, J.; Lugtenburg, J.; Griffin, R.G.; Herzfeld, J. Solid State 13C and 15N NMR Investigations of the N Intermediate of Bacteriorhodopsin. *Biochemistry* **1994**, *33*, 8853–8857. [CrossRef]
27. Hu, J.G.; Sun, B.Q.; Petkova, A.T.; Griffin, R.G.; Herzfeld, J. The predischarge chromophore in bacteriorhodopsin: A 15N sol-id-state NMR study of the L photointermediate. *Biochemistry* **1997**, *36*, 9316–9322. [CrossRef]
28. Harbison, G.S.; Smith, S.O.; Pardoen, J.A.; Winkel, C.; Lugtenburg, J.; Herzfeld, J.; Mathies, R.; Griffin, R.G. Dark-adapted bacteriorhodopsin contains 13-cis, 15-syn and all-trans, 15-anti retinal Schiff bases. *Proc. Natl. Acad. Sci. USA* **1984**, *81*, 1706–1709. [CrossRef]
29. Harbison, G.S.; Smith, S.O.; Pardoen, J.A.; Mulder, P.P.; Lugtenburg, J.; Herzfeld, J.; Mathies, R.; Griffin, R.G. Solid-state 13C NMR studies of retinal in bacteriorhodopsin. *Biochemistry* **1984**, *23*, 2662–2667. [CrossRef]
30. Smith, S.O.; de Groot, H.; Gebhard, R.; Courtin, J.M.L.; Lugtenburg, J.; Herzfeld, J.; Griffin, R.G. Structure and protein environment of the retinal chromophore in light- and dark-adapted bacteriorhodopsin studied by solid-state NMR. *Biochemistry* **1989**, *28*, 8897–8904. [CrossRef]
31. Hu, J.G.; Sun, B.Q.; Bizounok, M.; Hatcher, M.E.; Lansing, J.C.; Raap, J.; Verdegem, P.J.E.; Lugtenburg, J.; Griffin, R.G.; Herzfeld, J. Early and Late M Intermediates in the Bacteriorhodopsin Photocycle: A Solid-State NMR Study. *Biochemistry* **1998**, *37*, 8088–8096. [CrossRef] [PubMed]
32. Bajaj, V.S.; Mak-Jurkauskas, M.L.; Belenky, M.; Herzfeld, J.; Griffin, R.G. Functional and shunt states of bacteriorhodopsin re-solved by 250 GHz dynamic nuclear polarization-enhanced solid-state NMR. *Proc. Natl. Acad. Sci. USA* **2009**, *106*, 9244–9249. [CrossRef] [PubMed]
33. Kawamura, I.; Kihara, N.; Ohmine, M.; Nishimura, K.; Tuzi, S.; Saitô, H.; Naito, A. Solid-state NMR studies of two backbone conformations at Tyr185 as a function of retinal configurations in the dark, light, and pressure adapted bacteriorhodopsins. *J. Am. Chem. Soc.* **2007**, *129*, 1016–1017. [CrossRef]
34. Petkova, A.T.; Hatanaka, M.; Jaroniec, C.P.; Hu, J.G.; Belenky, M.; Verhoeven, M.; Lugtenburg, J.; Griffin, R.G.; Herzfeld, J. Tryp-tophan interaction in bacteriorhodopsin: A heteronuclear solid-state NMR study. *Biochemistry* **2002**, *41*, 2429–2437. [CrossRef]
35. Mak-Jurkauskas, M.L.; Bajaj, V.S.; Hornstein, M.K.; Belenky, M.; Griffin, R.G.; Herzfeld, J. Energy transformations early in the bacteriorhodopsin photocycle revealed by DNP-enhanced solid-state NMR. *Proc. Natl. Acad. Sci. USA* **2008**, *105*, 883–888. [CrossRef]
36. Becker-Baldus, J.; Bamann, C.; Saxena, K.; Gustmann, H.; Brown, L.J.; Brown, R.C.D.; Reiter, C.; Bamberg, E.; Wachtveitl, J.; Schwalbe, H.; et al. Enlightening the photoactive site of channelrhodopsin-2 by DNP-enhanced solid-state NMR spectroscopy. *Proc. Natl. Acad. Sci. USA* **2015**, *112*, 9896–9901. [CrossRef] [PubMed]
37. Tomonaga, Y.; Hidaka, T.; Kawamura, I.; Nishio, T.; Ohsawa, K.; Okitsu, T.; Wada, A.; Sudo, Y.; Kamo, N.; Ramamoorthy, A.; et al. An Active Photoreceptor Intermediate Revealed by In Situ Photoirradiated Solid-State NMR Spectroscopy. *Biophys. J.* **2011**, *101*, L50–L52. [CrossRef]
38. Yomoda, H.; Makino, Y.; Tomonaga, Y.; Hidaka, T.; Kawamura, I.; Okitsu, T.; Wada, A.; Sudo, Y.; Naito, A. Color-Discriminating Retinal Configurations of Sensory Rhodopsin I by Photo-Irradiation Solid-State NMR Spectroscopy. *Angew. Chem. Int. Ed.* **2014**, *53*, 6960–6964. [CrossRef]
39. Makino, Y.; Kawamura, I.; Okitsu, T.; Wada, A.; Kamo, N.; Sudo, Y.; Ueda, K.; Naito, A. Retinal Configuration of ppR Intermediates Revealed by Photoirradiation Solid-State NMR and DFT. *Biophys. J.* **2018**, *115*, 72–83. [CrossRef]
40. Naito, A.; Kawamura, I. *Photo-Activated Structural Changes in Photoreceptor Membrane Proteins as Revealed by In Situ Pho-Toirradiation Solid-State NMR*; Advances in Biological Solid-State NMR: Proteins and Membrane Active peptides; Separovic, F., Naito, A., Eds.; Royal Society of Chemistry: London, UK, 2014; pp. 387–404.

41. Naito, A.; Kawamura, I.; Javkhlantugs, N. Recent Solid-State NMR Studies of Membrane-Bound Peptides and Proteins. *Annu. Rep. NMR Spectrosc.* **2015**, *86*, 333–411. [CrossRef]
42. Naito, A.; Makino, Y.; Kawamura, I. In Situ Photo Irradiation Solid-State NMR Spectroscopy Applied to Retinal-Binding Membrane Proteins. In *Modern Magnetic Resonance*, 2nd ed.; Springer: Berlin, Germany, 2018; pp. 537–557.
43. Naito, A.; Makino, Y.; Tasei, Y.; Kawamura, I. Photoirradiation and Microwave Irradiation NMR Spectroscopy. *Exp. Approaches NMR Spectrosc.* **2018**, *5*, 135–170. [CrossRef]
44. Oshima, K.; Shigeta, A.; Makino, Y.; Kawamura, I.; Okitsu, T.; Wada, A.; Tuzi, S.; Iwasa, T.; Naito, A. Characterization of pho-to-intermediates in the photo-reaction pathways of a bacteriorhodopsin Y185F mutant using in situ photo-irradiation solid-state NMR spectroscopy. *Photochem. Photobiol. Sci.* **2015**, *14*, 1694–1702. [CrossRef]
45. Naito, A.; Makino, Y.; Shigeta, A.; Kawamura, I. Photoreaction pathways and photointermediates of retinal-binding photore-ceptor proteins as revealed by in situ photoirradiation solid-state NMR spectroscopy. *Biophys. Rev.* **2019**, *11*, 167–181. [CrossRef] [PubMed]
46. Oesterhelt, D.; Stoeckenious, W. Isolation of the cell membrane of Halobacterium halobium and its fractionation into red and purple membrane. *Methods Enzymol.* **1974**, *31*, 667–678. [PubMed]
47. Kawamura, I.; Tanabe, J.; Ohmine, M.; Yamaguchi, S.; Tuzi, S.; Naito, A. Participation of the BC Loop in the Correct Folding of Bacteriorhodopsin as Revealed by Solid-state NMR. *Photochem. Photobiol.* **2009**, *85*, 624–630. [CrossRef]
48. Bennett, A.E.; Rienstra, C.M.; Auger, M.; Lakshmi, K.V.; Griffin, R.G. Heteronuclear decoupling in rotating solids. *J. Chem. Phys.* **1995**, *103*, 6951–6958. [CrossRef]
49. Druckmann, S.; Friedman, N.; Lanyi, J.K.; Needleman, R.; Ottolenghi, M.; Sheves, M. The back photoreaction of the M intermediate in the photocycle of bacteriorhodopsin: Mechanism and evidence for two M species. *Photochem. Photobiol.* **1992**, *56*, 1041–1047. [CrossRef]

membranes

MDPI

Article

The Poly-Glutamate Motif of GmMATE4 Regulates Its Isoflavone Transport Activity

Yee-Shan Ku [1,*,†], Sau-Shan Cheng [1,†], Ming-Yan Cheung [1], Yongchao Niu [1], Ailin Liu [1], Gyuhwa Chung [2] and Hon-Ming Lam [1,*]

[1] School of Life Sciences and Centre for Soybean Research of the State Key Laboratory of Agrobiotechnology, The Chinese University of Hong Kong, Hong Kong, China; chengsaushan@yahoo.com (S.-S.C.); cheungmy@cuhk.edu.hk (M.-Y.C.); niuyongchao@link.cuhk.edu.hk (Y.N.); merestarry@gmail.com (A.L.)

[2] Department of Biotechnology, Chonnam National University, Yeosu 59626, Korea; chung@chonnam.ac.kr

* Correspondence: ysamyku@cuhk.edu.hk (Y.-S.K.); honming@cuhk.edu.hk (H.-M.L.)

† These authors contributed equally.

Abstract: Multidrug and toxic compound extrusion (MATE) transporters in eukaryotes have been characterized to be antiporters that mediate the transport of substrates in exchange for protons. In plants, alkaloids, phytohormones, ion chelators, and flavonoids have been reported to be the substrates of MATE transporters. Structural analyses have been conducted to dissect the functional significance of various motifs of MATE proteins. However, an understanding of the functions of the N- and C-termini has been inadequate. Here, by performing phylogenetic analyses and protein sequence alignment of 14 representative plant species, we identified a distinctive N-terminal poly-glutamate motif among a cluster of MATE proteins in soybean. Amongst them, GmMATE4 has the most consecutive glutamate residues at the N-terminus. A subcellular localization study showed that GmMATE4 was localized at the vacuolar membrane-like structure. Protein charge prediction showed that the mutation of the glutamate residues to alanine would reduce the negative charge at the N-terminus. Using yeast as the model, we showed that GmMATE4 mediated the transport of daidzein, genistein, glycitein, and glycitin. In addition, the glutamate-to-alanine mutation reduced the isoflavone transport capacity of GmMATE4. Altogether, we demonstrated GmMATE4 as an isoflavone transporter and the functional significance of the N-terminal poly-glutamate motif of GmMATE4 for regulating the isoflavone transport activity.

Keywords: multidrug and toxic compound extrusion (MATE) transporter; proton motive force; proton gradient; poly-glutamate motif; acidic amino acid; isoflavone; legume; soybean

1. Introduction

Multidrug and toxic compound extrusion (MATE) transporters are a family of active membrane transporters generally consisting of 12 transmembrane domains (TMDs) [1]. The 1st to the 6th TMDs are usually termed as the N-lobe while the 7th to the 12th TMDs are usually termed as the C-lobe [2,3]. MATE transporters have been reported to be localized at membranes including plasma, vacuolar, and mitochondrial membranes, the chloroplast envelope, and the surface of small vesicles [1]. In prokaryotes, MATE transporters transport the substrate across the membrane in exchange for Na^+ or H^+, while in eukaryotes, they tend to transport the substrate in exchange for H^+ only [4]. The electrochemical gradient across the membrane drives the transport of the substrate. Structurally, MATE proteins are generally characterized by the typical 12 TMDs [5]. However, the different amino acid residues of different MATE proteins enable different substrate specificities [1,6,7]. In plants, MATE transporters have been reported to be responsible for the movement of alkaloids, phytohormones, ion chelators, and flavonoids including anthocyanins and isoflavones across membranes [1,6–9]. The transport direction is largely associated with the nature of the substrate to be transported. For example, it has been reported that GmMATE1 and

GmMATE2 mediate the transport of isoflavones into the vacuole for storage [7]. In plant genomes, *MATE* genes usually compose a big family. For example, in studies reporting genome-wide identification of *MATE* genes, 117, 67, 53, 56, and 40 *MATE* genes were identified in *Glycine max*, [10], *Solanum lycopersicum* [11], *Oryza sativa* [12], *Arabidopsis thaliana* [13], and *Medicago truncatula* [9], respectively. The need to transport various metabolites by MATE proteins with different substrate specificities may be the possible reason behind the big MATE families in plants.

Research has been done to understand the protein structure of MATE transporters and the roles of the various domains within that structure. For example, it was found that the protonation of acidic amino acid residues at the C-lobe of the transmembrane domain of AtDTX14 (*Arabidopsis thaliana* detoxification efflux carrier 14) resulted in a structural change in the transmembrane helix 7, and thus, regulated the transport mechanism [2]. It was also suggested that the binding of protons at these acidic residues played a role in the coupling of the proton-motive force to regulate the substrate transport [2]. Previous studies also suggested the importance of the N-terminal domain (NTD) in substrate and ion binding [3,14]. Based on the crystal structure of the MATE protein from *Bacillus halodurans* (DinF-BH), it has been reported that D40 in TMD1 was the key cation-binding residue, which also bound to the substrate to be transported [3]. The D40N mutation led to drastically reduced binding of the drug to be transported, and the ability of the MATE protein to release H^+ upon drug binding was abolished [3].

Most research has focused on studying the structural importance of the TMDs. However, the non-transmembrane domains of these transporters, i.e., the N-terminal tail and the C-terminal tail, also play significant regulatory roles. It has been reviewed that the N- and the C-terminal tails of transporters are involved in other aspects of the regulation of the transporter proteins, such as the sorting to organelle, ubiquitination and turnover, and transport activity and substrate specificity [15]. Several transporters have been reported to have conserved N- ad C-termini among a group of organisms [15]. For example, nucleobase-ascorbic transporters (NATs) have a nearly absolutely conserved sequence in the N-terminus among fungi [15]. This conserved motif is not found in other organisms such as prokaryotes, slime molds, protists, plants, or metazoans [15,16]. Another example is the C-terminus of the yeast amino acid polyamine-organocation (APC) superfamily. It was found that the C-termini of yeast APCs were more conserved than the N-termini and were predicted to be involved in the sorting, turnover, palmitoylation, and the formation of secondary structures of the transporter proteins [15]. However, the degrees of sequence conservation and the functions of the N- and C-termini of MATE transporters in various organisms including plants are largely unknown.

Through phylogenetic analyses of MATE proteins from 14 representative plant species, including those in the Leguminosae family (*Arachis ipaensis, Cajanus cajan, Cicer arietinum, Lotus japonicus, Medicago truncatula, Phaseolus vulgaris, Vigna radiata* and *Glycine max*) and other non-Leguminosae species (*Arabidopsis thaliana, Gossypium hirsutum, Helianthus annuus, Oryza sativa, Solanum lycopersicum,* and *Solanum tuberosum*), we found clades enriched with MATE proteins from Leguminosae and a cluster consisting solely of MATE proteins from *Glycine max*. MATE proteins from this latter cluster, including GmMATE4 which was previously identified to be in the overlapping quantitative trait loci (QTLs) regulating the contents of antioxidants, phenolics, and flavonoids in soybean seeds [17], have a conserved poly-glutamate motif at the domain predicted to be inside the vacuolar compartment (vacuolar domain) of the N-terminus. Here, we demonstrated GmMATE4 as an isoflavone transporter and the role of the poly-glutamate motif in regulating the transport activity of GmMATE4.

2. Materials and Methods

2.1. Phylogenetic Analysis, Protein Sequence Alignment, Topology Prediction, and Protein Charge Prediction

The protein sequences of MATE family representatives from 14 plant species, including Leguminosae (*Arachis ipaensis, Cajanus cajan, Cicer arietinum, Lotus japonicus, Medicago truncatula, Phaseolus vulgaris, Vigna radiata,* and *Glycine max*) and non-Leguminosae (*Arabidopsis thaliana, Gossypium hirsutum, Helianthus annuus, Oryza sativa, Solanum lycopersicum,* and *Solanum tuberosum*), were retrieved from the NCBI or Phytozome database and used for gene family clustering. The longest protein sequences of each gene were retained and all-against-all BLASTP [18] was used to calculate pairwise sequence similarities with a *p*-value cutoff of 1×10^{-5}. Then, the homologous genes were identified by OrthoMCL [19] with the use of default parameters. The MATE protein sequences were aligned by ClustalW in MEGA11 [20]. The topology prediction of MATE proteins was done with the use of Protter [21]. Then, the N-termini of the MATE proteins were subjected to charge prediction by Prot-pi [22] using the pKa values based on ProMoST [23] for calculating the isoelectric point.

2.2. Subcellular Localization Study

GmMATE4, *GmMATE4Δ3ala*, and *GmMATE4Δ7ala* were amplified from pGBKT7ΔBD-*GmMATE4*, pGBKT7ΔBD-*GmMATE4Δ3ala*, and pGBKT7ΔBD-*GmMATE4Δ7ala* as the templates, respectively, with the use of PrimeSTAR GXL DNA Polymerase (R050B, TaKaRa). The DNA fragments were subjected to restriction digestion, and then, fused to the 5′ end of *green fluorescent protein* (*GFP*). The fusion constructs were cloned downstream of the CaMV 35S promoter of the vector V7. The primers and restriction enzymes used are listed in Supplementary Materials Table S1. Restriction enzymes and T4 DNA ligase used for cloning were from New England Biolabs.

The plasmids were, then, coated onto gold particles and bombarded into onion epidermis with the use of the Biolistic® PDS-1000/He Particle Delivery System (Bio-Rad, Hercules, CA, USA), according to manufacturer's protocol. Each 3 mg of gold particles (1.0 µm diameter, #1652263, Bio-Rad, Hercules, CA, USA) was washed with 1 mL 70% EtOH by vortexing for 5 min. After that, the gold particles were allowed to be incubated in the 70% EtOH before being fast spun for 5 s. Then, the supernatant was removed, the particles were added with 1 mL sterile water by vortexing, and then, the gold particles were allowed to be incubated in water for 1 min before being spun for 2 s. Then, the supernatant was removed. The water-wash steps were performed three times. After removing the supernatant, the washed particles were added with 50 µL sterile 50% glycerol. Then, 5 µL plasmid DNA (1 mg/mL), 50 µL $CaCl_2$ (2.5 M), and 20 µL spermidine (0.1 M) were mixed with the gold particle by vortexing. After the mixing, the gold particles were allowed to be incubated in the mixture for 1 min before being fast spun for 2 s. Then, the supernatant was removed. Without being disturbed, the gold particles were washed using 140 µL 70% EtOH, followed by being washed with 140 µL absolute EtOH. After removing the supernatant, 48 µL EtOH was added to the gold particles. The gold particles were resuspended by pipetting before the bombardment. All steps were performed at room temperature. For each bombardment under 1100 psi, 0.75 mg of the gold particles coated with 1.25 µg of plasmid DNA were used. The onion epidermis was left at room temperature overnight before being observed using a confocal microscope (Olympus FV1000, excitation 488 nm, the emission signal was collected between 500 and 545 nm). The images were processed by a FV10-ASW 4.2 Viewer. The experiment was performed twice. All cells showing the green, fluorescent signal ($\geq$11 cells) from the two experiments showed similar patterns.

2.3. Yeast Isoflavone Uptake Assay

The coding sequence (CDS) of *GmMATE4* (*Glyma.19G120900*) was cloned from a mixture of cDNAs from the pod and the seed of cultivated soybean (*Glycine max*) accession C08, using PrimeSTAR GXL DNA Polymerase (R050B, TaKaRa, Shiga, Japan). The poly-glutamate motif of GmMATE4 was converted from EEEEEEE to AAAEEEE (GmMATE4Δ3ala) or from EEEEEEE to AAAAAAA (GmMATE4Δ7ala) with the use of specially designed primers. Then, the DNA fragments were cloned into the vector pGBKT7ΔBD [7]. The primers and restriction enzymes used are listed in Table S1. Restriction enzymes and T4 DNA ligase used for cloning were from New England Biolabs. The plasmids were transformed into the yeast strain Y2HGold (630489, Takara, Shiga, Japan) by the LiAc/PEG method [24]. To prepare competent yeast cells, overnight culture of the yeast strain Y2HGold at stationary phase, grown in YPDA (yeast peptone dextrose adenine) broth at 30 °C with shaking at 200 rpm, was inoculated to a fresh broth of YPDA at a 1:10 ratio. Then, the inoculated broth was shaken at 30 °C, 200 rpm until OD600 reached 0.4–0.6. After that, the yeast cells were pelleted at 1000× g for 5 min. The supernatant was removed. Then, the pelleted cells were washed by resuspending with sterile water, pelleted again at 1000× g for 5 min, with the supernatant removed after centrifugation. The cells were washed twice with sterile water. After that, the washed cells were resuspended in TE/LiAc buffer (0.01 M Tris-HCl, 1 mM EDTA, 0.1 M lithium acetate, pH 7.5). The yeast cells resuspended in TE/LiAc were used as competent yeast cells for transformation. Each 1 mL competent yeast cells resulted from 20 mL overnight yeast culture at stationary phase. To transform the competent yeast cells, for each construct, 0.1 µg of plasmid was mixed with 0.1 mg denatured carrier DNA (630440, TaKaRa, Shiga, Japan) and 0.1 mL competent yeast cells by vortexing. After that, the mixture of plasmid and competent yeast cells was mixed with 0.6 mL filter-sterile PEG/LiAc solution (40% PEG 4000, 0.01 M Tris-HCl, 1 mM EDTA, 0.1 M lithium acetate, pH 7.5) by vortexing. Then, the mixture was shaken at 30 °C, 200 rpm, for 30 min. After that, 70 µL DMSO was added to each transformation mixture by gentle inversion. Then, the mixture was incubated in 30 °C for 15 min, and then, on ice for 2 min. After that, the yeast cells were pelleted at room temperature, 14,000 rpm, for 5 s. After removing the supernatant, the yeast cells were resuspended in 0.1 mL sterile TE buffer (0.01 M Tris-HCl, 1 mM EDTA, 0.1 M lithium acetate, pH 7.5) and spread on 1.5% agar in synthetic dropout medium without tryptophan (SD-Trp) for selection at 30 °C. Two days after the incubation at 30 °C, the yeast colonies were picked and streaked on a fresh plate of 1.5% agar in SD-Trp to confirm the selection.

The yeast isoflavone uptake assays were performed, as described previously [7,25] with slight modifications. Untransformed Y2HGold was compared to GmMATE4-transformed yeast to rule out that the uptakes of isoflavones in GmMATE4 expressing yeast cells were solely due to the endogenous yeast proteins. Yeast cells transformed with *GmMATE4*, *GmMATE4Δ3ala*, and *GmMATE4Δ7ala* were used to demonstrate the functional significance of the poly-glutamate motif. Untransformed Y2HGold (wildtype) was grown in SD-Trp broth (complement with Trp) to compensate the inability to synthesize tryptophan, while the transformed yeast cells were grown in SD-Trp broth. The yeast was grown in SD-Trp broth at 30 °C with shaking at 200 rpm until OD600 of the cells reached 2.0. The cells were pelleted at 1000× g for 3 min. The cell pellet was washed twice with phosphate-buffered saline (PBS, pH 7.4) and resuspended in 15 mL PBS (pH 7.4). Each 1 mL aliquot of cells was treated overnight with 200 µM isoflavone or DMSO only as the mock control at 30 °C with shaking at 200 rpm. Then, the treated yeast cells were pelleted at 1000× g for 3 min and washed twice with 50 mM potassium phosphate buffer (pH 7.8). Intracellular metabolites in yeast cells were extracted with pure methanol by the freeze-thaw method, as previously described [7,26]. In brief, the cell pellet was mixed with 0.1 g glass powder and 500 µL methanol by vortexing. Fifty-micromolar formononetin was added as a spike. Then, the mixture was frozen in liquid nitrogen and thawed on ice. The freeze-thaw cycle was performed twice. After that, the mixture was centrifuged at 14,000× g for 5 min. The supernatant was filtered through an Acrodisc 13 mm mini spike 0.2 µm PVDF filter (4450, Pall Corporation, Port Washington, NY, USA). The filtered supernatant was analyzed using a Vanquish Flex UHPLC System with a UV-Diode array detector (Thermo Scientific, Waltham, MA, USA) and with the use of Acquity BEH C18 reverse-phase column (1 mm × 100 mm, 1.7 µm) (186002346, Waters Corporation, Milford, MA, USA). Three microliters of the filtered supernatant were injected into the column with the flow rate maintained at 0.1 mL min^{-1}. A mixture of ultra-pure water with 0.1% formic acid (v/v) was used as solvent A while absolute acetonitrile was used as solvent B. Solvent A and solvent B at different percentages were used to make up 100% composition of the mobile phase. The sample was eluted with a gradient set at 0–5 min, 10–20% solvent B; 5–10 min, 20–90% solvent B; 10–20 min, 90% solvent B; 20–20.1 min, 90–10% solvent B; 20.1–25 min, 10% solvent B.

Different isoflavone species were separated by gradient elution. Since different isoflavone species have the best solubility at different compositions of solvent A and solvent B, the elution strategy was to, first, elute isoflavones which are comparatively more soluble in aqueous solvent then other isoflavone species (0–5 min), and then, elute isoflavones by progressively increasing the percentage of solvent B (5–10 min). In the first 10 min, most of the isoflavone species are eluted (Figure S3). After that, the composition of solvent B was increased to the maximum amount to 90% to elute any molecules remaining in the stationary phase (10–20 min). The final step was to increase the composition of solvent A to 90% (i.e., 10% solvent B) to prepare the column for the next cycle of elution. The identification of compounds was achieved by comparing the retention times and UV spectra at 260 nm [27]. Vials filled with either standard solutions or samples were kept at 10 °C in the autosampler. The standards, daidzein (D-101), glycitein (GL-001), genistein (G-103), daidzin (021096), glycitin (GL-002), genistin (021050), and malonylgenistin (06–1454) were purchased from Indofine Chemical Company, Inc (Hillsborough, NJ, USA). Malonyldaidzin (PS1633–0010) and malonylglycitin (139–13831) were purchased from Chengdu Push Bio-Technology Co., Ltd (Chengdo, China) and Fujifilm Wako Pure Chemical Corporation (Chuo-Ku, Japan), respectively. The isoflavones were dissolved in DMSO and stored at −80 °C before use.

3. Results and Discussion

3.1. Phylogenetic and Protein Sequence Analyses of MATE Proteins

In order to gain insights into the structures of MATE proteins, we performed a phylogenetic analysis of MATE proteins from 14 representative plant species including some in the Leguminosae group (*Arachis ipaensis, Cajanus cajan, Cicer arietinum, Lotus japonicus, Medicago truncatula, Phaseolus vulgaris, Vigna radiata,* and *Glycine max*) and others in the non-Leguminosae group (*Arabidopsis thaliana, Gossypium hirsutum, Helianthus annuus, Oryza sativa, Solanum lycopersicum,* and *Solanum tuberosum*) which have a Benchmarking sets of Universal Single-Copy Orthologs (BUSCO) index [28] of their genomes higher than 85 (Figure 1). The results show that some clades (clades 1–4) are enriched with MATE proteins from Leguminosae (Figure 1). Within clade 2, there is a cluster consisting solely of MATE proteins from *Glycine max* (Glyma.03G005200, Glyma.03G005400, Glyma.19G120900, Glyma.03G005300, and Glyma.08G244400). This cluster is indicated by a black bar in clade 2 in Figure 1.

3.2. Identification of the Poly-Glutamate Motif in GmMATE4 by Sequence Alignment

In the cluster within clade 2 that consists solely of proteins from *G. max*, besides Glyma.08G244400 being predicted to be a truncated protein (Figure S1), Glyma.03G005200, Glyma.03G005300, and Glyma.03G005400 were all predicted to have 12 transmembrane domains (Figure S1). In this cluster, *Glyma.19G120900* has been previously identified to encode a MATE-type protein and was located in the overlapping QTLs regulating the contents of antioxidants, phenolics, and flavonoids in soybean seeds [17]. By aligning the protein sequences of these MATE proteins, a conserved poly-glutamate motif with a varied number of glutamate residues was found in the N-terminus (Figure 2). As a comparison, GmMATE1 and GmMATE2, which have also been reported to be located in the same overlapping QTLs with GmMATE4 [17], were also included in the protein sequence alignment. Among these MATE proteins, GmMATE4 has the most (seven consecutive) glutamate residues. Although Glyma.03G005400 has only two glutamate residues in the motif, it also has an aspartate residue, which is an acidic amino acid as well.

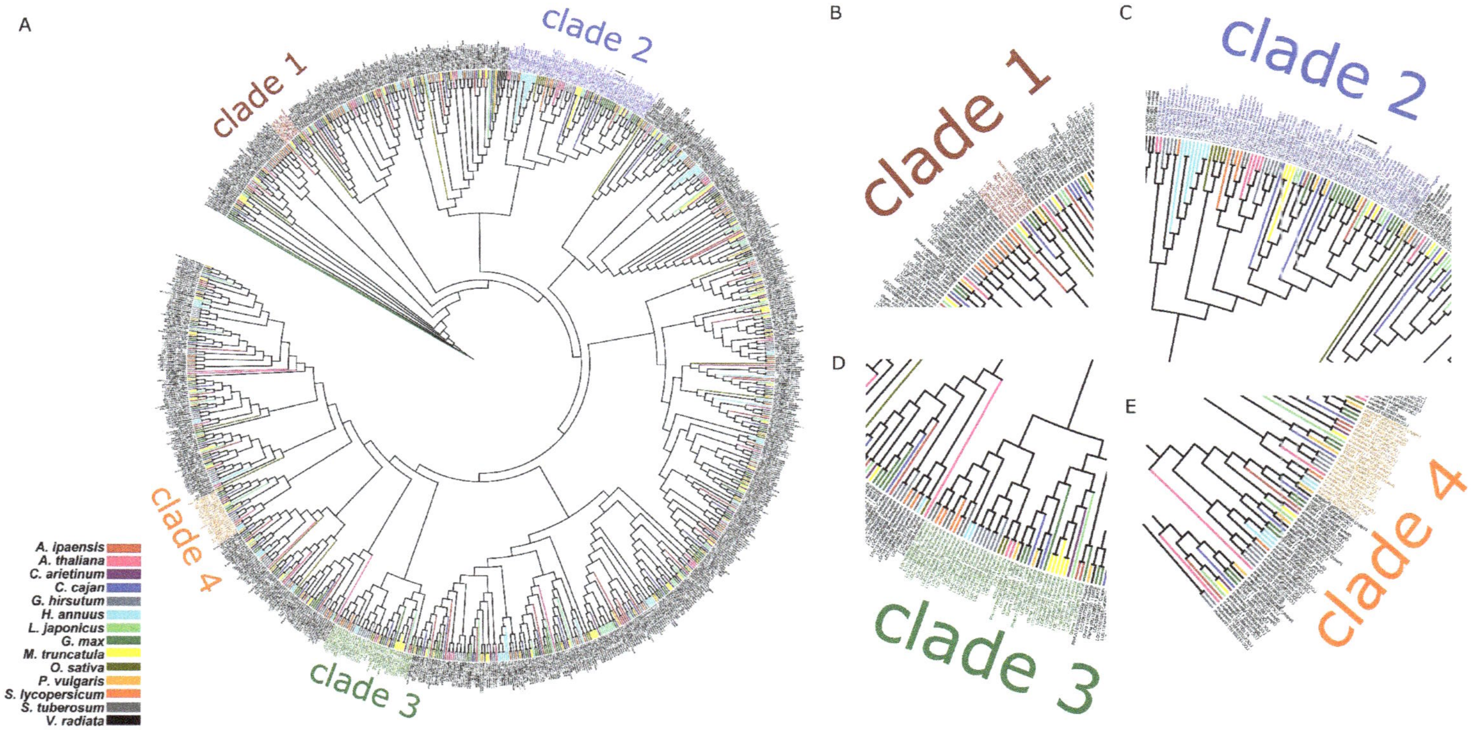

Figure 1. Phylogenetic analysis of the MATE family of proteins from 14 Leguminosae species (*Arachis ipaensis*, *Cajanus cajan*, *Cicer arietinum*, *Lotus japonicus*, *Medicago truncatula*, *Phaseolus vulgaris*, *Vigna radiata*, and *Glycine max*) and non-Leguminosae plants (*Arabidopsis thaliana*, *Gossypium hirsutum*, *Helianthus annuus*, *Oryza sativa*, *Solanum lycopersicum*, and *Solanum tuberosum*). The cluster consisting solely of MATE proteins from *Glycine max* is indicated by a black bar in clade 2. (**A**) The whole phylogenetic tree; (**B**) the enlarged portion of the phylogenetic tree showing clade 1; (**C**) the enlarged portion of the phylogenetic tree showing clade 2; (**D**) the enlarged portion of the phylogenetic tress showing clade 3; (**E**) the enlarged portion of the phylogenetic tree showing clade 4.

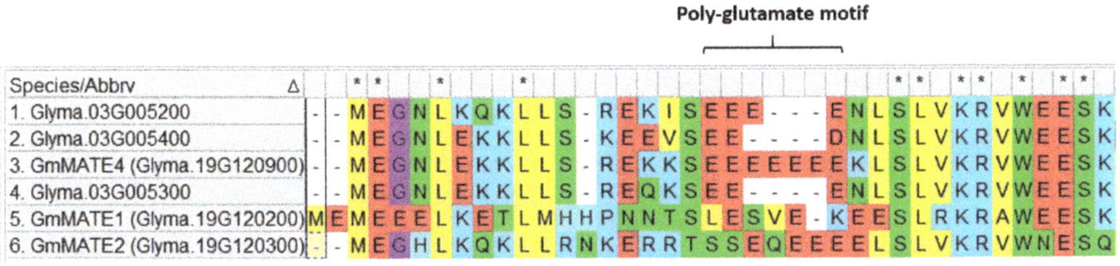

Figure 2. Alignment of the N-terminal amino acid sequences of Glyma.03G005200, Glyma.03G005300, Glyma.03G005400, GmMATE4, GmMATE1, and GmMATE2. The alignment of the full protein sequences is shown in Figure S2. The sequence alignment was done by ClustalW using MEGA11 [20].

3.3. The Effect of the Poly-Glutamate Motif on the Protein Net Charge

Next, we investigated the functional significance of this poly-glutamate motif. Based on the topology prediction (Figure S2) and using Prot-pi [22], the N-termini of the MATE proteins (before the first transmembrane domain) were subjected to net charge prediction (Table 1). Since the close homologs, GmMATE1 and GmMATE2, were previously characterized to be localized at the vacuolar membrane [7], the net charge prediction was done at pH 5 or pH 5.5, which is the vacuolar pH [29] (Table 1). To assess the significance of the long string of consecutive glutamate residues in GmMATE4, mutations were created to produce a fragment in which three of the glutamate residues were substituted by alanine (GmMATE4Δ3ala) and a fragment in which all the glutamate residues were substituted by alanine (GmMATE4Δ7ala). The results showed that, among the MATE proteins in the cluster, the vacuolar domain of GmMATE4 was the most negatively charged at pH 5 or pH 5.5 (Table 1). Furthermore, the substitutions of glutamate residues by alanine (GmMATE4Δ3ala and GmMATE4Δ7ala) rendered the N-termini positively charged (Table 1). Next, we wanted to find out whether GmMATE4 is an isoflavone transporter, like its homolog GmMATE1 and GmMATE2, and whether the alanine substitutions of this poly-glutamate motif affect the function of GmMATE4.

Table 1. Net charge prediction of the N-termini of MATE-type proteins.

	Net Charge of the Vacuolar N-Terminal Region		% of Acidic Amino Acids (No. of Acidic Amino Acids/Total Amino Acids)
	pH 5	pH 5.5	
Glyma.03G005200	+0.773	−0.37	24.2% (8/33)
Glyma.03G005400	−1.881	−3.245	29.4% (10/34)
Glyma.03G005300	+0.794	−0.363	23.5% (8/34)
GmMATE1 (Glyma.19G120200)	+0.3	−1.534	27.5% (11/40)
GmMATE2 (Glyma.19G120300)	+2.708	+1.426	21.6% (8/37)
GmMATE4 (Glyma.19G120900)	−0.348	−2.043	33.3% (12/36)
GmMATE4Δ3ala	+1.992	+0.712	25% (9/36)
GmMATE4Δ7ala	+5.113	+4.385	13.9% (5/36)

3.4. Subcellular Localization Study

The close homologs of GmMATE4, GmMATE1, and GmMATE2 were reported to be localized in the vacuolar membrane [7]. As a confirmation of the subcellular localization of GmMATE4, GmMATE4Δ3ala, and GmMATE4Δ7ala, the genes encoding these proteins were fused with *green fluorescent protein* (*GFP*). The fusion constructs, *GmMATE4-GFP*, *GmMATE4Δ3ala-GFP*, or *GmMATE4Δ7ala-GFP* were coated on gold particles and bombarded into onion epidermal cells. The green fluorescence signal was, then, detected using a confocal microscope. The results showed that GmMATE4-GFP, GmMATE4Δ3ala-GFP, and GmMATE4Δ7ala-GFP were preferably localized in the vacuolar membrane (Figure 3). Based on the topology prediction and the subcellular localization study, the N-termini used for the net charge prediction of GmMATE4, GmMATE4Δ3ala, and GmMATE4Δ7ala (Table 1) were confirmed to be a vacuolar domain.

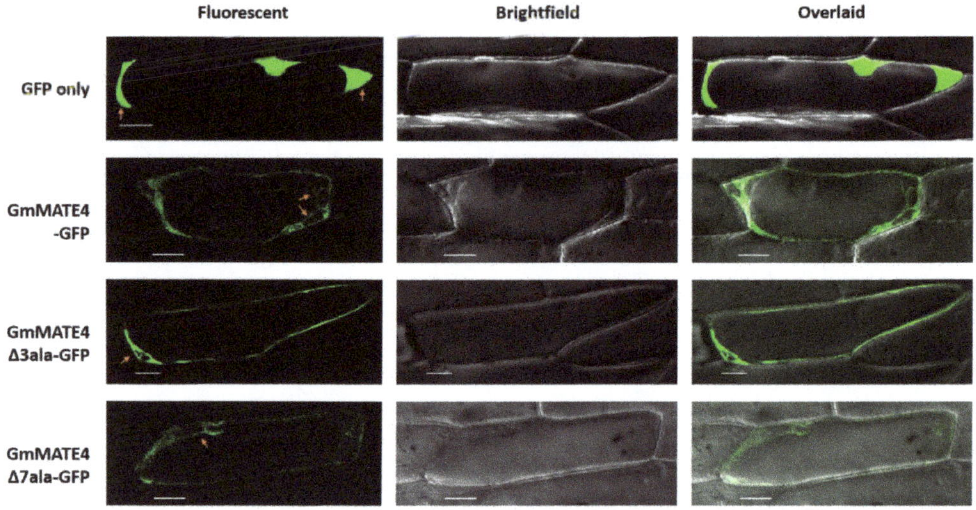

Figure 3. Subcellular localization study of GmMATE4, GmMATE4Δ3ala, and GmMATE4Δ7ala. GFP was fused to the C-termini of GmMATE4, GmMATE4Δ3ala, and GmMATE4Δ7ala. The fusion construct was cloned downstream of a CaMV 35S promoter in the plasmid V7. The plasmid was coated onto gold particles, bombarded into onion epidermal cells, and then, observed using a confocal microscope. Scale bar, 50 μm; excitation 488 nm; and the emission signal was collected between 500 and 545 nm. All cells having the green, fluorescent signal ($\geq$11 cells) from two biological repeats showed the same patterns.

3.5. The Transporter Activity of GmMATE4

The close homologs of GmMATE4, GmMATE1, and GmMATE2 were reported to be isoflavone transporters [7]. To test whether GmMATE4 is also an isoflavone transporter, using yeast as the model, *GmMATE4* was ectopically expressed in yeast cells. Untransformed yeast cells (wildtype) and *GmMATE4* expressing yeast cells were subjected to isoflavone uptake assays. The yeast cells were treated with individual isoflavones, including aglycones (daidzein, genistein, and glycitein), isoflavone glucosides (daidzin, genistin, and glycitin), and malonylated (M-) isoflavones (M-daidzin, M-genistin, and M-glycitin). The results of the uptake assays showed that, as compared with the wildtype, *GmMATE4* expressing yeast cells had significantly higher amounts of daidzein (Figure 4A), genistein (Figure 4B), and glycitein (Figure 4C) after the respective isoflavone treatments. When the yeast cells were treated with glycitin, a significantly higher amount of glycitein was found in *GmMATE4* expressing yeast cells as compared with the wildtype (Figure 4D). The chromatograms showing the detection of isoflavones after all the treatments are shown in Supplementary Figures S4–S6. The chromatograms of the mock treatment, in which only DMSO but not any isoflavone was added to the yeast cells, are shown in Supplementary Figure S7.

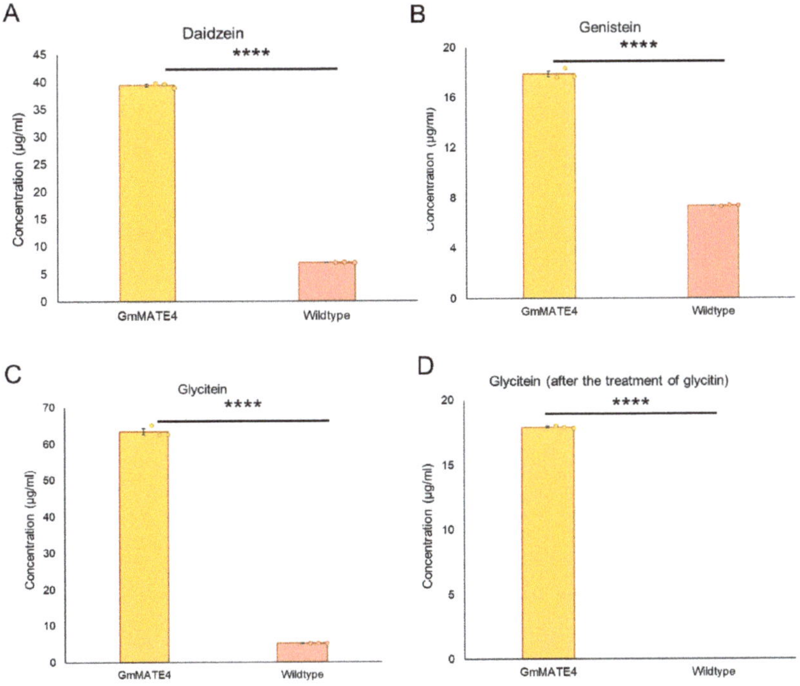

Figure 4. Uptake assays of isoflavones. The isoflavones detected in the yeast cells after the treatments of: (**A**) daidzein; (**B**) genistein; (**C**) glycitein; or (**D**) glycitin. Error bar: standard error from three technical repeats. Student's *t* test (two-tailed) was performed to compare the isoflavone uptake capacity between the constructs. **** indicates $p < 0.0001$. The experiments were performed twice with similar results.

3.6. The Functional Significance of the Poly-Glutamate Motif on the Transporter Activity of GmMATE4

The above results suggested that GmMATE4 mediates the uptake of daidzein, genistein, glycitein, and glycitin (Figure 4). Then, we tested the functional significance of the poly-glutamate motif on the activity of GmMATE4 to transport these isoflavone species (daidzein, genistein, glycitein, and glycitin). *GmMATE4*, *GmMATE4Δ3ala*, or *GmMATE4Δ7ala* was ectopically expressed in yeast cells for isoflavone uptake assays. The yeast cells were treated with daidzein, genistein, glycitein, or glycitin. The results of the uptake assays showed that, as compared with GmMATE4, GmMATE4Δ3ala and GmMATE4Δ7ala had weakened capacities to mediate the transport of daidzein (Figure 5A) and genistein (Figure 5B). GmMATE4Δ3ala and GmMATE4Δ7ala had similar capacities to mediate the transport of genistein (Figure 5B) but GmMATE4Δ7ala had even weaker capacity to mediate the transport of daidzein as compared with GmMATE4Δ3ala (Figure 5A). As compared with GmMATE4, GmMATE4Δ3ala had weakened capacity to transport glycitein (Figure 5C). However, as compared with GmMATE4, GmMATE4Δ7ala had similar capacity transport glycitein (Figure 5C). When the cells were treated with glycitin, a significantly higher amount of glycitein was found in both *GmMATE4Δ3ala* expressing yeast cells and *GmMATE4Δ7ala* expressing yeast cells as compared with *GmMATE4* expressing yeast cells (Figure 5D). The chromatograms showing the detection of isoflavones after the treatment of daidzein, genistein, glycitein, and glycitin in *GmMATE4*, *GmMATE4Δ3ala*, or *GmMATE4Δ7ala* expressing yeast cells are shown in Supplementary Figure S8.

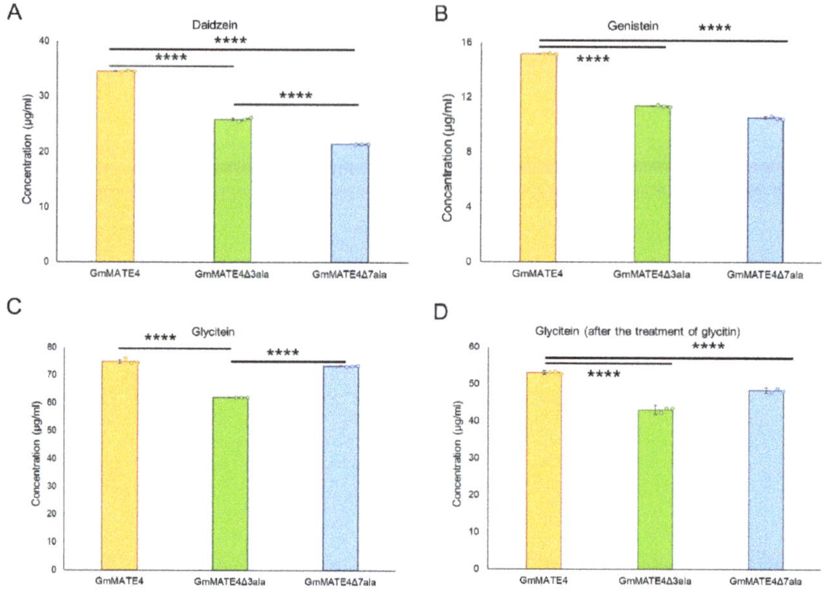

Figure 5. Uptake assays of isoflavones. The isoflavones detected in the yeast cells after the treatments of (**A**) daidzein, (**B**) genistein, (**C**) glycitein, and (**D**) glycitin. Error bar: standard error from three technical repeats. Student's *t* test (two-tailed) was performed to compare the isoflavone uptake capacity between the constructs. **** indicates $p < 0.0001$. The experiments were performed twice with similar results.

4. Discussion

Structural analyses have been done to understand the functional significance of each domain of MATE transporters. However, previous studies have largely been biased towards understanding the TMDs rather than the N- and C-termini [1–3]. In addition, studies on other transporter proteins have revealed that the N- and C-termini could also play important regulatory roles [15].

To improve the understanding of the protein domains of MATE transporters in plants, MATE transporters from 14 representative plant species, including Leguminosae and non-Leguminosae, were subjected to phylogenetic analyses (Figure 1). Among the clades enriched with MATE proteins from Leguminosae, there was a cluster consisting solely of MATE proteins from *G. max* (Figure 1). Protein sequence alignment of members in this cluster showed a motif rich in glutamate residues (Figure 2). This motif is, thus, named a poly-glutamate motif. The substitution of the negatively charged glutamate residues by alanine reduced the overall negative charge of this domain as expected (Table 1). Glutamate residues have been suggested to be involved in proton transport. For example, the D35, D38, and D144 in the TMDs of AtPHT1;1 (*Arabidopsis thaliana* phosphate transporter 1;1) have been reported to play an essential role in proton transfer by forming a conduit that facilitates proton transport across membranes [30]. The mutation of any of these residues to alanine hampered the phosphate transport activity [30]. In addition, the glutamate residue, named gating glutamate (Glu_{gate}) and conserved in the CLC (chloride channels) family of proteins, has been suggested to serve as a gate for proton transit in CLC antiporters [31–34]. The mutation of this glutamate residue to alanine (E148A) in EcCLC (*Escherichia coli* CLC) hampered proton transport as compared with the wild type [34]. The mutation of threonine to glutamate (T269E) in CmCLC (CLC from red algae) promoted proton transport as compared with the wild type [34]. The above examples show the importance of glutamate residues in TMDs for promoting proton transport. Despite these findings, the importance of glutamate residues in the N- or C-terminus, and the functional significance of consecutive glutamate residues, remain to be unclear. Nevertheless, based on the above examples, it could be deduced that glutamate residues play an important role in promoting proton affinity. In addition, the attraction of proton by the production of polyglutamic acid on the cell surface has been reported in *Bacillus pseudofirmus* OF4 cells [35,36]. It was suggested that such attraction of protons to the cell surface was associated with the formation of proton motive force across the membrane [36]. It is possible that the polyglutamate motif of GmMATE4 plays a similar role in attracting protons to the vacuolar membrane surface. Previous studies have shown that MATE transporters in eukaryotes are antiporters that transport their substrates in exchange for H^+ [4]. The proton gradient across the membrane drives the transport of the substrate [4]. Therefore, it is logical to deduce that the poly-glutamate motif of GmMATE4 at the vacuolar domain is associated with the attraction of protons to the membrane surface, and thus, promote the proton motive force and the transport of the substrates into the vacuole.

The results of the isoflavone uptake assays were in line with the above deduction. Similar to GmMATE1 and GmMATE2, which are close homologs of GmMATE4, GmMATE4 could mediate the transport of isoflavones including daidzein, genistein, glycitein, and glycitin (Figure 4). The mutation of glutamate residues to alanine (GmMATE4Δ3ala or GmMATE4Δ7ala) reduced the isoflavone transport activity of the mutant proteins as compared with the native GmMATE4 (Figure 5). Among the substrates being transported, GmMATE4Δ3ala or GmMATE4Δ7ala showed differential capacity to transport daidzein, with GmMATE4Δ3ala having a higher transport activity as compared with GmMATE4Δ7ala (Figure 5A). It is possible that a further reduction in the net charge of the N-terminus of GmMATE4Δ7ala (Table 1) led to a further reduction in the transport activity. However, such phenomenon was not observed when the cells were treated with genistein (Figure 5B) or glycitein (Figure 5C). In addition, when the cells were treated with glycitein, GmMATE4Δ3ala had reduced transport capacity as compared with the native GmMATE4, as expected (Figure 5C), while GmMATE4Δ7ala and the native GmMATE4 had a similar capacity to transport glycitein (Figure 5C). Such observations are possibly due to the different mechanics of the transport of different substrates and the potential effects of the mutation on the interaction with other protein motifs. A detailed mechanistic study would be needed to delineate the regulation.

Similar to GmMATE1 and GmMATE2 [7], which are the close homologs of GmMATE4, in this study, we showed that GmMATE4 localizes at the vacuolar membrane-like structure (Figure 3). Together with the isoflavone uptake assays (Figure 4), the results indicated that GmMATE4 has similar functions as GmMATE1 and GmMATE2, which were previously demonstrated to be involved in the storage of isoflavones in the vacuole. However, the longer consecutive glutamate residues at the vacuolar domain of GmMATE4 as compared with GmMATE1 and GmMATE2 (Figure 2) possibly allows further regulation of the transporter function.

5. Conclusions

In this study, by the phylogenetic analysis of MATE proteins from 14 plant species, we identified a cluster consisting of four full-length MATE proteins in soybean. By aligning the four full-length MATE proteins, a poly-glutamate motif was identified. Among the four MATE proteins, GmMATE4 has the most consecutive glutamate residues at the N-terminus, which is the vacuolar domain. Net charge prediction suggested the importance of the poly-glutamate motif to the negative charge of the N-termini of GmMATE4. In this study, GmMATE4 was shown to be localized at the vacuolar membrane-like structure. GmMATE4 was also found to mediate the uptake of isoflavones including daidzein, genistein, glycitein, and glycitin. The mutagenesis and isoflavone uptake assays show that the poly-glutamate motif of GmMATE4 regulates its activity to transport daidzein, genistein, glycitein, and glycitin. To the best of our knowledge, this is the first study reporting the isoflavone transport activity of GmMATE4 and the functional significance of poly-glutamate motif on the transporter activity of MATE protein.

Supplementary Materials: The following supporting information can be downloaded at https://www.mdpi.com/article/10.3390/membranes12020206/s1. Figure S1: The predicted transmembrane topologies of GmMATE transporters, Figure S2: Alignment of the full protein sequences of Glyam.03G005200, Glyma.03G005300, Glyma.03G005400, GmMATE4 (Glyma.19G120900), GmMATE1 (Glyma.19G120200), and GmMATE2 (Glyma.19G120300), Figure S3: A UHPLC chromatogram showing the absorption peaks of the nine isoflavone standards, Figure S4: Representative UHPLC chromatograms of the detection of isoflavones in the wildtype and *GmMATE4* expressing yeast cells after the treatment with daidzein, genistein, and glycitein, Figure S5: Representative UHPLC chromatograms of the detection of isoflavones in the wildtype and *GmMATE4* expressing yeast cells after the treatment with daidzin, genistin, and glycitin, Figure S6: Representative UHPLC chromatograms of the detection of isoflavones in the wildtype and *GmMATE4* expressing yeast cells after the treatment with malonyl-daidzin, malonyl-genistin, and malonyl-glycitin, Figure S7: Representative UHPLC chromatograms of the detection of isoflavones after the mock treatment, Figure S8: Representative UHPLC chromatograms of the detection of isoflavones in GmMATE4, GmMATE4Δ3ala, or GmMATE4Δ7ala expressing yeast cells after the treatment with daidzein, genistein, glycitein, and glycitin, Table S1: List of primers and restriction enzymes used for cloning.

Author Contributions: Conceptualization, Y.K. and H.L.; methodology, Y.K., S.C., M.C., Y.N. and A.L.; formal analysis, Y.K., S.C., M.C. and A.L.; investigation, Y.K., S.C., M.C., A.L. and G.C.; data curation, Y.K., S.C., M.C., Y.N. and A.L., writing—original draft preparation, Y.K., S.C., M.C. and Y.N.; writing—review and editing, Y.K. and H.L.; supervision, H.L.; project administration, Y.K. and H.L.; funding acquisition, H.L. All authors have read and agreed to the published version of the manuscript.

Funding: This research was funded by the Hong Kong Research Grants Council General Research Fund (14143916) to H.L.

Institutional Review Board Statement: Not applicable.

Informed Consent Statement: Not applicable.

Data Availability Statement: All data generated in this study are available within this manuscript and companion Supplementary Materials.

Acknowledgments: J.Y. Chu copy-edited this manuscript. Any opinions, findings, conclusions, or recommendations expressed in this publication do not reflect the views of the Government of the Hong Kong Special Administrative Region or the Innovation and Technology Commission.

Conflicts of Interest: The authors declare no conflict of interest.

References

1. Takanashi, K.; Shitan, N.; Yazaki, K. The multidrug and toxic compound extrusion (MATE) family in plants. *Plant Biotechnol.* **2014**, *31*, 417–430. [CrossRef]
2. Miyauchi, H.; Moriyama, S.; Kusakizako, T.; Kumazaki, K.; Nakane, T.; Yamashita, K.; Hirata, K.; Dohmae, N.; Nishizawa, T.; Ito, K.; et al. Structural basis for xenobiotic extrusion by eukaryotic MATE transporter. *Nat. Commun.* **2017**, *8*, 1633. [CrossRef]
3. Lu, M.; Radchenko, M.; Symersky, J.; Nie, R.; Guo, Y. Structural insights into H^+-coupled multidrug extrusion by a MATE transporter. *Nat. Struct. Mol. Biol.* **2013**, *20*, 1310–1317. [CrossRef] [PubMed]
4. Meyer, S.; De Angeli, A.; Fernie, A.R.; Martinoia, E. Intra- and extra-cellular excretion of carboxylates. *Trends Plant Sci.* **2009**, *15*, 40–47. [CrossRef]
5. Shoji, T.; Inai, K.; Yazaki, Y.; Sato, Y.; Takase, H.; Shitan, N.; Yazaki, K.; Goto, Y.; Toyooka, K.; Matsuoka, K.; et al. Multidrug and toxic compound extrusion-type transporters implicated in vacuolar sequestration of nicotine in tobacco roots. *Plant Physiol.* **2008**, *149*, 708–718. [CrossRef]
6. Kusakizako, T.; Miyauchi, H.; Ishitani, R.; Nureki, O. Structural biology of the multidrug and toxic compound extrusion superfamily transporters. *Biochim. Biophys. Acta-Biomembr.* **2020**, *1862*, 183154. [CrossRef] [PubMed]
7. Ng, M.-S.; Ku, Y.-S.; Yung, W.-S.; Cheng, S.-S.; Man, C.-K.; Yang, L.; Song, S.; Chung, G.; Lam, H.-M. MATE-type proteins are responsible for isoflavone transportation and accumulation in soybean seeds. *Int. J. Mol. Sci.* **2021**, *22*, 12017. [CrossRef] [PubMed]
8. Debeaujon, I.; Peeters, A.J.M.; Léon-Kloosterziel, K.M.; Koornneef, M. The *TRANSPARENT TESTA12* gene of Arabidopsis encodes a multidrug secondary transporter-like protein required for flavonoid sequestration in vacuoles of the seed coat endothelium. *Plant Cell* **2001**, *13*, 853–872. [CrossRef]

9. Zhao, J.; Dixon, R.A. MATE transporters facilitate vacuolar uptake of epicatechin 3′-O-glucoside for proanthocyanidin biosynthesis in Medicago truncatula and Arabidopsis. *Plant Cell* **2009**, *21*, 2323–2340. [CrossRef] [PubMed]
10. Liu, J.; Li, Y.; Wang, W.; Gai, J.; Li, Y. Genome-wide analysis of MATE transporters and expression patterns of a subgroup of *MATE* genes in response to aluminum toxicity in soybean. *BMC Genom.* **2016**, *17*, 1–15. [CrossRef]
11. dos Santos, A.L.; Chaves-Silva, S.; Yang, L.; Maia, L.G.S.; Chalfun-Júnior, A.; Sinharoy, S.; Zhao, J.; Benedito, V.A. Global analysis of the MATE gene family of metabolite transporters in tomato. *BMC Plant Biol.* **2017**, *17*, 1–13. [CrossRef]
12. Tiwari, M.; Sharma, D.; Singh, M.; Tripathi, R.D.; Trivedi, P.K. Expression of OsMATE1 and OsMATE2 alters development, stress responses and pathogen susceptibility in *Arabidopsis*. *Sci. Rep.* **2014**, *4*, 1–12. [CrossRef]
13. Li, L.; He, Z.; Pandey, G.K.; Tsuchiya, T.; Luan, S. Functional cloning and characterization of a plant efflux carrier for multidrug and heavy metal detoxification. *J. Biol. Chem.* **2002**, *277*, 5360–5368. [CrossRef] [PubMed]
14. Jagessar, K.L.; McHaourab, H.S.; Claxton, D.P. The N-terminal domain of an archaeal multidrug and toxin extrusion (MATE) transporter mediates proton coupling required for prokaryotic drug resistance. *J. Biol. Chem.* **2019**, *294*, 12807–12814. [CrossRef]
15. Mikros, E.; Diallinas, G. Tales of tails in transporters. *Open Biol.* **2019**, *9*, 190083. [CrossRef] [PubMed]
16. Diallinas, G. Dissection of transporter function: From genetics to structure. *Trends Genet.* **2016**, *32*, 576–590. [CrossRef] [PubMed]
17. Li, M.W.; Muñoz, N.B.; Wong, C.F.; Wong, F.L.; Wong, K.-S.; Wong, J.W.-H.; Qi, X.; Li, K.-P.; Ng, M.-S.; Lam, H.-M. QTLs regulating the contents of antioxidants, phenolics, and flavonoids in soybean seeds share a common genomic region. *Front. Plant Sci.* **2016**, *7*, 854. [CrossRef] [PubMed]
18. Camacho, C.; Coulouris, G.; Avagyan, V.; Ma, N.; Papadopoulos, J.; Bealer, K.; Madden, T.L. BLAST+: Architecture and applications. *BMC Bioinform.* **2009**, *10*, 421. [CrossRef] [PubMed]
19. Li, L.; Stoeckert, C.J.J.; Roos, D.S. OrthoMCL: Identification of ortholog groups for eukaryotic genomes. *Genome Res.* **2003**, *13*, 2178–2189. [CrossRef] [PubMed]
20. Tamura, K.; Stecher, G.; Kumar, S. MEGA11: Molecular evolutionary genetics analysis version 11. *Mol. Biol. Evol.* **2021**, *38*, 3022–3027. [CrossRef] [PubMed]
21. Omasits, U.; Ahrens, C.H.; Müller, S.; Wollscheid, B. Protter: Interactive protein feature visualization and integration with experimental proteomic data. *Bioinformatics* **2014**, *30*, 884–886. [CrossRef]
22. Prot-pi. Available online: https://www.protpi.ch/Calculator/ProteinTool (accessed on 22 December 2021).
23. Halligan, B.D.; Ruotti, V.; Jin, W.; Laffoon, S.; Twigger, S.N.; Dratz, E.A. ProMoST (Protein Modification Screening Tool): A web-based tool for mapping protein modifications on two-dimensional gels. *Nucleic Acids Res.* **2004**, *32*, W638–W644. [CrossRef] [PubMed]
24. Gietz, R.D.; TriggsRaine, B.; Robbins, A.; Graham, K.C.; Woods, R.A. Identification of proteins that interact with a protein of interest: Applications of the yeast two-hybrid system. *Mol. Cell. Biochem.* **1997**, *172*, 67–79. [CrossRef] [PubMed]
25. Slatnar, A.; Jakopic, J.; Stampar, F.; Veberic, R.; Jamnik, P. The effect of bioactive compounds on *in vitro* and *in vivo* antioxidant activity of different berry juices. *PLoS ONE* **2012**, *7*, e47880. [CrossRef]
26. Canelas, A.B.; ten Pierick, A.; Ras, C.; Seifar, R.M.; van Dam, J.C.; van Gulik, W.M.; Heijnen, J.J. Quantitative Evaluation of Intracellular Metabolite Extraction Techniques for Yeast Metabolomics. *Anal. Chem.* **2009**, *81*, 7379–7389. [CrossRef] [PubMed]
27. Toro-Funes, N.; Odriozola-Serrano, I.; Bosch-Fusté, J.; Latorre-Moratalla, M.L.; Veciana-Nogués, M.T.; Izquierdo-Pulido, M.; Vidal-Carou, M.C. Fast simultaneous determination of free and conjugated isoflavones in soy milk by UHPLC–UV. *Food Chem.* **2012**, *135*, 2832–2838. [CrossRef]
28. Griesmann, M.; Chang, Y.; Liu, X.; Song, Y.; Haberer, G.; Crook, M.B.; Billault-Penneteau, B.; Lauressergues, D.; Keller, J.; Imanishi, L.; et al. Phylogenomics reveals multiple losses of nitrogen-fixing root nodule symbiosis. *Science* **2018**, *361*, eaat1743. [CrossRef]
29. Taiz, L. The plant vacuole. *J. Exp. Biol.* **1992**, *172*, 113–122. [CrossRef]
30. Liao, Y.-Y.; Li, J.-L.; Pan, R.-L.; Chiou, T.-J. Structure–function analysis reveals amino acid residues of Arabidopsis phosphate transporter AtPHT1;1 crucial for its activity. *Front. Plant Sci.* **2019**, *10*, 1158. [CrossRef]
31. Miller, C. ClC chloride channels viewed through a transporter lens. *Nature* **2006**, *440*, 484–489. [CrossRef]
32. Miller, C. ClC channels: Reading eukaryotic function through prokaryotic spectacles. *J. Gen. Physiol.* **2003**, *122*, 129–131. [CrossRef] [PubMed]
33. Park, K.; Lee, B.-C.; Lim, H.-H. Mutation of external glutamate residue reveals a new intermediate transport state and anion binding site in a CLC Cl^-/H^+ antiporter. *Proc. Natl. Acad. Sci. USA* **2019**, *116*, 17345–17354. [CrossRef] [PubMed]
34. Feng, L.; Campbell, E.B.; MacKinnon, R. Molecular mechanism of proton transport in CLC Cl^-/H^+ exchange transporters. *Proc. Natl. Acad. Sci. USA* **2012**, *109*, 11699–11704. [CrossRef] [PubMed]
35. Gilmour, R.; Messner, P.; Guffanti, A.A.; Kent, R.; Scheberl, A.; Kendrick, N.; Krulwich, T.A. Two-dimensional gel electrophoresis analyses of pH-dependent protein expression in facultatively alkaliphilic *Bacillus pseudofirmus* OF4 lead to characterization of an S-layer protein with a role in alkaliphily. *J. Bacteriol.* **2000**, *182*, 5969–5981. [CrossRef]
36. Matsuno, T.; Goto, T.; Ogami, S.; Morimoto, H.; Yamazaki, K.; Inoue, N.; Matsuyama, H.; Yoshimune, K.; Yumoto, I. Formation of proton motive force under low-aeration alkaline conditions in Alkaliphilic bacteria. *Front. Microbiol.* **2018**, *9*, 2331. [CrossRef]

membranes

MDPI

Review

Endothelial Cell Plasma Membrane Biomechanics Mediates Effects of Pro-Inflammatory Factors on Endothelial Mechanosensors: Vicious Circle Formation in Atherogenic Inflammation

Nadezhda Barvitenko [1,*], Mohammad Ashrafuzzaman [2,†], Alfons Lawen [3,†], Elisaveta Skverchinskaya [4], Carlota Saldanha [5], Alessia Manca [6], Giuseppe Uras [7], Muhammad Aslam [8,‡] and Antonella Pantaleo [6,*,‡]

1 Independent Researcher, 191014 Saint-Petersburg, Russia
2 Department of Biochemistry, College of Science, King Saud University, Riyadh 11451, Saudi Arabia; mashrafuzzaman@ksu.edu.sa
3 Department of Biochemistry and Molecular Biology, School of Biomedical Sciences, Monash University, Melbourne, VIC 3800, Australia; alfons.lawen@monash.edu
4 Sechenov Institute of Evolutionary Physiology and Biochemistry, 194223 Saint-Petersburg, Russia; lisarafail@mail.ru
5 Institute of Biochemistry, Institute of Molecular Medicine, Faculty of Medicine, University of Lisbon, 1649-028 Lisboa, Portugal; carlotasaldanha@fm.ul.pt
6 Department of Biomedical Science, University of Sassari, 07100 Sassari, Italy; alessia_manca@hotmail.it
7 Department of Clinical and Movement Neurosciences, Institute of Neurology, University College London, London NW3 2PF, UK; g.uras@ucl.ac.uk
8 Experimental Cardiology, Department of Cardiology/Angiology, Justus Liebig University, 35392 Giessen, Germany; muhammad.aslam@physiomed.jlug.de
* Correspondence: nbarvitenko@mail.ru (N.B.); apantaleo@uniss.it (A.P.)
† These authors contributed equally to this work.
‡ Two senior co-authors.

Abstract: Chronic low-grade vascular inflammation and endothelial dysfunction significantly contribute to the pathogenesis of cardiovascular diseases. In endothelial cells (ECs), anti-inflammatory or pro-inflammatory signaling can be induced by different patterns of the fluid shear stress (SS) exerted by blood flow on ECs. Laminar blood flow with high magnitude is anti-inflammatory, while disturbed flow and laminar flow with low magnitude is pro-inflammatory. Endothelial mechanosensors are the key upstream signaling proteins in SS-induced pro- and anti-inflammatory responses. Being transmembrane proteins, mechanosensors, not only experience fluid SS but also become regulated by the biomechanical properties of the lipid bilayer and the cytoskeleton. We review the apparent effects of pro-inflammatory factors (hypoxia, oxidative stress, hypercholesterolemia, and cytokines) on the biomechanics of the lipid bilayer and the cytoskeleton. An analysis of the available data suggests that the formation of a vicious circle may occur, in which pro-inflammatory cytokines enhance and attenuate SS-induced pro-inflammatory and anti-inflammatory signaling, respectively.

Keywords: shear stress; inflammation; endothelial cell; lipid bilayer; cytoskeleton; mechanosensor; atherosclerosis; oxidative stress

1. Introduction

1.1. Blood Flow Patterns and the Pro-Inflammatory Response of Endothelial Cells

Inflammation is a key mechanism affecting endothelial cells (ECs) and leading to atherosclerosis [1–4]. ECs are subjected to three main hemodynamic forces: hydrostatic pressure, cyclic stretch, and shear stress (SS). Fluid SS is a frictional force, produced by blood flow, acting tangentially to the surface of the ECs. Depending on the patterns of fluid SS acting on ECs, either anti- or pro-inflammatory mechanisms can be triggered in

ECs. High laminar SS induces an anti-inflammatory response [2,5,6], while low laminar SS and disturbed patterns of SS, on the other hand, activate pro-inflammatory mechanisms in ECs [2,3]. The key transcription factors responsible for the low flow-induced inflammatory response of ECs are activator protein 1 (AP-1), nuclear factor κB (NF-κB) [3], and yes-associated protein/transcriptional coactivators with a PDZ-binding motif (YAP/TAZ) [7]. The transcription factors Krüppel-like factor 2 (KLF2), myocyte enhancer factor 2 (MEF2) and nuclear factor erythroid 2-related factor 2 (NRF2) are known to activate an anti-inflammatory response [3,6].

The molecular pathways leading to inflammation in ECs begin at the endothelial mechanosensors, one of which is the platelet endothelial cell adhesion molecule-1 (PECAM-1), which forms a mechanosensory complex with vascular endothelial cadherin (VE-cadherin) and vascular endothelial growth factor receptor 2 (VEGFR2). The PECAM-1/VE-cadherin/VEGFR2 complex triggers the activation of the nuclear factor κB (NF-κB) transcription factor [8,9]. In this complex, PECAM-1 transmits the mechanical signal, VE-cadherin is an adaptor protein, and VEGFR2 activates phosphatidylinositol-3-OH kinase (PI3K) [8]. The activation of the PECAM-1/VE-cadherin/VEGFR2 complex by SS (12 dyn/cm^2) occurs within 15 s, leading to the activation of integrin and integrin-dependent EC alignment in the direction of flow, and the transient activation of NF-κB [8]. It should be noted that the signaling triggered by VEGF via its receptors can exhibit both pro-inflammatory and vasculo-protective anti-inflammatory actions. For example, VEGF participates in angiotensin II (Ang II)-induced vascular inflammation [10], while the local application of recombinant VEGF or the local VEGF gene transfer to arterial wall ECs in the hind-limb ischemia model enhanced vascularization and collateral circulation [11,12]. In mouse aortic ECs (MAECs) and bovine aortic ECs (BAECs), high laminar SS (24 dyn/cm^2) activates integrins and NF-κB [9]. The seemingly paradoxical nature of that high laminar SS-activated pro-inflammatory NF-κB can be explained via transient NF-κB activation (in the case of high laminar SS) versus sustained NF-κB activation (in the case of oscillatory or disturbed SS) [9]. In human aortic ECs (HAECs), the activation of αvβ3 integrins by oscillatory flow promotes NF-κB activation and atherogenesis [13]. Activated NF-κB induces the expression of pro-inflammatory cytokines, chemokines, and cell adhesion molecules [14,15]. Likewise, atheroprotective laminar SS induces the inactivation of pro-inflammatory YAP/TAZ via their phosphorylation-mediated cytoplasmic retention in ECs [7]. On the other hand, oscillatory SS results in the hyperactivation of YAP/TAZ via RhoA/Rock-dependent actin stress fiber formation, leading to the enhanced expression of inflammatory ICAM1 and VCAM1 [7,16]. The endothelial specific over-expression of YAP exacerbates, while CRISPR-mediated YAP-knockdown retards, plaque formation in ApoE$^{-/-}$ mice [16,17].

These links between blood flow patterns, the mechanosensitivity of ECs, and inflammation evoked deep interest in the study of endothelial mechanosensors [18–27], which include primary cilia, glycocalyx, integrins, caveolae, ion channels, heterotrimeric G proteins, G protein-coupled receptors (GPCRs), PECAM-1, VE-cadherin, VEGFR2, and the Tie family of receptor tyrosine kinases [20]. Moreover, Notch1 and guidance receptor plexin D1 can also function as endothelial mechanosensors [28,29]. Likewise, in human pulmonary aortic ECs (HPAECs), mitochondria respond to fluid SS with increased oxidative phosphorylation and elevated ATP production [30,31], suggesting their possible role as mechanosensors.

1.2. Forces and Plasma Membrane Mechanosensors

Two models have been proposed for the mechanical force-induced activation of mechanosensitive ion channels (Figure 1): "force-from-lipids" and "force-from-filament" [32]. The "force-from-lipid" model suggests the gating of mechanosensitive ion channels with inputs from the lipid bilayer, while the "force-from-filament" model suggests primary roles of the extracellular matrix and intracellular cytoskeleton (CSK) in the mechanical activation of ion channels [32]. Great efforts have been devoted to deciphering the multiple effects exerted by membrane lipids on the structure and function of transmembrane proteins [33–44].

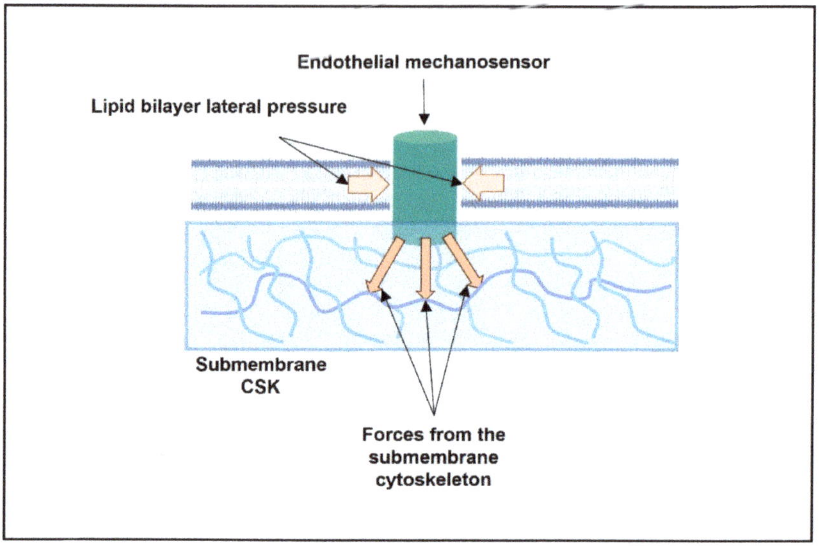

Figure 1. Scheme illustrating the actions of forces arising from the lipid bilayer and from the submembrane cytoskeleton and converging on a single mechanosensitive transmembrane protein.

Two major types of physical effects, integral membrane proteins or ion channels, are expected to draw from the hosting lipid bilayer due to its profiles of electrical charges and mechanical properties [43]. The former one arises due to the consideration that the distribution of the charges of the lipids on either monolayer may or may not show any net charges, but due to their coupling with the integral membrane proteins (MPs) or channel proteins, the charges on both lipids and proteins become redistributed or polarized. As a result, any MP-lipid coupling appears with a distinctive type of physical phenomenon, many charges interactions, as explained using screened Coulomb interactions (SCIs) in ref. [43]. Considering the mechanical properties (bilayer elasticity and lipid intrinsic curvature) of lipid layers, only the bilayer regulation of the integral membrane protein function has long been addressed (see refs. [45–52]). However, in refs. [43,44], it is clearly shown that the SCI model that considers the charge-based interactions among integral MPs and hosting bilayer lipids can correctly address the bilayer regulation of MP or channel functions, and that the charge-based interactions appear to be primary regulators of channel functions. The bilayer mechanical property-based regulation of MP functions also appears in SCI treatment, but only to produce a secondary effect on MP functions. In the elastic bilayer model, the primary effect due to charge-based interactions was totally ignored [45–52]. The charge-based effects have later been consistently found to be appearing as primary molecular mechanisms, especially when using molecular dynamics (MD) simulations, on varieties of membrane-adsorbed peptides and drugs cases (see refs. [53,54]).

Recently, the plasma membrane of immune cells was suggested to integrate multiple biophysical and biochemical stimuli (such as cholesterol content, negatively charged lipids, electrical potential) in order to regulate immune receptor function [55].

Both the composition of the lipid bilayer [22] and the NMMII-generated basal tension of EC [56] can be altered by pro-inflammatory stimuli. Endothelial mechanosensors, which are embedded in the plasma membrane, not only influence the blood flow, but also regulate the mechanical properties of the lipid bilayer and CSK. Here, we discuss if and how pro-inflammatory factors may change the mechanical properties of the lipid bilayer and CSK. The rigidification of the lipid bilayer and the increase in the cytoskeletal NMMII-generated tension can increase the energy barrier for the activation of endothelial mechanosensors by SS. As a result, high SS can be perceived by ECs as low SS. Since low SS induces the release of pro-inflammatory mediators [2,4] the *vicious circle* can be formed. It keeps up the low-grade vascular inflammation and promotes the development of endothelial dysfunction and atherosclerosis. Recently, the primary cilia on ECs were suggested to amplify low unidirectional SS signaling, resulting in the activation of NRF2 and the protection of ECs from oxidative damage [6]. We discuss the data suggesting that, under inflammatory conditions, the lipid bilayer and the NMMII-generated tension could dampen high laminar SS.

2. What Are the Intracellular Forces Acting on Any Single Transmembrane Endothelial Mechanosensor?

2.1. From Stiffness of the Whole EC to the Mapping of Intracellular Forces Acting on Single Transmembrane Mechanosensor: From Cell- to Protein-Scale Studies

Blood flow exerts extracellular forces, such as hydrodynamic pressure, cyclic stretch, and fluid SS. In addition to these forces, the substrate stiffness is also sensed by ECs [57,58]. The dependence of the mechanical properties of the cortical CSK in HAECs and human umbilical vein ECs (HUVECs) on laminar SS strength was investigated using acoustic force spectroscopy [59]. The exposure of HAECs and HUVECs to laminar SS (6 dyn/cm^2 for up to 48 h) was found to evoke an increase in the membrane cortex stiffness [59].

The stiffness of the whole EC, or its plasma membrane with the underlying submembrane actin-based CSK (smACSK), is an integral parameter, evaluating the behavior of the whole cell. At the cellular level, the stiffness of bovine pulmonary arterial ECs (BPAECs) depends on the basal isometric tension, which is determined by NMMII contractility [56]. Thrombin induces a rapid increase in basal isometric tension in BPAECs via the MLCK- and RhoA-mediated activation of NMMII [56]. Further studies on the biomechanics of whole ECs should lead to an analysis of the spectrum of intracellularly generated forces that converge on any single transmembrane mechanosensor [60]. Evidence is accumulating that forces generated within the cell regulate mechanical tension across the transmembrane, cytoskeletal, and scaffolding proteins [61–63].

Endothelial mechanosensors are subjected to forces generated within the cell, in particular within the lipid bilayer of the plasma membrane, and forces arising from smACSK (Figure 2). Energy inputs from the lipid bilayer and the smACSK are likely to increase, or decrease, the activation energy required for the stimulation of mechanosensors by fluid SS (Figure 2). Earlier, we proposed that the gradients in hydrostatic pressures across the plasma membrane induced by changes in cell volume are actively probed by cells via the pulling activity of non-muscle myosin II (NMMII) and the pushing activity of smACSK [64]. Considering the forces arising from the lipid bilayer (see Section 1.2), and the forces generated by smACSK, the energy input (E(intracellular)) that is received—in addition to extracellular mechanical forces—by any single mechanosensory can be presented as the sum of the following energies:

$$E(\text{intracellular}) = E(\text{lipid bilayer}) + E(\text{NMMII}) + E(\text{protrusion}) + E(\text{resistance}) + E(\text{smACSK spring}) \quad (1)$$

where E(lipid bilayer) is an energy from the lipid bilayer, E(NMMII) is an energy of NMMII-generated pulling (directed into the cell) force, E(protrusion) is an energy of pushing (directed out of the cell) force due to the actin-based assembly of lamellipodia and filopodia, E(resistance) is an energy generated by the lipid bilayer together with smACSK, and E(smACSK spring) is an energy stored by smACSK during its mechanical deformation.

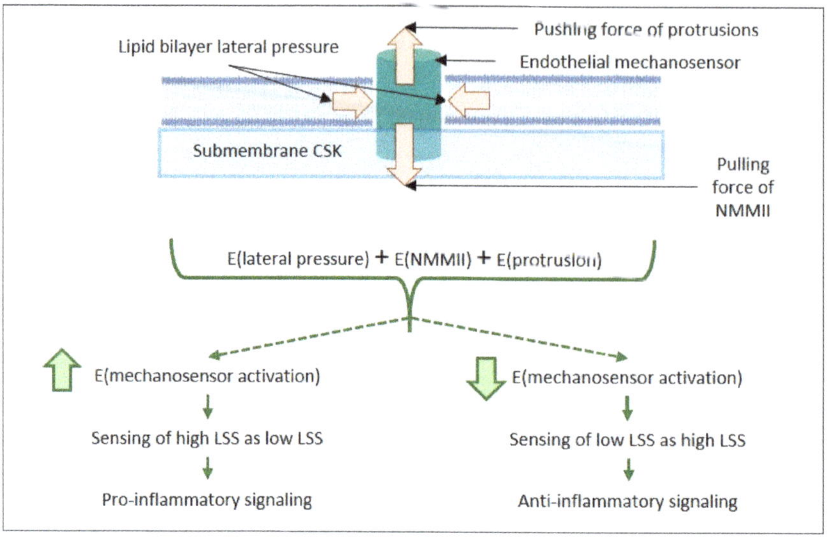

Figure 2. Scheme illustrating the actions of intracellular forces on any single endothelial mechanosensor. The transmembrane domain(s) of the mechanosensor is sensing the lipid bilayer lateral pressure. A pulling force from NMMII-generated tension and a pushing force from an actin-based protrusion are also sensed by the mechanosensor, meaning that the energy required for mechanosensor activation can be increased; in this case, the high LSS can be perceived as low LSS, or decreased, in which case the low LSS can be perceived as high LSS. Thus, the interplay of intracellular forces can shift LSS-induced signaling to be either pro-atherogenic or anti-atherogenic. Abbreviations: LSS, laminar shear stress; NMMII, non-muscle myosin II; smACSK, submembrane actin-based cytoskeleton.

Thus, the energies received from within the cell would also influence the activation of an endothelial mechanosensor:

$$E(\text{activation of endothelial mechanosensor}) = E(\text{extracellular}) + E(\text{intracellular}) \quad (2)$$

In this paper, we only pay attention to three intracellular forces: the biomechanics of the lipid bilayer, the pulling force of NMMII, and the pushing force of the lamellipodia. NMMII and smACSK are controlled by many intracellular signaling mechanisms; however, we only consider two functionally antagonistic signaling proteins, RhoA and Rac1. The convergence of several pro-inflammatory stimuli on the two functionally antagonistic small GTPases RhoA and Rac1 is discussed in Section 3.

2.2. Biomechanics of the Lipid Bilayer and the Activation Energy of Mechanosensors

There are two mechanisms for the regulation of transmembrane proteins by lipids: a ligand-like mechanism, when the direct high-affinity binding of lipids to proteins occurs, and a solvent-like mechanism, when the addition or removal of a lipid changes the biomechanics of the lipid bilayer [65,66]. For example, many transmembrane proteins contain specific motifs for cholesterol binding: a cholesterol recognition/interaction amino acid consensus (CRAC, R/K-X_5-Y-X_5-L/V), a reversed CRAC motif named CARC (L/V-X_5-Y-X_5-R/K), and a cholesterol consensus motif (CCM) [65–69]. In this paper, we mainly pay attention to the solvent-like scenario, which deals with alterations in plasma membrane biomechanics.

There is convincing evidence that the lipid composition, which determines the lipid bilayer's fluidity, influences both basal and SS-induced GTPase activities of $G_{\alpha q}$ and $G_{\alpha i3}$ subunits of heterotrimeric G proteins in phospholipid vesicles [70]. The incorporation of lysophosphatidylcholine into liposomes increases the fluidity of the lipid bilayer and elevates the basal activity of $G_{\alpha q}$ and $G_{\alpha i3}$ proteins from 0.47 to 1.35 pmol/min per μg of protein [70]. The incorporation of benzyl alcohol, another fluidizing agent, increases the basal activity of G proteins from 0.47 to 2.37 pmol/min per μg of proteins. On the other hand, the incorporation of cholesterol, which decreases bilayer fluidity, diminishes the basal activity of Gαq and Gαi3 proteins from 0.47 to 0.113 pmol/min per μg of protein and reduces the activation of G proteins by SS [70].

In HUVECs, fluid SS (from 0.7 to 33 dyn/cm^2) induces an increase in membrane fluidity [71]. The addition of benzyl alcohol also increases the membrane fluidity [71]. The exposure of BAECs to fluid SS increases the plasma membrane fluidity [72]. Additionally, BAECs' membrane fluidity is increased and decreased by benzyl alcohol (a fluidizing agent) and cholesterol (a rigidifying agent), respectively [72]. In HPAECs, fluid SS increases membrane fluidity [30,31]. Furthermore, the plasma membranes in HPAECs discriminate between cyclic stretch and fluid SS, in that cyclic stretch increases the lipid bilayer order and decreases fluidity, while fluid SS decreases the lipid bilayer order and increases fluidity [73].

Due to all the changes in the membrane composition, which have been explained above, a membrane's two major physical properties, namely, the charge profiles and mechanical properties, may especially become altered. Consequently, as explained earlier, the membrane regulation of integral MP functions also changes [43,45–52]. However, the exact energy (generally refereed as the 'free energy of bilayer–integral protein coupling') that plays important roles in such membrane regulation of integral protein functions has been correctly calculated using the SCI models that consider charge-based interactions [43,53,54]. The SCI model explains all the parameters behind calculating the free energy of bilayer–integral protein coupling, and that this energy consists of both components drawn from the charge properties and mechanical properties of the bilayer and MPs. Thus, it appears to be a universal mechanism which also raises some universal probability functions related to any bilayer-MP coupling energetics (see details in ref. [54]). MD simulations on lipid–drug pair interactions in the bilayer environment have especially demonstrated these universal probability functions to be primarily relying on two major types of charge-based lipid–drug interactions, namely the electrostatic and van der Waals interactions.

2.3. *Mechanosensors and Force Generated by NMMII*

2.3.1. Control of NMMII Contractility

NMMII is an actin-based heterohexameric molecular motor consisting of two heavy chains (HCs), two essential light chains (ELCs), and two regulatory light chains (RLCs) [74–78]. There are three isoforms of NMMII depending on HC paralog: NMMIIA, NMMIIB, and NMMIIC [74–77], which apparently have partially distinctive roles [79]. In ECs, mainly NMMIIA and NMMIIB are expressed [80,81].

The phosphorylation of the Ser19/Thr18 of RLC by Ca^{2+}/calmodulin-dependent myosin light chain kinase (MLCK) activates NMMII contractility (Figure 3) [82], whereas the dephosphorylation of this site by myosin light chain phosphatase (MLCP) inhibits NMMII [83,84]. MLCK can itself be activated by Ca^{2+}/calmodulin, protein tyrosine kinases (PTKs), which phosphorylate Tyr464 and Tyr471, and protein kinase C (PKC), while phosphorylation by protein kinase A (PKA) inhibits MLCK [85,86].

The small GTPases Rho, Rac, and Cdc42, govern the formation of actin stress fibers, lamellipodia, and filopodia, respectively [87]. In humans, there are 20 members in the Rho family which are subdivided into subfamilies: Rho, Rac, Cdc42, RhoU/V, RhoD/F, Rnd, RhoH, and RhoBTB [88]. The Rho subfamily consists of RhoA, RhoB, and RhoC, while the Rac subfamily includes Rac1, Rac2, Rac3, and RhoG [88]. In the vasculature, the small GTPases Rho, Rac, and Cdc42, control a number of functions, including the maintenance of the endothelial barrier, the response to SS, the regulation of endothelial nitric oxide synthase, migration, and apoptosis [89]. There is reciprocal regulation of RhoA and Rac1. For example, in BAECs, the activation of integrins by laminar SS (12 dynes/cm^2) transiently inhibits Rho, but activates Rac1 [90,91]. The small GTPases, RhoA and Rac1, are

HUVECs [102]. It should be noted that NMMII contractility is required for lamellipodia formation [102,103].

3. Effects of Pro-Inflammatory Stimuli on the Biomechanics of the Lipid Bilayer and Submembrane Cytoskeleton; Focus on Counterbalance between RhoA and Rac1

3.1. Pro-Inflammatory Stimuli and the Lipid Bilayer Biomechanics

Hypoxia itself can be a mechanical signal for ECs, as a decrease in the number of dioxygen molecules dissolved in the lipid bilayer of the plasma membrane is a mechanical stimulus that can influence the mechanosensitive transmembrane proteins and, thus, participate in the hypoxia response [104].

However, hypoxia can also lead to the increased production of reactive oxygen species (ROS) [105], which may increase membrane lipid peroxidation, with effects on the mechanical properties of the lipid bilayer. Oxysterols, products of cholesterol oxidation, promote the development of atherosclerosis [106]. Connections between dyslipidemia and the mechanical properties of ECs were reviewed elsewhere [22]. Hypercholesterolemia leads to the accumulation of cholesterol in ECs, and promotes the development of inflammation and atherosclerosis [22,107,108]. An increase in low-density lipoprotein (LDL) cholesterol levels and a decrease in high-density lipoprotein (HDL) cholesterol levels in blood plasma are among the key risk factors for atherogenesis [107,109].

Oxidative stress leads to an accumulation of oxidized phospholipids in the EC plasma membrane [110]. The peroxidation of membrane lipids decreases the lipid bilayer thickness [111,112]. On the other hand, long-chain polyunsaturated fatty acids (PUFAs)—such as eicosapentaenoic acid (EPA, C20:5, n-3), docosahexaenoic acid (DHA, C22:6, n-3), and docosapentaenoic acid (DPA, C22:5, n-3)—trigger anti-inflammatory anti-atherogenic responses in ECs, as can be exemplified with docosahexaenoic acid (DHA) (22:6ω-3) [113–115].

3.2. Pro-Inflammatory Stimuli in RhoA and Rac1 Regulation in ECs

3.2.1. Hypoxia and Oxidative Stress in the Regulation of RhoA and Rac1 in ECs

In cultured porcine aortic ECs (PAECs), hypoxia induces the activation of RhoA and the inhibition of Rac1 [93] (Table 2). In piglet PAECs, hypoxia activates RhoA and inhibits Rac1 [116]. In rat PAECs, hypoxia activates RhoA [117]. In rat PAECs, oxidative stress (H_2O_2) activates RhoA [117]. In BAECs, H_2O_2 induces the activation of Rac1 [118].

Table 2. Hypoxia and oxidative stress in RhoA and Rac1 regulation in ECs. ↑—activation, ↓—inhibition.

Stress Factor	Effect on RhoA or Rac1	EC Type	Reference
Hypoxia	↑ RhoA	Porcine AECs [1]	[93]
	↑ RhoA	Piglet PAEC	[116]
	↑ RhoA	Rat PAECs	[117]
	↓ Rac1	Porcine -AECs	[93]
	↓ Rac1	Piglet PAECs	[116]
Oxidative stress	↑ RhoA	Rat PAECs	[117]
	↑ Rac1	BAECs	[118]

[1] Abbreviations: AECs, aortic endothelial cells; BAECs, bovine aortic endothelial cells; PAECs, pulmonary artery endothelial cells.

3.2.2. Pro-Inflammatory Cytokines in RhoA and Rac1 Regulation in ECs

As discussed above, RhoA-ROCK activation via the induction of actomyosin contractility may increase the tension experienced by EC mechanosensors, leading to an increase in the pro-inflammatory response in ECs. Pro-inflammatory mediators unbalance RhoA–Rac1 activities and homeostasis, resulting in changes to the intracellularly generated tension and mechanosensors' activation threshold.

Ang II can induce vascular inflammation and remodeling [10]. Ang II acts via two types of Ang receptors, type 1 (AT_1R) and type 2 receptor (AT_2R), which significantly differ in their physiological effects [119]. Signaling through AT_1R leads to vasoconstriction, oxidative stress, and inflammation, while signaling through AT_2R mediates anti-inflammatory effects [120,121] and prevents the development of hypertension in animal models of hypertension [119]. In BAECs, the stimulation of AT_1R leads to the sequential activation of $G_{\alpha 12/13}$ and RhoA [122]. In BAECs, Ang II acting via AT_1R activates Rac1, and elevates focal adhesion complexes and actin fiber formation [118]. In contrast, the activation of AT_2R has been linked with the negative regulation of RhoA activity in vascular smooth muscle cells [123]. In BAECs, C-reactive protein (CRP) activates the RhoA–ROCK pathway to induce the expression of plasminogen activator inhibitor-1 [124].

Sphingosine 1-phosphate (S1P) may exert both anti-inflammatory anti-atherogenic effects, when acting through S1P receptor type 1 ($S1P_1$ receptor), and pro-inflammatory pro-atherogenic effects, when acting through $S1P_2$ and $S1P_3$ receptors [125–127]. The stimulation of $S1P_1$ receptor signaling activates $G_{\alpha i}$ and Rac1, leading to the suppression of the pro-inflammatory response [126]. Both $S1P_2$ and $S1P_3$ receptors are coupled to $G_{\alpha i/o}$, $G_{\alpha q}$, and $G_{\alpha 12/13}$ and their stimulation activates the RhoA–ROCK axis and destabilizes endothelial barrier [125–127].

Thrombin, an important regulator of acute and chronic vascular inflammation, acts via protease-activated receptors (PARs), of which there are four isoforms (PAR-1, -2, -3, and -4) [128]. In HUVECs, thrombin activates RhoA, suppresses Rac1 activity, and induces actomyosin contractility [129,130]. Intermedin, a member of the calcitonin gene-related peptide family, and acting via calcitonin receptor-like receptors, antagonizes thrombin-induced endothelial hyperpermeability via the activation of Rac1 [129].

4. Some Mechanosensors Are Located in the Plasma Membrane of the EC; Their Sensitivity to the Lipid Bilayer and the CSK Biomechanics

4.1. Piezo1

Piezo1 is a transmembrane cation channel [131–133] that is gated by membrane tension and SS [134,135]. Full-length *Piezo1* was cloned from the mouse neuroblastoma N2A cell line and expressed in several other cell lines [136]. Piezo1 exhibits activation by stretching [136,137]. Fluorescent Piezo1 constructs were expressed in HEK293 cells [134]. Blebs formed in these transfected cells were deficient in the cytoskeletal proteins, and the basal Piezo1 activity in the bleb-attached patches was higher than in whole-cell-attached patches, suggesting that the membrane tension is a main driver of mechanosensitive gating of Piezo1, whereas the CSK has a mechanoprotective role [134].

Piezo1 performs multiple roles in the cardiovascular system, and links mechanical stimuli to the triggering of both pro- and anti-atherogenic responses in ECs [138]. Piezo1 is expressed in many cell types participating in the development of atherosclerosis, including ECs, vascular smooth muscles cells, T and B cells, and monocytes, which undergo sequential transition into macrophages and lipid-engorged foam cells [138]. The endothelial-specific deletion of Piezo1 in mice impairs SS-mediated vascular development [131], sprouting angiogenesis, and vascular lumen formation [139]. Moreover, EC Piezo1 can sense disturbed blood flow and is linked to inflammatory signaling [140]. In human umbilical arterial ECs (HUAECs), the fluid SS-induced activation of Piezo1 leads to ATP release and autocrine stimulation of the purinergic $P2Y_2$ receptor and its downstream effectors, $G_{\alpha q}$ and $G_{\alpha 11}$, which triggers the activation of eNOS, NO release, and vasodilatation [141]. Depending on blood flow patterns, Piezo1 activation, together with the activation of purinergic $P2Y_2$ receptor and $G_{\alpha q/11}$, may lead to atheroprotective signaling or atherogenic signaling in response to laminar SS or disturbed SS, respectively [140]. In the case of disturbed flow, the induction of atherogenic signaling proceeds via the activation of integrin by SS sensors Piezo1 and $P2Y_2$–$G_{\alpha q/11}$ [140].

4.2. Mechanosensory PECAM-1/VE-Cadherin/VEGFR2 Complex

VE-cadherin is both a key player in the regulation of the endothelial barrier function [94,95] and an element in the PECAM-1/VE-cadherin/VEGFR2 mechanosensory complex [8]. Interestingly, a direct association between PECAM-1 and $G_{\alpha q/11}$ was reported in HUVECs [142,143]. In HUVECs, the dissociation of $G_{\alpha q/11}$ from PECAM-1 is dependent on $G_{\alpha q/11}$ activation by fluid SS, and on the patterns of fluid SS. Here, impulse and oscillatory, but not ramped-transient flow induced $G_{\alpha q/11}$ dissociation from PECAM-1 [142]. VEGFR2 is localized in lipid rafts, and interference in the lipid raft structure may impair its activation by SS [144]. The transmembrane domains of VE-cadherin and VEGFR2 and VEGFR3 can interact with each other, and this interaction is important for SS signal transduction onto and the activation of VEGFR2/3 [145]. The linking of VE-cadherin to the actin CSK seems to be mediated by β- and α-catenins, as it takes place in epithelial cells containing epithelial (E)-cadherin [146]. E-cadherin binds to β-catenin, β-catenin interacts with α-catenin, and α-catenin binds to actin filaments [146].

The SS response of ECs includes re-arrangement in the EC cytoskeleton and the redistribution of the intracellular forces acting on PECAM-1 and VE-cadherin [63]. The onset of the flow elicited an increase in the tension on PECAM-1 and a decrease in the tension on VE-cadherin [63]. Vimentin, an intermediate filament, appears to transmit NMMII tension onto PECAM-1 [63]. In our hypothesis, RhoA- and Rac1-dependent intracellular forces can tune the mechanosensitivity of the PECAM-1/VE-cadherin/VEGFR2 complex (Figure 5). Indeed, in HUVECs, the VE-cadherin located in the lamellipodia undergoes exhibits both pushing and pulling forces [102]. S1P, via the $S1P_1$ receptor, activates Rac1 and suppresses pro-inflammatory elevation in endothelial permeability [126,127] (Figure 5). Rac1 inhibits the RhoA–ROCK pathway that reduces NMMII-generated tension on VE-cadherin, resulting in the stabilization of VE-cadherin [101]. On the other hand, NMMII contractility also contributes to lamellipodia formation [102,103].

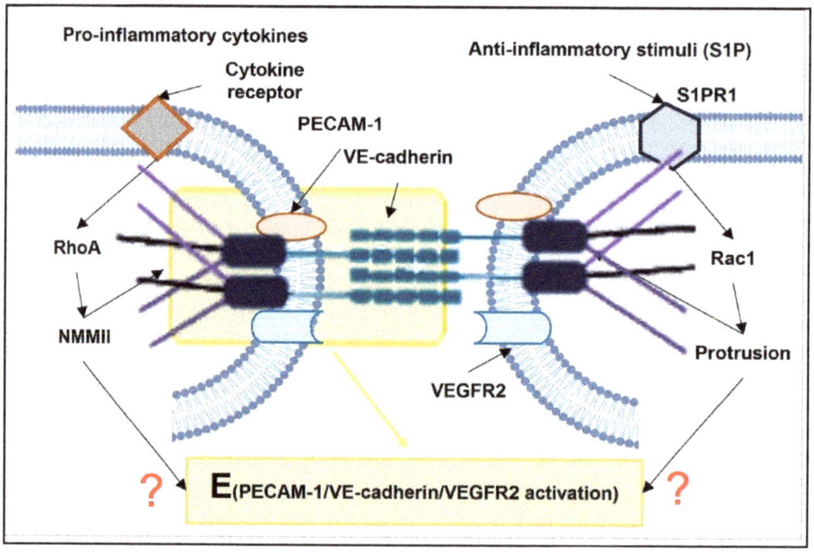

Figure 5. Scheme illustrating the actions of RhoA- and Rac1-dependent intracellular forces on the PECAM-1/VE-cadherin/VEGFR2 complex.

4.3. Heterotrimeric G Proteins and GPCRs

There is experimental evidence for the activation of GPCRs and heterotrimeric G proteins by fluid SS in the absence of GPCR agonists. The exposure of HUVECs to fluid SS (10 dyne/cm^2) induces the rapid (within 1 s) activation of $G_{\alpha q}$ and $G_{\alpha i3}$ [147]. The reconstitution of purified $G_{\alpha q}$ and $G_{\alpha i3}$ in phospholipid liposomes reveals that the lipid composition significantly influences the activation of G proteins in response to fluid SS (0–30 dynes/cm^2) [70]. In human coronary artery ECs (HCAECs), $G_{\alpha q/11}$ proteins are activated by fluid SS independently of upstream GPCRs [148]. In addition, in HCAECs, the stimulation of $G_{\alpha q/11}$ proteins by fluid SS is independent of the Piezo1 channel [143].

The human bradykinin type 2 receptor (B_2R), with an inserted yellow fluorescent protein and fused to a cyan fluorescent protein, was expressed in BAECs [149]. The recombinant B_2R was activated by fluid SS, hypotonic stress, and benzyl alcohol in the absence of a B_2R agonist [149]. Several long-chain polyunsaturated fatty acids (eicosapentaenoic acid, docosahexaenoic acid, docosadienoic acid, and dihomo-γ-linoleic acid) activated B_2R in BAECs and in B_2R-transfected HEK293 cells in a ligand-independent manner [150]. The search for a structural motif(s) responsible for the mechanosensitivity of the human histamine type 1 (H_1) receptor in HUVECs revealed that helix 8, located in the cytoplasmic C-terminus of the H_1 receptor, is essential [151]. Whether intracellular forces, such as NMMII-dependent contractions, influence the activities of heterotrimeric G proteins and GPCRs remains poorly understood.

4.4. Integrins

Integrins, heterodimeric adhesion proteins linking extracellular matrix proteins to the cytoskeleton, comprise 18 α- and 8 β-subunits, which form 24 heterodimers [152]. Integrins sense various mechanical stresses and convert a mechanical force into biochemical signaling within the cell [58]. The multiple roles of integrins—expressed on ECs, leukocytes, monocytes/macrophages, vascular smooth muscle cells, and platelets—in the pathogenesis of atherosclerosis are reviewed in depth elsewhere [153,154].

Integrins are sensitive to the lipid bilayer ordering and thickness [155–158]. Moreover, in PAECs and mouse embryonic fibroblasts, integrins themselves seem to increase the lipid bilayer order [155]. The exposure of HUVECs to pro-inflammatory oscillatory SS (0.5 ± 4 dyn/cm^2) or anti-inflammatory pulsatile SS (12 ± 4 dyn/cm^2) induced the opposite effects on α5β1 integrin localization in lipid rafts [156]. Oscillatory SS and pulsatile SS increase or decrease the levels of α5β1 integrins in lipid raft regions, respectively [156]. The exposure of HAECs to three non-lipid amphiphiles (vitamin E, Triton X-100, and benzyl alcohol) was used for the study of β1 integrin dependence on lipid bilayer order and domain thickness [157]. Only benzyl alcohol partitioned into the liquid-disordered domains and thinned these domains, enhancing β1 integrin affinity and valency, and inducing β1 integrin clustering [157]. Lietha and Izard suggested that mechanical stress-induced β integrin activation is mediated by membrane thinning [158].

Integrins are linked to the actin-based cytoskeleton via numerous scaffolding proteins, such as talin, vinculin, filamin A, and zyxin [58,152]. Integrins themselves regulate NMMII [153]. NMMII-generated contractility is required for the maintenance of integrin adhesion complexes [159]. In the stationary membranes of BAECs, the tension across vinculin was generated by NMMII and its upstream regulator RhoA, and was about 2.5 pN (0.25 μdyn) [62].

5. Conclusions

The experimental data considered here suggest that internal cellular forces—mainly NMMII contractility and actin-based lamellipodial protrusions, together with pro-inflammatory cholesterol- and oxidative stress-induced changes in the biomechanics of the lipid bilayer—tune the mechanosensitivity of the endothelial mechanosensors. Since RhoA and Rac1, key regulators of NMMII contractility and actin filaments assembly, are themselves critically regulated by pro-inflammatory agents, the biomechanics of the plasma membrane of ECs can be softened or stiffened upon inflammation, with subsequent alteration in ECs response to fluid SS. Fluid SS, depending on its patterns, is among the main controllers of endothelial inflammation. We propose that the altered—by first exposure of ECs to pro-inflammatory agents—mechanosensitivity of mechanosensors leads to their aberrant activation, and to the formation of a vicious cycle where pro-inflammatory signaling is sustained. Low-grade vascular inflammation then promotes atherogenesis.

Author Contributions: Literature review, N.B., M.A. (Mohammad Ashrafuzzaman), M.A. (Muhammad Aslam) and A.L.; methodology, N.B., C.S., A.M. and G.U.; investigation, N.B., A.L., M.A. (Mohammad Ashrafuzzaman), M.A. (Muhammad Aslam), E.S., C.S., G.U., A.M. and A.P.; writing—original draft preparation, N.B., E.S., A.L., M.A. (Mohammad Ashrafuzzaman), M.A. (Muhammad Aslam), C.S. and A.P.; writing—review and editing, E.S., A.L., M.A. (Mohammad Ashrafuzzaman), M.A. (Muhammad Aslam), A.M., G.U., C.S. and A.P.; visualization, E.S., A.M. and G.U.; supervision, N.B.; project administration, A.P.; funding acquisition, A.P. All authors have read and agreed to the published version of the manuscript.

Funding: This research was funded by grant from "Fondo di Ateneo per la ricerca 2020", University of Sassari, Italy.

Institutional Review Board Statement: Not applicable.

Informed Consent Statement: Not applicable.

Data Availability Statement: Not applicable.

Acknowledgments: We thank Owen P. Hamill for kind sharing of ideas and inspiration.

Conflicts of Interest: The authors declare no conflict of interest.

Abbreviations

Ang II, angiotensin; AT_1R, angiotensin II type 1 receptor; AP-1, activator protein 1; BAECs, bovine aorta endothelial cells; CSK, cytoskeleton; BPAECs, bovine pulmonary arterial endothelial cells; ECs, endothelial cells; ELC, essential light chain; FAK, focal adhesion kinase; GPCRs, G protein-coupled receptors; HAECs, human aortic endothelial cells; HC, heavy chain; HCAECs, human coronary artery endothelial cells; HUAECs, human umbilical arterial endothelial cells; HUVECs, human umbilical vein endothelial cells; KLF2, Krüppel-like factor 2; LSS, laminar shear stress; MAECs, mouse aortic endothelial cells; MD, molecular dynamics; MEF2, myocyte enhancer factor 2; MLCK, myosin light chain kinase; MLCP, myosin light chain phosphatase; MP, membrane protein; NF-κB, nuclear factor κB; NMMII, non-muscle myosin II; NRF2, nuclear factor erythroid 2-related factor 2; PAK, p21-activated kinase; PECAM-1, platelet endothelial cell adhesion molecule-1; PI3K, phosphatidylinositol-3-OH kinase; PKA, protein kinase A; PKC, protein kinase C; PTKs, protein tyrosine kinases; RLC, regulatory light chain; ROCK, Rho-associated coiled-coil-containing kinase; ROS, reactive oxygen species; SCI, screened Coulomb interactions; smACSK, submembrane actin-based cytoskeleton; SS, shear stress; TRP, transient receptor potential; VE-cadherin, vascular endothelial cadherin; VEGFR2, vascular endothelial growth factor receptor 2.

References

1. Libby, P. Inflammation in Atherosclerosis. *Arterioscler. Thromb. Vasc. Biol.* **2012**, *32*, 2045–2051. [CrossRef] [PubMed]
2. Chiu, J.-J.; Chien, S. Effects of Disturbed Flow on Vascular Endothelium: Pathophysiological Basis and Clinical Perspectives. *Physiol. Rev.* **2011**, *91*, 327–387. [CrossRef] [PubMed]
3. Bryan, M.T.; Duckles, H.; Feng, S.; Hsiao, S.T.; Kim, H.R.; Serbanovic-Canic, J.; Evans, P.C. Mechanoresponsive Networks Controlling Vascular Inflammation. *Arterioscler. Thromb. Vasc. Biol.* **2014**, *34*, 2199–2205. [CrossRef] [PubMed]
4. Ramji, D.P.; Davies, T.S. Cytokines in Atherosclerosis: Key Players in All Stages of Disease and Promising Therapeutic Targets. *Cytokine Growth Factor Rev.* **2015**, *26*, 673–685. [CrossRef]
5. Hirata, T.; Yamamoto, K.; Ikeda, K.; Arita, M. Functional Lipidomics of Vascular Endothelial Cells in Response to Laminar Shear Stress. *FASEB J.* **2021**, *35*, e21301. [CrossRef]
6. Ishii, T.; Warabi, E.; Mann, G.E. Mechanisms Underlying Unidirectional Laminar Shear Stress-Mediated Nrf2 Activation in Endothelial Cells: Amplification of Low Shear Stress Signaling by Primary Cilia. *Redox Biol.* **2021**, *46*, 102103. [CrossRef]
7. Wang, K.-C.; Yeh, Y.-T.; Nguyen, P.; Limqueco, E.; Lopez, J.; Thorossian, S.; Guan, K.-L.; Li, Y.-S.J.; Chien, S. Flow-Dependent YAP/TAZ Activities Regulate Endothelial Phenotypes and Atherosclerosis. *Proc. Natl. Acad. Sci. USA* **2016**, *113*, 11525–11530. [CrossRef]
8. Tzima, E.; Irani-Tehrani, M.; Kiosses, W.B.; Dejana, E.; Schultz, D.A.; Engelhardt, B.; Cao, G.; DeLisser, H.; Schwartz, M.A. A Mechanosensory Complex That Mediates the Endothelial Cell Response to Fluid Shear Stress. *Nature* **2005**, *437*, 426–431. [CrossRef]
9. Petzold, T.; Orr, A.W.; Hahn, C.; Jhaveri, K.A.; Parsons, J.T.; Schwartz, M.A. Focal Adhesion Kinase Modulates Activation of NF-KB by Flow in Endothelial Cells. *Am. J. Physiol. Cell Physiol.* **2009**, *297*, C814–C822. [CrossRef]

10. Zhao, Q.; Ishibashi, M.; Hiasa, K.; Tan, C.; Takeshita, A.; Egashira, K. Essential Role of Vascular Endothelial Growth Factor in Angiotensin II–Induced Vascular Inflammation and Remodeling. *Hypertension* **2004**, *44*, 264–270. [CrossRef]
11. Takeshita, S.; Zheng, L.P.; Brogi, E.; Kearney, M.; Pu, L.Q.; Bunting, S.; Ferrara, N.; Symes, J.F.; Isner, J.M. Therapeutic Angiogenesis. A Single Intraarterial Bolus of Vascular Endothelial Growth Factor Augments Revascularization in a Rabbit Ischemic Hind Limb Model. *J. Clin. Investig.* **1994**, *93*, 662–670. [CrossRef] [PubMed]
12. Riessen, R.; Rahimizadeh, H.; Blessing, E.; Takeshita, S.; Barry, J.J.; Isner, J.M. Arterial Gene Transfer Using Pure DNA Applied Directly to a Hydrogel-Coated Angioplasty Balloon. *Hum. Gene* **1993**, *4*, 749–758. [CrossRef]
13. Chen, J.; Green, J.; Yurdagul, A.; Albert, P.; McInnis, M.C.; Orr, A.W. Avβ3 Integrins Mediate Flow-Induced NF-KB Activation, Proinflammatory Gene Expression, and Early Atherogenic Inflammation. *Am. J. Pathol.* **2015**, *185*, 2575–2589. [CrossRef] [PubMed]
14. Oeckinghaus, A.; Ghosh, S. The NF-KB Family of Transcription Factors and Its Regulation. *Cold Spring Harb. Perspect. Biol.* **2009**, *1*, a000034. [CrossRef] [PubMed]
15. Liu, T.; Zhang, L.; Joo, D.; Sun, S.-C. NF-KB Signaling in Inflammation. *Signal Transduct. Target. Ther.* **2017**, *2*, 17023. [CrossRef] [PubMed]
16. Wang, L.; Luo, J.-Y.; Li, B.; Tian, X.Y.; Chen, L.-J.; Huang, Y.; Liu, J.; Deng, D.; Lau, C.W.; Wan, S.; et al. Integrin-YAP/TAZ-JNK Cascade Mediates Atheroprotective Effect of Unidirectional Shear Flow. *Nature* **2016**, *540*, 579–582. [CrossRef] [PubMed]
17. Li, B.; He, J.; Lv, H.; Liu, Y.; Lv, X.; Zhang, C.; Zhu, Y.; Ai, D. C-Abl Regulates YAPY357 Phosphorylation to Activate Endothelial Atherogenic Responses to Disturbed Flow. *J. Clin. Investig.* **2019**, *129*, 1167–1179. [CrossRef]
18. Conway, D.E.; Schwartz, M.A. Flow-Dependent Cellular Mechanotransduction in Atherosclerosis. *J. Cell Sci.* **2013**, *126*, 5101–5109. [CrossRef]
19. Davies, P.F.; Civelek, M.; Fang, Y.; Fleming, I. The Atherosusceptible Endothelium: Endothelial Phenotypes in Complex Haemodynamic Shear Stress Regions in Vivo. *Cardiovasc. Res.* **2013**, *99*, 315–327. [CrossRef]
20. Givens, C.; Tzima, E. Endothelial Mechanosignaling: Does One Sensor Fit All? *Antioxid. Redox Signal.* **2016**, *25*, 373–388. [CrossRef]
21. Baratchi, S.; Khoshmanesh, K.; Woodman, O.L.; Potocnik, S.; Peter, K.; McIntyre, P. Molecular Sensors of Blood Flow in Endothelial Cells. *Trends Mol. Med.* **2017**, *23*, 850–868. [CrossRef] [PubMed]
22. Le Master, E.; Ahn, S.J.; Levitan, I. Chapter Five—Mechanisms of Endothelial Stiffening in Dyslipidemia and Aging: Oxidized Lipids and Shear Stress. In *Current Topics in Membranes*; Levitan, I., Trache, A., Eds.; Academic Press: London, UK, 2020; Volume 86, pp. 185–215. ISBN 1063-5823.
23. Mahmoudi, M.; Farghadan, A.; McConnell, D.R.; Barker, A.J.; Wentzel, J.J.; Budoff, M.J.; Arzani, A. The Story of Wall Shear Stress in Coronary Artery Atherosclerosis: Biochemical Transport and Mechanotransduction. *J. Biomech. Eng.* **2020**, *143*, 041002. [CrossRef] [PubMed]
24. Roux, E.; Bougaran, P.; Dufourcq, P.; Couffinhal, T. Fluid Shear Stress Sensing by the Endothelial Layer. *Front. Physiol.* **2020**, *11*, 861. [CrossRef] [PubMed]
25. Tanaka, K.; Joshi, D.; Timalsina, S.; Schwartz, M.A. Early Events in Endothelial Flow Sensing. *Cytoskeleton* **2021**, *78*, 217–231. [CrossRef]
26. Mylvaganam, S.; Yusuf, B.; Li, R.; Lu, C.-Y.; Robinson, L.A.; Freeman, S.A.; Grinstein, S. Endothelial Integration of Mechanosensory Signals by the Spectrin Cytoskeleton. *bioRxiv* **2021**. [CrossRef]
27. Fang, Y.; Wu, D.; Birukov, K.G. Mechanosensing and Mechanoregulation of Endothelial Cell Functions. *Compr. Physiol.* **2019**, *9*, 873.
28. Mack, J.J.; Mosqueiro, T.S.; Archer, B.J.; Jones, W.M.; Sunshine, H.; Faas, G.C.; Briot, A.; Aragón, R.L.; Su, T.; Romay, M.C.; et al. NOTCH1 Is a Mechanosensor in Adult Arteries. *Nat. Commun.* **2017**, *8*, 1620. [CrossRef]
29. Mehta, V.; Pang, K.-L.; Rozbesky, D.; Nather, K.; Keen, A.; Lachowski, D.; Kong, Y.; Karia, D.; Ameismeier, M.; Huang, J.; et al. The Guidance Receptor Plexin D1 Is a Mechanosensor in Endothelial Cells. *Nature* **2020**, *578*, 290–295. [CrossRef]
30. Yamamoto, K.; Imamura, H.; Ando, J. Shear Stress Augments Mitochondrial ATP Generation That Triggers ATP Release and Ca2+ Signaling in Vascular Endothelial Cells. *Am. J. Physiol. Heart Circ. Physiol.* **2018**, *315*, H1477–H1485. [CrossRef]
31. Yamamoto, K.; Nogimori, Y.; Imamura, H.; Ando, J. Shear Stress Activates Mitochondrial Oxidative Phosphorylation by Reducing Plasma Membrane Cholesterol in Vascular Endothelial Cells. *Proc. Natl. Acad. Sci. USA* **2020**, *117*, 33660. [CrossRef]
32. Cox, C.D.; Bavi, N.; Martinac, B. Biophysical Principles of Ion-Channel-Mediated Mechanosensory Transduction. *Cell Rep.* **2019**, *29*, 1–12. [CrossRef]
33. Hamill, O.P.; Martinac, B. Molecular Basis of Mechanotransduction in Living Cells. *Physiol. Rev.* **2001**, *81*, 685. [CrossRef]
34. Hamill, O.P. Twenty Odd Years of Stretch-Sensitive Channels. *Pflügers Arch.* **2006**, *453*, 333–351. [CrossRef] [PubMed]
35. Balleza, D. Mechanical Properties of Lipid Bilayers and Regulation of Mechanosensitive Function. *Channels* **2012**, *6*, 220–233. [CrossRef] [PubMed]
36. Lee, A.G. How Lipids Affect the Activities of Integral Membrane Proteins. *Biochim. Biophys. Acta Biomembr.* **2004**, *1666*, 62–87. [CrossRef] [PubMed]
37. Corradi, V.; Sejdiu, B.I.; Mesa-Galloso, H.; Abdizadeh, H.; Noskov, S.Y.; Marrink, S.J.; Tieleman, D.P. Emerging Diversity in Lipid–Protein Interactions. *Chem. Rev.* **2019**, *119*, 5775–5848. [CrossRef]

38. Muller, M.P.; Jiang, T.; Sun, C.; Lihan, M.; Pant, S.; Mahinthichaichan, P.; Trifan, A.; Tajkhorshid, E. Characterization of Lipid–Protein Interactions and Lipid-Mediated Modulation of Membrane Protein Function through Molecular Simulation. *Chem. Rev.* **2019**, *119*, 6086–6161. [CrossRef]
39. Jodaitis, L.; van Oene, T.; Martens, C. Assessing the Role of Lipids in the Molecular Mechanism of Membrane Proteins. *Int. J. Mol. Sci.* **2021**, *22*, 7267. [CrossRef]
40. Overduin, M.; Trieber, C.; Prosser, R.S.; Picard, L.-P.; Sheff, J.G. Structures and Dynamics of Native-State Transmembrane Protein Targets and Bound Lipids. *Membranes* **2021**, *11*, 451. [CrossRef]
41. Majeed, S.; Ahmad, A.B.; Sehar, U.; Georgieva, E.R. Lipid Membrane Mimetics in Functional and Structural Studies of Integral Membrane Proteins. *Membranes* **2021**, *11*, 685. [CrossRef]
42. Renard, K.; Byrne, B. Insights into the Role of Membrane Lipids in the Structure, Function and Regulation of Integral Membrane Proteins. *Int. J. Mol. Sci.* **2021**, *22*, 9026. [CrossRef]
43. Ashrafuzzaman, M.; Tuszynski, J. Regulation of Channel Function Due to Coupling with a Lipid Bilayer. *J. Comput. Theor. Nanosci.* **2012**, *9*, 564–570. [CrossRef]
44. Ashrafuzzaman, M.; Tuszynski, J. Lipid Bilayer-Membrane Protein Coupling. In *Membrane Biophysics*; Springer: Berlin/Heidelberg, Germany, 2012; pp. 75–125.
45. Huang, H.W. Deformation Free Energy of Bilayer Membrane and Its Effect on Gramicidin Channel Lifetime. *Biophys. J.* **1986**, *50*, 1061–1070. [CrossRef]
46. Gruner, S.M. Intrinsic Curvature Hypothesis for Biomembrane Lipid Composition: A Role for Nonbilayer Lipids. *Proc. Natl. Acad. Sci. USA* **1985**, *82*, 3665–3669. [CrossRef] [PubMed]
47. Helfrich, P.; Jakobsson, E. Calculation of Deformation Energies and Conformations in Lipid Membranes Containing Gramicidin Channels. *Biophys. J.* **1990**, *57*, 1075–1084. [CrossRef]
48. Durkin, J.T.; Providence, L.L.; Koeppe, R.E., II; Andersen, O.S.; Koeppe, R.E., II; Andersen, O.S. Energetics of Heterodimer Formation among Gramicidin Analogues with an NH2-Terminal Addition or Deletion: Consequences of Missing a Residue at the Join in the Channel. *J. Mol. Biol.* **1993**, *231*, 1102–1121. [CrossRef]
49. Nielsen, C.; Goulian, M.; Andersen, O.S. Energetics of Inclusion-Induced Bilayer Deformations. *Biophys. J.* **1998**, *74*, 1966–1983. [CrossRef]
50. Nielsen, C.; Andersen, O.S. Inclusion-Induced Bilayer Deformations: Effects of Monolayer Equilibrium Curvature. *Biophys. J.* **2000**, *79*, 2583–2604. [CrossRef]
51. Lundbæk, J.A.; Birn, P.; Hansen, A.J.; Søgaard, R.; Nielsen, C.; Girshman, J.; Bruno, M.J.; Tape, S.E.; Egebjerg, J.; Greathouse, D.V. Regulation of Sodium Channel Function by Bilayer Elasticity The Importance of Hydrophobic Coupling. Effects of Micelle-Forming Amphiphiles and Cholesterol. *J. Gen. Physiol.* **2004**, *123*, 599–621. [CrossRef]
52. Lundbaek, J.; Birn, P.; Tape, S.; Toombes, G.E.; Søgaard, R.; Koeppe, R.E.; Gruner, S.M.; Hansen, A.J.; Andersen, O.S. Capsaicin Regulates Voltage-Dependent Sodium Channels by Altering Lipid Bilayer Elasticity. *Mol. Pharmacol.* **2005**, *68*, 680–689. [CrossRef]
53. Ashrafuzzaman, M.; Tseng, C.-Y.; Tuszynski, J. Regulation of Channel Function Due to Physical Energetic Coupling with a Lipid Bilayer. *Biochem. Biophys. Res. Commun.* **2014**, *445*, 463–468. [CrossRef]
54. Ashrafuzzaman, M.; Tseng, C.-Y.; Tuszynski, J. Charge-Based Interactions of Antimicrobial Peptides and General Drugs with Lipid Bilayers. *J. Mol. Graph. Model.* **2020**, *95*, 107502. [CrossRef] [PubMed]
55. Zhang, T.; Hu, W.; Chen, W. Plasma Membrane Integrates Biophysical and Biochemical Regulation to Trigger Immune Receptor Functions. *Front. Immunol.* **2021**, *12*, 53. [CrossRef] [PubMed]
56. Goeckeler, Z.M.; Bridgman, P.C.; Wysolmerski, R.B. Nonmuscle Myosin II Is Responsible for Maintaining Endothelial Cell Basal Tone and Stress Fiber Integrity. *Am. J. Physiol. Cell Physiol.* **2008**, *295*, C994–C1006. [CrossRef] [PubMed]
57. Gordon, E.; Schimmel, L.; Frye, M. The Importance of Mechanical Forces for in Vitro Endothelial Cell Biology. *Front. Physiol.* **2020**, *11*, 684. [CrossRef]
58. Ross, T.D.; Coon, B.G.; Yun, S.; Baeyens, N.; Tanaka, K.; Ouyang, M.; Schwartz, M.A. Integrins in Mechanotransduction. *Curr. Opin. Cell Biol.* **2013**, *25*, 613–618. [CrossRef]
59. Silvani, G.; Romanov, V.; Cox, C.D.; Martinac, B. Biomechanical Characterization of Endothelial Cells Exposed to Shear Stress Using Acoustic Force Spectroscopy. *Front. Bioeng. Biotechnol.* **2021**, *9*, 21. [CrossRef]
60. Engler, A.J.; Wang, Y. Editorial: Understanding Molecular Interactions That Underpin Vascular Mechanobiology. *APL Bioeng.* **2021**, *5*, 030401. [CrossRef]
61. Katsumi, A.; Milanini, J.; Kiosses, W.B.; del Pozo, M.A.; Kaunas, R.; Chien, S.; Hahn, K.M.; Schwartz, M.A. Effects of Cell Tension on the Small GTPase Rac. *J. Cell Biol.* **2002**, *158*, 153–164. [CrossRef]
62. Grashoff, C.; Hoffman, B.D.; Brenner, M.D.; Zhou, R.; Parsons, M.; Yang, M.T.; McLean, M.A.; Sligar, S.G.; Chen, C.S.; Ha, T.; et al. Measuring Mechanical Tension across Vinculin Reveals Regulation of Focal Adhesion Dynamics. *Nature* **2010**, *466*, 263–266. [CrossRef]
63. Conway, D.E.; Breckenridge, M.T.; Hinde, E.; Gratton, E.; Chen, C.S.; Schwartz, M.A. Fluid Shear Stress on Endothelial Cells Modulates Mechanical Tension across VE-Cadherin and PECAM-1. *Curr. Biol.* **2013**, *23*, 1024–1030. [CrossRef]
64. Barvitenko, N.; Aslam, M.; Lawen, A.; Saldanha, C.; Skverchinskaya, E.; Uras, G.; Manca, A.; Pantaleo, A. Two Motors and One Spring: Hypothetic Roles of Non-Muscle Myosin II and Submembrane Actin-Based Cytoskeleton in Cell Volume Sensing. *Int. J. Mol. Sci.* **2021**, *22*, 7967. [CrossRef]

65. Song, Y.; Kenworthy, A.K.; Sanders, C.R. Cholesterol as a Co-Solvent and a Ligand for Membrane Proteins. *Protein Sci.* **2014**, *23*, 1–22. [CrossRef]
66. Salas-Estrada, L.A.; Leioatts, N.; Romo, T.D.; Grossfield, A. Lipids Alter Rhodopsin Function via Ligand-like and Solvent-like Interactions. *Biophys. J.* **2018**, *114*, 355–367. [CrossRef]
67. Fantini, J.; Barrantes, F. How Cholesterol Interacts with Membrane Proteins: An Exploration of Cholesterol-Binding Sites Including CRAC, CARC, and Tilted Domains. *Front. Physiol.* **2013**, *4*, 31. [CrossRef]
68. Lee, A.G. Interfacial Binding Sites for Cholesterol on G Protein-Coupled Receptors. *Biophys. J.* **2019**, *116*, 1586–1597. [CrossRef] [PubMed]
69. Jakubík, J.; El-Fakahany, E.E. Allosteric Modulation of GPCRs of Class A by Cholesterol. *Int. J. Mol. Sci.* **2021**, *22*, 1953. [CrossRef] [PubMed]
70. Gudi, S.; Nolan, J.P.; Frangos, J.A. Modulation of GTPase Activity of G Proteins by Fluid Shear Stress and Phospholipid Composition. *Proc. Natl. Acad. Sci. USA* **1998**, *95*, 2515–2519. [CrossRef] [PubMed]
71. Haidekker, M.A.; L'Heureux, N.; Frangos, J.A. Fluid Shear Stress Increases Membrane Fluidity in Endothelial Cells: A Study with DCVJ Fluorescence. *Am. J. Physiol. Heart Circ. Physiol.* **2000**, *278*, H1401–H1406. [CrossRef] [PubMed]
72. Butler, P.J.; Norwich, G.; Weinbaum, S.; Chien, S. Shear Stress Induces a Time- and Position-Dependent Increase in Endothelial Cell Membrane Fluidity. *Am. J. Physiol. Cell Physiol.* **2001**, *280*, C962–C969. [CrossRef]
73. Yamamoto, K.; Ando, J. Vascular Endothelial Cell Membranes Differentiate between Stretch and Shear Stress through Transitions in Their Lipid Phases. *Am. J. Physiol. Heart Circ. Physiol.* **2015**, *309*, H1178–H1185. [CrossRef]
74. Billington, N.; Wang, A.; Mao, J.; Adelstein, R.S.; Sellers, J.R. Characterization of Three Full-Length Human Nonmuscle Myosin II Paralogs. *J. Biol. Chem.* **2013**, *288*, 33398–33410. [CrossRef] [PubMed]
75. Dulyaninova, N.G.; Bresnick, A.R. The Heavy Chain Has Its Day. *BioArchitecture* **2013**, *3*, 77–85. [CrossRef] [PubMed]
76. Shutova, M.S.; Svitkina, T.M. Common and Specific Functions of Nonmuscle Myosin II Paralogs in Cells. *Biochemistry* **2018**, *83*, 1459–1468. [CrossRef] [PubMed]
77. Sellers, J.R.; Heissler, S.M. Nonmuscle Myosin-2 Isoforms. *Curr. Biol.* **2019**, *29*, R275–R278. [CrossRef] [PubMed]
78. Garrido-Casado, M.; Asensio-Juárez, G.; Vicente-Manzanares, M. Nonmuscle Myosin II Regulation Directs Its Multiple Roles in Cell Migration and Division. *Annu. Rev. Cell Dev. Biol.* **2021**, *37*, 285–310. [CrossRef] [PubMed]
79. Wang, A.; Ma, X.; Conti, M.A.; Adelstein, R.S. Distinct and Redundant Roles of the Non-Muscle Myosin II Isoforms and Functional Domains. *Biochem. Soc. Trans.* **2011**, *39*, 1131–1135. [CrossRef]
80. Kolega, J. Asymmetric Distribution of Myosin IIB in Migrating Endothelial Cells Is Regulated by a Rho-Dependent Kinase and Contributes to Tail Retraction. *Mol. Biol. Cell* **2003**, *14*, 4745–4757. [CrossRef]
81. Kolega, J. The Role of Myosin II Motor Activity in Distributing Myosin Asymmetrically and Coupling Protrusive Activity to Cell Translocation. *Mol. Biol. Cell* **2006**, *17*, 4435–4445. [CrossRef]
82. Kamm, K.E.; Stull, J.T. Dedicated Myosin Light Chain Kinases with Diverse Cellular Functions. *J. Biol. Chem.* **2001**, *276*, 4527–4530. [CrossRef]
83. Hartshorne, D.J.; Ito, M.; Erdodi, F. Myosin Light Chain Phosphatase: Subunit Composition, Interactions and Regulation. *J. Muscle Res. Cell Motil.* **1998**, *19*, 325–341. [CrossRef]
84. Kiss, A.; Erdődi, F.; Lontay, B. Myosin Phosphatase: Unexpected Functions of a Long-Known Enzyme. *Biochim. Biophys. Acta Mol. Cell Res.* **2019**, *1866*, 2–15. [CrossRef] [PubMed]
85. Mehta, D.; Malik, A.B. Signaling Mechanisms Regulating Endothelial Permeability. *Physiol. Rev.* **2006**, *86*, 279–367. [CrossRef] [PubMed]
86. Rigor, R.R.; Shen, Q.; Pivetti, C.D.; Wu, M.H.; Yuan, S.Y. Myosin Light Chain Kinase Signaling in Endothelial Barrier Dysfunction. *Med. Res. Rev.* **2013**, *33*, 911–933. [CrossRef]
87. Nobes, C.D.; Hall, A. Rho, Rac, and Cdc42 GTPases Regulate the Assembly of Multimolecular Focal Complexes Associated with Actin Stress Fibers, Lamellipodia, and Filopodia. *Cell* **1995**, *81*, 53–62. [CrossRef]
88. Ridley, A.J. Rho GTPase Signalling in Cell Migration. *Curr. Opin. Cell Biol.* **2015**, *36*, 103–112. [CrossRef] [PubMed]
89. Flentje, A.; Kalsi, R.; Monahan, T.S. Small GTPases and Their Role in Vascular Disease. *Int. J. Mol. Sci.* **2019**, *20*, 917. [CrossRef]
90. Tzima, E.; del Pozo, M.A.; Shattil, S.J.; Chien, S.; Schwartz, M.A. Activation of Integrins in Endothelial Cells by Fluid Shear Stress Mediates Rho-Dependent Cytoskeletal Alignment. *EMBO J.* **2001**, *20*, 4639–4647. [CrossRef] [PubMed]
91. Tzima, E.; Del Pozo, M.A.; Kiosses, W.B.; Mohamed, S.A.; Li, S.; Chien, S.; Schwartz, M.A. Activation of Rac1 by Shear Stress in Endothelial Cells Mediates Both Cytoskeletal Reorganization and Effects on Gene Expression. *EMBO J.* **2002**, *21*, 6791–6800. [CrossRef]
92. Birukov, K.G. Small GTPases in Mechanosensitive Regulation of Endothelial Barrier. *Microvasc. Res.* **2009**, *77*, 46–52. [CrossRef]
93. Aslam, M.; Schluter, K.-D.; Rohrbach, S.; Rafiq, A.; Nazli, S.; Piper, H.M.; Noll, T.; Schulz, R.; Gündüz, D. Hypoxia–Reoxygenation-Induced Endothelial Barrier Failure: Role of RhoA, Rac1 and Myosin Light Chain Kinase. *J. Physiol.* **2013**, *591*, 461–473. [CrossRef]
94. Komarova, Y.A.; Kruse, K.; Mehta, D.; Malik, A.B. Protein Interactions at Endothelial Junctions and Signaling Mechanisms Regulating Endothelial Permeability. *Circ. Res.* **2017**, *120*, 179–206. [CrossRef]
95. Rho, S.-S.; Ando, K.; Fukuhara, S. Dynamic Regulation of Vascular Permeability by Vascular Endothelial Cadherin-Mediated Endothelial Cell-Cell Junctions. *J. Nippon Med. Sch.* **2017**, *84*, 148–159. [CrossRef] [PubMed]

96. Amano, M.; Ito, M.; Kimura, K.; Fukata, Y.; Chihara, K.; Nakano, T.; Matsuura, Y.; Kaibuchi, K. Phosphorylation and Activation of Myosin by Rho-Associated Kinase (Rho-Kinase). *J. Biol. Chem.* **1996**, *271*, 20246–20249. [CrossRef] [PubMed]
97. Kimura, K.; Ito, M.; Amano, M.; Chihara, K.; Fukata, Y.; Nakafuku, M.; Yamamori, B.; Feng, J.; Nakano, T.; Okawa, K. Regulation of Myosin Phosphatase by Rho and Rho-Associated Kinase (Rho-Kinase). *Science* **1996**, *273*, 245–248. [CrossRef] [PubMed]
98. Noma, K.; Oyama, N.; Liao, J.K. Physiological Role of ROCKs in the Cardiovascular System. *Am. J. Physiol. Cell Physiol.* **2006**, *290*, C661–C668. [CrossRef] [PubMed]
99. Sanders, L.C.; Matsumura, F.; Bokoch, G.M.; De Lanerolle, P. Inhibition of Myosin Light Chain Kinase by P21-Activated Kinase. *Science* **1999**, *283*, 2083–2085. [CrossRef] [PubMed]
100. Borghi, N.; Sorokina, M.; Shcherbakova, O.G.; Weis, W.I.; Pruitt, B.L.; Nelson, W.J.; Dunn, A.R. E-Cadherin Is under Constitutive Actomyosin-Generated Tension That Is Increased at Cell–Cell Contacts upon Externally Applied Stretch. *Proc. Natl. Acad. Sci. USA* **2012**, *109*, 12568. [CrossRef]
101. Daneshjou, N.; Sieracki, N.; van Nieuw Amerongen, G.P.; Conway, D.E.; Schwartz, M.A.; Komarova, Y.A.; Malik, A.B. Rac1 Functions as a Reversible Tension Modulator to Stabilize VE-Cadherin Trans-Interaction. *J. Cell Biol.* **2015**, *208*, 23–32. [CrossRef]
102. Efimova, N.; Svitkina, T.M. Branched Actin Networks Push against Each Other at Adherens Junctions to Maintain Cell–Cell Adhesion. *J. Cell Biol.* **2018**, *217*, 1827–1845. [CrossRef]
103. Morimura, S.; Suzuki, K.; Takahashi, K. Nonmuscle Myosin IIA Is Required for Lamellipodia Formation through Binding to WAVE2 and Phosphatidylinositol 3, 4, 5-Triphosphate. *Biochem. Biophys. Res. Commun.* **2011**, *404*, 834–840. [CrossRef]
104. Barvitenko, N.N.; Aslam, M.; Lawen, A.; Pantaleo, A.; Saldanha, C.; Matteucci, E. Effects of Oxygen Depletion on Transmembrane Protein Activities. *Curr. Org. Chem.* **2015**, *19*, 2002–2010. [CrossRef]
105. Chandel, N.S.; McClintock, D.S.; Feliciano, C.E.; Wood, T.M.; Melendez, J.A.; Rodriguez, A.M.; Schumacker, P.T. Reactive Oxygen Species Generated at Mitochondrial Complex III Stabilize Hypoxia-Inducible Factor-1α during Hypoxia: A Mechanism of O2 Sensing. *J. Biol. Chem.* **2000**, *275*, 25130–25138. [CrossRef] [PubMed]
106. Zmysłowski, A.; Szterk, A. Current Knowledge on the Mechanism of Atherosclerosis and Pro-Atherosclerotic Properties of Oxysterols. *Lipids Health Dis.* **2017**, *16*, 188. [CrossRef] [PubMed]
107. Tall, A.R.; Yvan-Charvet, L. Cholesterol, Inflammation and Innate Immunity. *Nat. Rev. Immunol.* **2015**, *15*, 104–116. [CrossRef] [PubMed]
108. Pederiva, C.; Capra, M.E.; Viggiano, C.; Rovelli, V.; Banderali, G.; Biasucci, G. Early Prevention of Atherosclerosis: Detection and Management of Hypercholesterolaemia in Children and Adolescents. *Life* **2021**, *11*, 345. [CrossRef]
109. Wang, H.H.; Garruti, G.; Liu, M.; Portincasa, P.; Wang, D.Q.-H. Cholesterol and Lipoprotein Metabolism and Atherosclerosis: Recent Advances in Reverse Cholesterol Transport. *Ann. Hepatol.* **2017**, *16*, S27–S42. [CrossRef]
110. Berliner, J.A.; Leitinger, N.; Tsimikas, S. The Role of Oxidized Phospholipids in Atherosclerosis. *J. Lipid Res.* **2009**, *50*, S207–S212. [CrossRef]
111. Jurkiewicz, P.; Olżyńska, A.; Cwiklik, L.; Conte, E.; Jungwirth, P.; Megli, F.M.; Hof, M. Biophysics of Lipid Bilayers Containing Oxidatively Modified Phospholipids: Insights from Fluorescence and EPR Experiments and from MD Simulations. *Biochim. Biophys. Acta Biomembr.* **2012**, *1818*, 2388–2402. [CrossRef]
112. Yadav, D.K.; Kumar, S.; Choi, E.-H.; Chaudhary, S.; Kim, M.-H. Molecular Dynamic Simulations of Oxidized Skin Lipid Bilayer and Permeability of Reactive Oxygen Species. *Sci. Rep.* **2019**, *9*, 4496. [CrossRef]
113. Mozaffarian, D.; Wu, J.H.Y. (N-3) Fatty Acids and Cardiovascular Health: Are Effects of EPA and DHA Shared or Complementary? *J. Nutr.* **2012**, *142*, 614S–625S. [CrossRef]
114. Yamagata, K. Docosahexaenoic Acid Regulates Vascular Endothelial Cell Function and Prevents Cardiovascular Disease. *Lipids Health Dis.* **2017**, *16*, 118. [CrossRef] [PubMed]
115. Innes, J.K.; Calder, P.C. Marine Omega-3 (N-3) Fatty Acids for Cardiovascular Health: An Update for 2020. *Int. J. Mol. Sci.* **2020**, *21*, 1362. [CrossRef] [PubMed]
116. Wojciak-Stothard, B.; Tsang, L.Y.F.; Paleolog, E.; Hall, S.M.; Haworth, S.G. Rac1 and RhoA as Regulators of Endothelial Phenotype and Barrier Function in Hypoxia-Induced Neonatal Pulmonary Hypertension. *Am. J. Physiol. Lung Cell. Mol. Physiol.* **2006**, *290*, L1173–L1182. [CrossRef] [PubMed]
117. Chi, A.Y.; Waypa, G.B.; Mungai, P.T.; Schumacker, P.T. Prolonged Hypoxia Increases ROS Signaling and RhoA Activation in Pulmonary Artery Smooth Muscle and Endothelial Cells. *Antioxid. Redox Signal.* **2010**, *12*, 603–610. [CrossRef] [PubMed]
118. Kalwa, H.; Sartoretto, J.L.; Sartoretto, S.M.; Michel, T. Angiotensin-II and MARCKS: A Hydrogen Peroxide- and Rac1-Dependent Signaling Pathway in Vascular Endothelium. *J. Biol. Chem.* **2012**, *287*, 29147–29158. [CrossRef]
119. Fatima, N.; Patel, S.N.; Hussain, T. Angiotensin II Type 2 Receptor: A Target for Protection Against Hypertension, Metabolic Dysfunction, and Organ Remodeling. *Hypertension* **2021**, *77*, 1845–1856. [CrossRef]
120. Terenzi, R.; Manetti, M.; Rosa, I.; Romano, E.; Galluccio, F.; Guiducci, S.; Ibba-Manneschi, L.; Matucci-Cerinic, M. Angiotensin II Type 2 Receptor (AT2R) as a Novel Modulator of Inflammation in Rheumatoid Arthritis Synovium. *Sci. Rep.* **2017**, *7*, 13293. [CrossRef]
121. Patel, S.N.; Fatima, N.; Ali, R.; Hussain, T. Emerging Role of Angiotensin AT2 Receptor in Anti-Inflammation: An Update. *Curr. Pharm. Des.* **2020**, *26*, 492–500. [CrossRef]

122. Shatanawi, A.; Romero, M.J.; Iddings, J.A.; Chandra, S.; Umapathy, N.S.; Verin, A.D.; Caldwell, R.B.; Caldwell, R.W. Angiotensin II-Induced Vascular Endothelial Dysfunction through RhoA/Rho Kinase/P38 Mitogen-Activated Protein Kinase/Arginase Pathway. *Am. J. Physiol. Cell Physiol.* **2011**, *300*, C1181–C1192. [CrossRef]
123. Savoia, C.; Tabet, F.; Yao, G.; Schiffrin, E.L.; Touyz, R.M. Negative Regulation of RhoA/Rho Kinase by Angiotensin II Type 2 Receptor in Vascular Smooth Muscle Cells: Role in Angiotensin II-Induced Vasodilation in Stroke-Prone Spontaneously Hypertensive Rats. *J. Hypertens.* **2005**, *23*, 1037–1045. [CrossRef]
124. Nakakuki, T.; Ito, M.; Iwasaki, H.; Kureishi, Y.; Okamoto, R.; Moriki, N.; Kongo, M.; Kato, S.; Yamada, N.; Isaka, N.; et al. Rho/Rho-Kinase Pathway Contributes to C-Reactive Protein–Induced Plasminogen Activator Inhibitor-1 Expression in Endothelial Cells. *Arterioscler. Thromb. Vasc. Biol.* **2005**, *25*, 2088–2093. [CrossRef] [PubMed]
125. Kurano, M.; Yatomi, Y. Sphingosine 1-Phosphate and Atherosclerosis. *J. Atheroscler. Thromb.* **2018**, *25*, 16–26. [CrossRef] [PubMed]
126. Anwar, M.; Mehta, D. Post-Translational Modifications of S1PR1 and Endothelial Barrier Regulation. *Biochim. Biophys. Acta Mol. Cell Biol. Lipids* **2020**, *1865*, 158760. [CrossRef]
127. Obinata, H.; Hla, T. Sphingosine 1-Phosphate and Inflammation. *Int. Immunol.* **2019**, *31*, 617–625. [CrossRef]
128. Chen, D.; Dorling, A. Critical Roles for Thrombin in Acute and Chronic Inflammation. *J. Thromb. Haemost.* **2009**, *7*, 122–126. [CrossRef] [PubMed]
129. Aslam, M.; Pfeil, U.; Gündüz, D.; Rafiq, A.; Kummer, W.; Piper, H.; Noll, T. Intermedin (Adrenomedullin2) Stabilizes the Endothelial Barrier and Antagonizes Thrombin-Induced Barrier Failure in Endothelial Cell Monolayers. *Br. J. Pharmacol.* **2012**, *165*, 208–222. [CrossRef] [PubMed]
130. Aslam, M.; Tanislav, C.; Troidl, C.; Schulz, R.; Hamm, C.; Gündüz, D. CAMP Controls the Restoration of Endothelial Barrier Function after Thrombin-Induced Hyperpermeability via Rac1 Activation. *Physiol Rep.* **2014**, *2*, e12175. [CrossRef]
131. Li, J.; Hou, B.; Tumova, S.; Muraki, K.; Bruns, A.; Ludlow, M.J.; Sedo, A.; Hyman, A.J.; McKeown, L.; Young, R.S.; et al. Piezo1 Integration of Vascular Architecture with Physiological Force. *Nature* **2014**, *515*, 279–282. [CrossRef]
132. Coste, B.; Xiao, B.; Santos, J.S.; Syeda, R.; Grandl, J.; Spencer, K.S.; Kim, S.E.; Schmidt, M.; Mathur, J.; Dubin, A.E.; et al. Piezo Proteins Are Pore-Forming Subunits of Mechanically Activated Channels. *Nature* **2012**, *483*, 176–181. [CrossRef]
133. Ge, J.; Li, W.; Zhao, Q.; Li, N.; Chen, M.; Zhi, P.; Li, R.; Gao, N.; Xiao, B.; Yang, M. Architecture of the Mammalian Mechanosensitive Piezo1 Channel. *Nature* **2015**, *527*, 64–69. [CrossRef]
134. Cox, C.D.; Bae, C.; Ziegler, L.; Hartley, S.; Nikolova-Krstevski, V.; Rohde, P.R.; Ng, C.-A.; Sachs, F.; Gottlieb, P.A.; Martinac, B. Removal of the Mechanoprotective Influence of the Cytoskeleton Reveals PIEZO1 Is Gated by Bilayer Tension. *Nat. Commun.* **2016**, *7*, 10366. [CrossRef] [PubMed]
135. Lewis, A.H.; Grandl, J. Mechanical Sensitivity of Piezo1 Ion Channels Can Be Tuned by Cellular Membrane Tension. *eLife* **2015**, *4*, e12088. [CrossRef] [PubMed]
136. Coste, B.; Mathur, J.; Schmidt, M.; Earley, T.J.; Ranade, S.; Petrus, M.J.; Dubin, A.E.; Patapoutian, A. Piezo1 and Piezo2 Are Essential Components of Distinct Mechanically Activated Cation Channels. *Science* **2010**, *330*, 55–60. [CrossRef] [PubMed]
137. Parpaite, T.; Coste, B. Piezo Channels. *Curr. Biol.* **2017**, *27*, R250–R252. [CrossRef] [PubMed]
138. Shinge, S.A.U.; Zhang, D.; Muluh, T.A.; Nie, Y.; Yu, F. Mechanosensitive Piezo1 Channel Evoked-Mechanical Signals in Atherosclerosis. *J. Inflamm. Res.* **2021**, *14*, 3621. [CrossRef] [PubMed]
139. Kang, H.; Hong, Z.; Zhong, M.; Klomp, J.; Bayless, K.J.; Mehta, D.; Karginov, A.V.; Hu, G.; Malik, A.B. Piezo1 Mediates Angiogenesis through Activation of MT1-MMP Signaling. *Am. J. Physiol. Cell Physiol.* **2019**, *316*, C92–C103. [CrossRef]
140. Albarrán-Juárez, J.; Iring, A.; Wang, S.; Joseph, S.; Grimm, M.; Strilic, B.; Wettschureck, N.; Althoff, T.F.; Offermanns, S. Piezo1 and Gq/G11 Promote Endothelial Inflammation Depending on Flow Pattern and Integrin Activation. *J. Exp. Med.* **2018**, *215*, 2655–2672. [CrossRef]
141. Wang, S.; Chennupati, R.; Kaur, H.; Iring, A.; Wettschureck, N.; Offermanns, S. Endothelial Cation Channel PIEZO1 Controls Blood Pressure by Mediating Flow-Induced ATP Release. *J. Clin. Investig.* **2016**, *126*, 4527–4536. [CrossRef]
142. Otte, L.A.; Bell, K.S.; Loufrani, L.; Yeh, J.-C.; Melchior, B.; Dao, D.N.; Stevens, H.Y.; White, C.R.; Frangos, J.A. Rapid Changes in Shear Stress Induce Dissociation of a Gαq/11–Platelet Endothelial Cell Adhesion Molecule-1 Complex. *J. Physiol.* **2009**, *587*, 2365–2373. [CrossRef]
143. Dela Paz, N.G.; Frangos, J.A. Rapid Flow-Induced Activation of Gαq/11 Is Independent of Piezo1 Activation. *Am. J. Physiol. Cell Physiol.* **2019**, *316*, C741–C752. [CrossRef]
144. Zabroski, I.O.; Nugent, M.A. Lipid Raft Association Stabilizes VEGF Receptor 2 in Endothelial Cells. *Int. J. Mol. Sci.* **2021**, *22*, 798. [CrossRef]
145. Coon, B.G.; Baeyens, N.; Han, J.; Budatha, M.; Ross, T.D.; Fang, J.S.; Yun, S.; Thomas, J.-L.; Schwartz, M.A. Intramembrane Binding of VE-Cadherin to VEGFR2 and VEGFR3 Assembles the Endothelial Mechanosensory Complex. *J. Cell Biol.* **2015**, *208*, 975–986. [CrossRef] [PubMed]
146. Ladoux, B.; Nelson, W.J.; Yan, J.; Mège, R.M. The Mechanotransduction Machinery at Work at Adherens Junctions. *Integr. Biol.* **2015**, *7*, 1109–1119. [CrossRef] [PubMed]
147. Gudi, S.R.P.; Clark, C.B.; Frangos, J.A. Fluid Flow Rapidly Activates G Proteins in Human Endothelial Cells. *Circ. Res.* **1996**, *79*, 834–839. [CrossRef] [PubMed]
148. Dela Paz, N.G.; Melchior, B.; Frangos, J.A. Shear Stress Induces Gαq/11 Activation Independently of G Protein-Coupled Receptor Activation in Endothelial Cells. *Am. J. Physiol. Cell Physiol.* **2017**, *312*, C428–C437. [CrossRef]

149. Chachisvilis, M.; Zhang, Y.-L.; Frangos, J.A. G Protein-Coupled Receptors Sense Fluid Shear Stress in Endothelial Cells. *Proc. Natl. Acad. Sci. USA* **2006**, *103*, 15463–15468. [CrossRef]
150. Candelario, J.; Chachisvilis, M. Activity of Bradykinin B2 Receptor Is Regulated by Long-Chain Polyunsaturated Fatty Acids. *PLoS ONE* **2013**, *8*, e68151. [CrossRef]
151. Erdogmus, S.; Storch, U.; Danner, L.; Becker, J.; Winter, M.; Ziegler, N.; Wirth, A.; Offermanns, S.; Hoffmann, C.; Gudermann, T.; et al. Helix 8 Is the Essential Structural Motif of Mechanosensitive GPCRs. *Nat. Commun.* **2019**, *10*, 5784. [CrossRef]
152. Hynes, R.O. Integrins: Bidirectional, Allosteric Signaling Machines. *Cell* **2002**, *110*, 673–687. [CrossRef]
153. Finney, A.C.; Stokes, K.Y.; Pattillo, C.B.; Orr, A.W. Integrin Signaling in Atherosclerosis. *Cell. Mol. Life Sci.* **2017**, *74*, 2263–2282. [CrossRef]
154. Mo, F.-E. Shear-Regulated Extracellular Microenvironments and Endothelial Cell Surface Integrin Receptors Intertwine in Atherosclerosis. *Front. Cell Dev. Biol.* **2021**, *9*, 832. [CrossRef] [PubMed]
155. Gaus, K.; Le Lay, S.; Balasubramanian, N.; Schwartz, M.A. Integrin-Mediated Adhesion Regulates Membrane Order. *J. Cell Biol.* **2006**, *174*, 725–734. [CrossRef] [PubMed]
156. Sun, X.; Fu, Y.; Gu, M.; Zhang, L.; Li, D.; Li, H.; Chien, S.; Shyy, J.Y.-J.; Zhu, Y. Activation of Integrin A5 Mediated by Flow Requires Its Translocation to Membrane Lipid Rafts in Vascular Endothelial Cells. *Proc. Natl. Acad. Sci. USA* **2016**, *113*, 769. [CrossRef] [PubMed]
157. Son, S.; Moroney, G.J.; Butler, P.J. B1-Integrin-Mediated Adhesion Is Lipid-Bilayer Dependent. *Biophys. J.* **2017**, *113*, 1080–1092. [CrossRef]
158. Lietha, D.; Izard, T. Roles of Membrane Domains in Integrin-Mediated Cell Adhesion. *Int. J. Mol. Sci.* **2020**, *21*, 5531. [CrossRef]
159. Parsons, J.T.; Horwitz, A.R.; Schwartz, M.A. Cell Adhesion: Integrating Cytoskeletal Dynamics and Cellular Tension. *Nat. Rev. Mol. Cell Biol.* **2010**, *11*, 633–643. [CrossRef]

membranes

MDPI

Article

Formation of β-Strand Oligomers of Antimicrobial Peptide Magainin 2 Contributes to Disruption of Phospholipid Membrane

Munehiro Kumashiro [1], Ryoga Tsuji [2], Shoma Suenaga [1] and Koichi Matsuo [1,2,3,*]

[1] Department of Physical Science, Graduate School of Science, Hiroshima University, 1-3-1 Kagamiyama, Higashi-Hiroshima, Hiroshima 739-8526, Japan; kumashiro@hiroshima-u.ac.jp (M.K.); barian1212@gmail.com (S.S.)
[2] Physics Program, Graduate School of Advanced Science and Engineering, Hiroshima University, 1-4-1 Kagamiyama, Higashi-Hiroshima, Hiroshima 739-8527, Japan; m211517@hiroshima-u.ac.jp
[3] Hiroshima Synchrotron Radiation Center, Hiroshima University, 2-313 Kagamiyama, Higashi-Hiroshima, Hiroshima 739-0046, Japan
* Correspondence: pika@hiroshima-u.ac.jp

Abstract: The antimicrobial peptide magainin 2 (M2) interacts with and induces structural damage in bacterial cell membranes. Although extensive biophysical studies have revealed the interaction mechanism between M2 and membranes, the mechanism of membrane-mediated oligomerization of M2 is controversial. Here, we measured the synchrotron-radiation circular dichroism and linear dichroism (LD) spectra of M2 in dipalmitoyl-phosphatidylglycerol lipid membranes in lipid-to-peptide (L/P) molar ratios from 0–26 to characterize the conformation and orientation of M2 on the membrane. The results showed that M2 changed from random coil to α-helix structures via an intermediate state with increasing L/P ratio. Singular value decomposition analysis supported the presence of the intermediate state, and global fitting analysis revealed that M2 monomers with an α-helix structure assembled and transformed into M2 oligomers with a β-strand-rich structure in the intermediate state. In addition, LD spectra showed the presence of β-strand structures in the intermediate state, disclosing their orientations on the membrane surface. Furthermore, fluorescence spectroscopy showed that the formation of β-strand oligomers destabilized the membrane structure and induced the leakage of calcein molecules entrapped in the membrane. These results suggest that the formation of β-strand oligomers of M2 plays a crucial role in the disruption of the cell membrane.

Keywords: amyloid; antimicrobial peptide; β-strand oligomers; fluorescence anisotropy; linear dichroism; magainin 2; peptide-membrane interaction; synchrotron-radiation circular dichroism

1. Introduction

Antimicrobial resistance (AMR) is one of the most serious global threats to healthcare and agriculture. The overuse and misuse of antimicrobial agents, such as antibiotics, can lead to the emergence of antimicrobial-resistant microorganisms, resulting in prolonged hospitalizations and increased medical costs [1]. Antimicrobial peptides (AMPs) are a potential solution to this problem because they have broad-spectrum antibacterial, antifungal, and antiviral activities and it is difficult for microorganisms to acquire AMR against them [2,3]. The antimicrobial mechanism of AMP is different from that of conventional antibiotics; antibiotics inhibit the synthesis of bacterial components, whereas AMPs interact with bacterial cell membranes and directly cause damage to the membrane structure [2,4]. Extensive efforts have been devoted to use AMP as a medicine; however, the clinical application of AMP is currently limited due to its short half-life and toxic side effects [5–7]. Hence, further understanding of the interaction mechanism between AMP and membranes at the molecular level are necessary to gain new insights into the design strategy of effective AMP [8].

Magainin 2 (M2), a cationic and amphipathic peptide composed of 23 amino acids (GIGKFLHSAKKFGKAFVGEIMNS), is an AMP present in the immune system of African clawed frog *Xenopus laevis* [9]. The interactions between M2 and the membrane have been investigated extensively through various biophysical techniques such as fluorescence spectroscopy, circular dichroism (CD) spectroscopy, nuclear magnetic resonance (NMR) spectroscopy, and isothermal titration calorimetry (ITC) [10–13]. Regarding the molecular mechanism of the antimicrobial activity of M2, Matsuzaki et al. suggested that the M2 peptides assembled and transformed the oligomers in phosphatidylglycerol (PG) membranes, and their oligomerization induced pore formation in the membrane, leading to membrane disruption by M2 [14,15]. Schümann et al. [16] and Gregory et al. [17] found that no peptide oligomerization occurs when M2 interacts with 1-palmitoyl-2-oleoylphosphatidylglycerol phospholipids (POPG) membranes and with the mixed membranes of 1-palmitoyl-2-oleoylphosphatidylcholine (POPC) and POPG, suggesting that in a peptide state (no oligomerization), M2 works in a chaotic or stochastic manner on the pores on these membranes when obtaining the antimicrobial activity [18]. These results indicate that the mechanisms of the membrane interaction or the antimicrobial activity of M2 might strongly depend on the constituents of the lipid membrane [19]. Although these studies commonly support that M2 peptides tightly interact with negatively charged lipid membranes, such as PG, compared to neutral lipid membranes, these interactions remain to be further explored [10,14]. As for the conformation of the M2 peptide on the membrane, most reports concluded that the membrane-bound conformation of M2 was an α-helix-rich structure, but some groups, including Hirsh et al. showed that there are two populations of M2 peptides in the membranes of dipalmitoyl-phosphatidylglycerol (DPPG) and in the mixed membranes of DPPG and dipalmitoyl-phosphatidylcholine (DPPC), with one population being completely α-helical and another being completely β-strand [20–22], as observed in other AMPs, such as bovine lactoferricin, protegrin 1, and human β-defensin-3 [23]. These results suggest that M2 could also form both α-helix and β-strand structures on the membrane, although the contribution of these M2 conformations to the antimicrobial activity is unclear and still controversial.

In this study, we characterized the unique and complicated conformations of M2 peptides on the negatively charged DPPG membrane and their contributions to the antimicrobial activity. We measured the synchrotron-radiation circular dichroism (SRCD) and linear dichroism (LD) spectra of M2 in the DPPG liposome in the lipid-to-peptide (L/P) molar ratio from 0 to 26 and analyzed the types of secondary structures (α-helix and β-strand) of M2 and their orientation on the membrane. SRCD has been applied to characterize the secondary structures of membrane-bound proteins in various liposomes [24–26] and can measure the CD spectra down to the vacuum-ultraviolet region (~160 nm), whose wide range spectra could realize the component analyses of the unique and complicated conformations of M2 induced by the interaction with the DPPG membrane. LD spectroscopy can provide important information on the secondary structure orientation, disclosing the direction of the helical and strand axes of M2 against the liposome surface [24]. Furthermore, the fluorescence anisotropy of DPPG liposomes in the presence of M2 was measured to elucidate the effect of M2-membrane binding on the structure and stability of lipid membranes. The combination of SRCD, LD, and fluorescence spectroscopy could be useful for characterizing the relationships between the unique conformations and activities of M2 on the DPPG membrane.

2. Materials and Methods

2.1. Materials

The M2 peptide was synthesized in GL Biochem (Shanghai, China), GenScript Biotech Corporation (Piscataway, NJ, USA), and Bio-Synthesis (Lewisville, TX, USA). This peptide was purified by HPLC (>95%) and its molecular weight was analyzed by mass spectroscopy. 1,2-dipalmitoyl-sn-glycero-3-phospho-(1′-rac-glycerol) (sodium salt) was purchased from Avanti (Alabaster, AL, USA). 1,2-dipalmitoyl-sn-glycero-3-phosphatidylethanolamine was obtained from Cayman Chemical (Ann Arbor, MI, USA). All other chemicals were analytical grade products purchased from Sigma (St. Louis, MO, USA).

2.2. Liposomes and Sample Preparation

DPPG liposomes with a diameter of 100 nm were prepared by an extrusion technique, as described previously [25,27]. The DPPG lipid molecule was dissolved in 10 mM phosphate buffer (pH 7.0), and the suspension was vortexed vigorously at least above the phase-transition temperature (~41 °C). The solution was alternately placed in liquid nitrogen and heated five times (freeze-thaw cycles). The lipid suspension was then passed through a 100 nm polycarbonate membrane (Whatman, Clifton, NJ, USA) 25 times using a Mini-Extruder (Avanti). This step was performed above the phase-transition temperature. The obtained liposome vesicles were mixed with the M2 solution at a final M2 concentration of 300 μM. The concentration of M2 was determined by measuring the absorption spectra of M2 in the absence of liposomes (molar extinction coefficient at 205 nm: 94,390 M^{-1} cm^{-1}). The concentration of DPPG phospholipid was changed from 0 to 8 mM (i.e., the L/P molar ratio of 0–26). The peptide-liposome mixtures were incubated overnight at 25 °C prior to the CD and LD measurements. For fluorescence anisotropy measurements, the following steps were performed before dissolving the lipids in the buffer: DPPG and DPH were dissolved in chloroform, and the stock solutions were then mixed at a DPPG/DPH molar ratio of 400/1. To remove the organic solvent, the solution was placed under a stream of nitrogen gas and then under vacuum for at least three hours [24]. The obtained DPPG film was dissolved in the buffer at final concentrations of 400 μM and 1 μM of DPPG and DPH, respectively.

2.3. Synchrotron-Radiation Circular Dichroism

The SRCD spectra of M2 in the presence and absence of DPPG liposomes were recorded from 260 to 178 nm using a vacuum-ultraviolet circular dichroism (VUVCD) spectrophotometer at Hiroshima Synchrotron Radiation Center (HiSOR) and an assembled optical cell with a 50 μm path length Teflon spacer. The details of the optical cell and spectrophotometer have been reported in previous studies [28,29]. The distance between the optical cell and the window of the photomultiplier tube was set to less than 10 mm to minimize the effect of light scattering from the liposome particles [24]. The temperature of the cell was controlled using a Peltier device. The actual temperature of the sample solution was obtained by a calibration curve of the actual temperature (in the optical cell) against the set temperature, which was investigated in advance. The SRCD spectra of each sample were measured four times and averaged. The SRCD spectra of the liposome solutions were also measured as a background and subtracted from the spectra of the peptide-liposome suspensions. The secondary structures of M2 in the presence and absence of DPPG liposomes were analyzed using the SELCON3 program [30,31] and the VUVCD dataset obtained by Matsuo et al. [32,33], as described previously [24,32,33].

2.4. Synchrotron-Radiation Linear Dichroism

The LD spectra of M2 in the presence of liposomes were measured from 300 to 185 nm using an LD spectrometer installed in the VUVCD spectrophotometer at HiSOR. The system of the LD spectrometer is described in detail elsewhere [24]. An LD flow cell with a light path length of 75 μm (Translume, MI, USA) was located within 10 mm of the photomultiplier tube. A shear flow with a flow velocity of 1.0 mL min^{-1} was applied using a dual-plunger parallel-flow pump (LC-20AD, Shimadzu, Kyoto, Japan), and the sample solution was circulated between the sample container and the LD flow cell. All LD spectra were recorded at room temperature (25 °C) with four accumulations. The scattering contribution to the LD signal was subtracted using the method described by Nordh et al. [34].

2.5. Fluorescence Anisotropy

The steady-state fluorescence anisotropy of DPH in DPPG liposomes was measured using a fluorescence spectrophotometer (FP-8300, Jasco, Japan) [35,36]. The excitation and emission wavelengths were 357 and 430 nm, respectively, with 5.0 nm band width. The measurements were conducted in the temperature range of 25–55 °C with 0.1 °C steps. The fluorescence anisotropy (r) was calculated using the following equation:

$$r = \frac{I_{vv} - GI_{vh}}{I_{vv} + 2GI_{vh}}, \tag{1}$$

where I_{vv} is the intensity of the detected light with both excitation and emission polarizers mounted vertically, I_{vh} is the intensity of the detected light with the excitation polarizer installed vertically and the emission polarizer mounted horizontally. The factor G in Equation (1) for each sample was determined by measuring the intensities of the detected light, I_{hv} and I_{hh}, with the emission polarizer installed horizontally and vertically, respectively, under the excitation polarizer mounted horizontally and calculating I_{hv}/I_{hh} (=G). The curves of the temperature dependence of anisotropy were fitted using:

$$r(T) = r_l + m_l(T - T_l) + \frac{r_g - r_l + m_g(T - T_g) - m_l(T - T_l)}{1 + \exp\left(-\frac{\Delta H}{RT}\left(1 - \frac{T}{T_m}\right)\right)}, \tag{2}$$

where T is the absolute temperature, T_l and T_g are the reference temperatures for DPPG liposome in the liquid-crystalline phase (55 °C) and the gel phase (25 °C), respectively; r_l and r_g are the values of fluorescence anisotropy at 55 °C (liquid-crystalline phase) and 25 °C (gel phase), respectively; T_m is the midpoint temperature of the phase transition of the DPPG membrane; ΔH is the molar enthalpy change of the phase transition at T_m; and R is the gas constant. m_l and m_g in Equation (2) are temperature dependences of the fluorescence anisotropy of lipid vesicles in the liquid-crystalline and gel phases, respectively, which are empirical parameters. The adjustable parameters in Equation (2) are ΔH, T_m, m_l, and m_g.

2.6. Calcein Leakage Assay

M2-induced leakage of calcein entrapped in DPPG liposomes was investigated at 25 °C, as described in a previous study [10]. The fluorescence intensity of calcein was monitored at 515 nm using a fluorescence spectrometer (FP-8300, Jasco, Tokyo, Japan) with an excitation wavelength of 490 nm. The extent of calcein release was calculated according to $F = (I_f - I_0)/(I_{max} - I_0)$, where I_0 and I_f are the initial and final (approximately 40 min after the initial manipulation) intensities of fluorescence, respectively, and I_{max} is the maximal fluorescence intensity obtained by adding Triton X-100 (all the entrapped calcein was released). Calcein-loaded DPPG liposomes were prepared by the extrusion method using a solution containing 60 mM calcein and 10 mM HEPES (pH 7.0). Untrapped calcein was separated from the liposomes by gel filtration using a bio-spin chromatography column (BIO-RAD, Hercules, CA, USA) with Sephadex G-75. This step was performed above the lipid phase-transition temperature. Calcein was purchased from Tokyo Chemical Industry (Tokyo, Japan). The lipid concentration (400 μM) was determined using a phosphorus assay [37].

2.7. Adsorption Model Fitting Procedure

The L/P dependence of the SRCD spectra of M2 in DPPG liposomes was analyzed using an adsorption model and a global fitting algorithm. The model of adsorption for a large self-associating ligand based on scaled particle theory was used as the model for fitting in this study [38]. The model can be written as [38]:

$$Kc_f = \Phi_1 \gamma_1(\Phi_1, \Phi_z), \tag{3}$$

$$\Phi_z = zK_{1z}\frac{(\gamma_1(\Phi_1,\ \Phi_z)^z)}{\gamma_z\ (\Phi_1,\ \Phi_z)}\ \Phi_1^z, \quad (4)$$

$$\ln\gamma_1 = -\ln(1-\Phi_1-\Phi_z) + \frac{3\Phi_1 + \left(\frac{2}{f}+\frac{1}{f^2}\right)\Phi_z}{1-\Phi_1-\Phi_z} + \left(\frac{\Phi_1+\frac{1}{f}\ \Phi_z}{1-\Phi_1-\Phi_z}\right)^2, \quad (5)$$

$$\ln\gamma_z = -\ln(1-\Phi_1-\Phi_z) + \frac{3\Phi_z + (2f+f^2)\Phi_1}{1-\Phi_1-\Phi_z} + \left(\frac{\Phi_z+f\Phi_1}{1-\Phi_1-\Phi_z}\right)^2, \quad (6)$$

where K is the association constant; c_f is the concentration of the peptide free in solution (or native state); Φ_1 ($=nc_1/c_L$) and Φ_z ($=nc_z/c_L$) are the fractions of the surface area occupied by the monomer and z-mer, respectively; c_1 and c_z are the concentrations of monomer and z-mer, respectively; c_L is the total lipid concentration; n is the number of lipid molecules covered by a single peptide; K_{1z} is the equilibrium constant for the formation of the z-mer; and f is the ratio of the radii of the circles representing the monomer and z-mer. When the area does not change due to self-association, f is equal to $z^{1/2}$.

To fit the adsorption model to the SRCD data, the thermodynamic parameters K, K_{1z}, z, and n were solved numerically using an iterative method. Their initial values were set, c_f, c_1, and c_z at each L/P were calculated according to Equations (3)–(6), and then the concentration matrix C was output. The matrix of the pure component spectra S for M2 in native, monomeric, and oligomeric states was calculated using the classical least-squares approach [39]:

$$S = \left(C^{\mathrm{T}}C\right)^{-1}CD, \quad (7)$$

where D is the matrix of measured SRCD data. The 2-norm of the error $E = |CS - D|_2$ was minimized by repeating the calculation with different initial values of K, K_{1z}, z, and n. The obtained solutions for the parameters were then used to calculate the fractions of M2 in the three states at each L/P using the least-squares method [39]:

$$C = D(S^{\mathrm{T}}\left(SS^{\mathrm{T}}\right)^{-1}). \quad (8)$$

The populations of M2 in the three states at each L/P obtained using Equation (8) were compared with the fitting curves calculated using the optimized thermodynamic parameters. The optimized component spectra of M2 in the three states were calculated using Equation (7) with the optimized parameters.

3. Results

3.1. Synchrotron-Radiation Circular Dichroism

The SRCD spectra of M2 in the presence of DPPG liposomes (pH 7.0 and 25 °C) were measured from 260 to 178 nm in the L/P ratios from 0 to 26, as shown in Figure 1. The spectrum at L/P = 0 exhibited a negative shoulder around 220 nm, a negative peak around 200 nm, and a negative sign around 178 nm, indicating that M2 formed a random coil structure in the native state. As the L/P ratio increased, the spectral intensities of the shoulder and peak observed in the native state decreased and increased, respectively. The spectral change was completely saturated around L/P = 15, and the spectrum at L/P = 26 had two successive negative peaks around 222 and 208 nm and a positive peak around 190 nm, indicating that M2 formed a helical structure in the membrane-bound state. Furthermore, we found two iso-dichroic points around 200 and 210 nm during the spectral change, implying that M2 formed a helical conformation in the DPPG membranes through an intermediate state. M2 in the mixed liposomes with a DPPE/DPPG molar ratio of 3/1, which is believed to mimic the cell membranes of bacteria [40,41], also exhibited the same spectral change at 25 °C as those of M2 in DPPG liposomes (Figure S1), suggesting that M2 also has an intermediate state when interacting with the bacterial cell membrane.

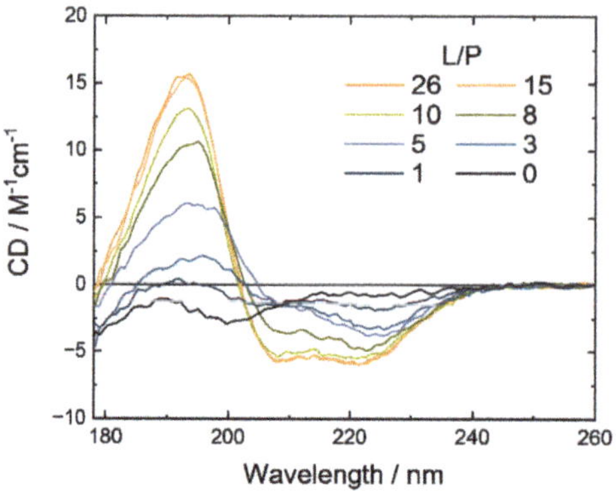

Figure 1. Synchrotron-radiation circular dichroism (SRCD) spectra of M2 in the presence of DPPG liposome in the L/P ratio from 0 to 26. All spectra were recorded at 25 °C, pH 7.0, and the M2 concentration of 300 μM. Experimental conditions are described in detail in Section 2.3.

We analyzed the spectral dataset in Figure 1 using the singular value decomposition (SVD) method. SVD, which is closely related to a principal component analysis, is a useful method for determining the dimension of a data matrix or the number of component spectra in the dataset [42,43]. As a result, the SVD analysis provided three singular values that were significantly larger than zero and estimated three component spectra that can construct the SRCD dataset within experimental error (Figures S2 and S3). This indicates that the SRCD spectra over the entire range of the L/P ratio can be explained by only three components, meaning that the spectral set in Figure 1 can be explained by the three-state model. However, the component spectra obtained from SVD might not correspond to that of M2 in each independent state (native, intermediate, and membrane-bound states) because the spectrum of M2, even in the native state, is composed of the components of secondary structures [43]. Hence, further analyses were conducted as follows:

Several physical models can explain the adsorption of peptides on membrane surfaces [38,44,45]. Among them, the adsorption model with a large self-associating ligand based on scaled particle theory successfully explained the sigmoidal shape of the adsorption isotherm of lysozyme (or ligand) on negatively charged membrane surfaces, showing that the lysozyme gathered upon the membrane association [44]. The sigmoidal shape of the adsorption isotherm is also a key behavior for interpreting the interaction between M2 and negatively charged PG because the same sigmoidal curve was observed in the tryptophan fluorescence experiments for M2 analogs and PG liposomes [14]. Hence, to characterize the intermediate state of M2, we analyzed the SRCD spectral data (Figure 1) using the adsorption model and the global fitting simulation (fitting procedures are provided in Section 2.7). This model assumes that peptides are in equilibrium between only three states: the native state in aqueous solution, the membrane-bound monomeric state, and the membrane-bound *z*-meric (oligomeric) state.

Figure 2a shows the L/P dependence of the fractional populations of M2 in the native, membrane-bound monomeric, and oligomeric states. The inset of Figure 2a shows the plots and fitting curve of CD at 193 nm against L/P. These experimental data were reproduced well from the adsorption model with a large self-associating ligand. We also found that the fitting error obtained using the adsorption model ($E = 9.5$) was much smaller than that obtained using the simplest monomodal adsorption model [44] ($E = 16.5$), which also supports the presence of the intermediate state. To understand the conformation of M2 in the intermediate state, we computed the spectra of M2 in the three states, as shown in Figure 2b. Native M2 exhibited a negative CD peak at 200 nm, which is a characteristic peak of the random coil structure [46], whereas M2 in the membrane-bound monomeric state showed two negative peaks at 208 and 222 nm, and a positive peak around 193 nm, which are characteristic of the α-helix structure [46]. Moreover, M2 in the oligomeric state exhibited a negative peak around 225 nm, and a positive peak at 200 nm, which are characteristic peaks of the β-strand structure [46], indicating that the α-helix monomers of M2 in DPPG membranes self-associate and transform to β-strand oligomers in the intermediate state.

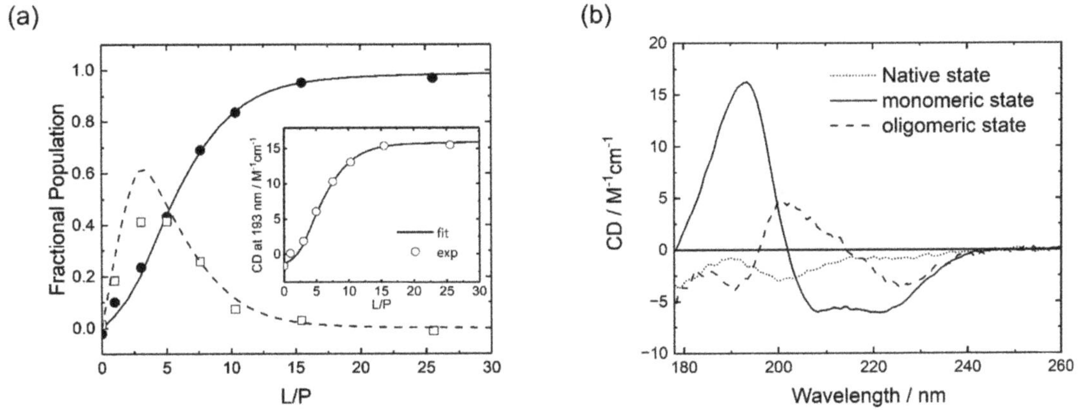

Figure 2. (**a**) Lipid-to-peptide (L/P) dependence of fractional populations of M2 in membrane-bound monomeric (closed circle: plots; solid line: fitting curve) and oligomeric states (open square: plots; dashed line: fitting curve) (see Section 2.7 for the details of calculations). The fitting curves in (**a**) were calculated using the following optimized parameters: $K = 17500$, $K_{1z} = 3$, $z = 5$, and $n = 2.2$. The inset of (**a**) shows the L/P dependence of the CD at 193 nm; (**b**) spectra of M2 at native, membrane-bound monomeric, and oligomeric states.

In this model, the membrane-bound monomeric state of the peptide is modeled as a spherical form [38]. However, the helix structure of M2 might need to be treated in a cylindrical form when considering the NMR structure of M2 bound to dodecylphosphocholine micelles [47]. In fact, when Zuckermann and Heimburg [45] described the equilibrium between the peptides adsorbed on membranes and the peptide oligomers inserted into the membranes, they proposed an SPT-based model in which the membrane-bound peptide was modeled as a cylindrical form. This model might improve the fitting error in our study, but it requires the detail structural parameters of cylindrical form such as the length and the radius. Since it was difficult to obtain these structural parameters of random coil, α-helix, and β-strand structures of M2 in this study, we used the spherical form. In fact, this form reproduced well the fractional population of the three states obtained from the global fitting simulation (Figure 2).

Previous research suggested that light scattering due to the self-association of peptides and liposomes induces the distortion of CD spectra [48], and the presence of an intermediate state in the research was attributed to the effect of light scattering distortion. We measured the optical density of the M2-DPPG liposome system at 450 nm using a commercial absorption spectrophotometer and a quartz cuvette with an optical path length of 1 cm, and observed slight light scattering in the system at an L/P ratio of 3 (Figure S4). Thus, we optimized the measurement system, such as the positioning of the sample holder and detector, because this optimized system realized the measurements of undistorted CD spectra of proteins even in the presence of large unilamellar vesicles [24,49]. We observed that the raw CD spectra of M2 exhibited no baseline shift at 260 nm, which is believed to be an artifact of light scattering [48] (data not shown). These results indicate that the effect of light scattering on the SRCD spectra was negligible. Therefore, the intermediate state of M2 in DPPG liposomes does not originate from scattering artifacts.

The spectra of M2 in the three states were analyzed using the SELCON3 program to estimate the secondary structures of M2 in the native, membrane-bound monomeric, and oligomeric states. The analytical results from the CDSSTR and BeStSel programs [30,50] were almost consistent with those of SELCON3 (Figure S5). Figure 3 shows the secondary structure contents of M2 in the three states. The SELCON3 analysis revealed that native M2 comprised 7.5% α-helix, 28.4% β-strand, and 64.1% turn and unordered structures (others), showing that M2 forms a random coil in an aqueous solution. In contrast, M2 in the membrane-bound monomeric and oligomeric states included 70.2% and 1.2% α-helix, 10.9% and 46.3% β-strand, and 18.9% and 52.5% others, respectively. These data suggest that the oligomeric state of M2 interacting with DPPG membranes induced a large number of β-strand structures.

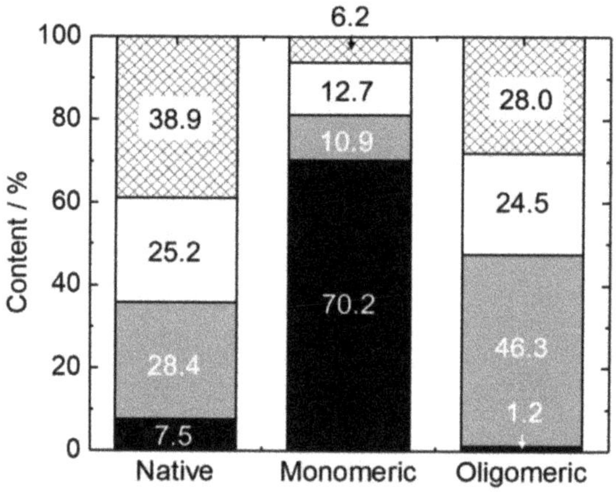

Figure 3. Bar graphs of secondary structures of M2 in native, membrane-bound monomeric, and oligomeric states analyzed using the SELCON3 program (black: α-helix; gray: β-strand; white: turn; meshed: unordered structure).

3.2. Synchrotron-Radiation Linear Dichroism

We then conducted LD measurements of M2 in the presence of DPPG liposomes around L/P ratios of 4 and 25, which correspond to the experimental conditions mainly occupied by the oligomeric state and the membrane-bound monomeric state, respectively, to estimate the orientation of the secondary structures (α-helices and β-strands) of M2 on the DPPG membrane surface. To obtain the LD spectra of M2, we used a liquid circulation system in which liposomes can deform from spherical to elliptical shapes in a shear flow environment. In the flow system, the average ratio of major axis/minor axis for the typical liposome, dimyristoyl phosphatidylcholine liposome, is approximately 1.7 at a flow velocity of 1.0 mL min^{-1}. [24,51]. Because the long axis of the elliptical liposome was oriented along the flow direction, the average orientations of secondary structures in the liposomes showed slight biases, providing the LD signals depending on the secondary structure orientations against the membrane surface. M2 in DPPG liposomes at a flow velocity of 1.0 mL min^{-1} showed a positive peak around 195 nm with a shoulder around 205 nm at L/P = 25 and a positive peak around 200 nm and a small shoulder around 220 nm at L/P = 4, as shown in Figure 4. Furthermore, both spectra depended on the flow velocity from 0 to 1.0 mL min^{-1} (the insets of Figure 4). The observations of LD signals mean that the helical and strand structure directly interact with the DPPG liposomes. Since M2 forms helical and strand structures around L/P = 25 and 4, respectively (Figure 1) and free peptides in aqueous solution show no LD signals, these results suggest that the observed LD was mainly affected by the oriented helical structure of M2 for L/P = 25 and the oriented strand structure for L/P = 4 on the elliptical liposomes.

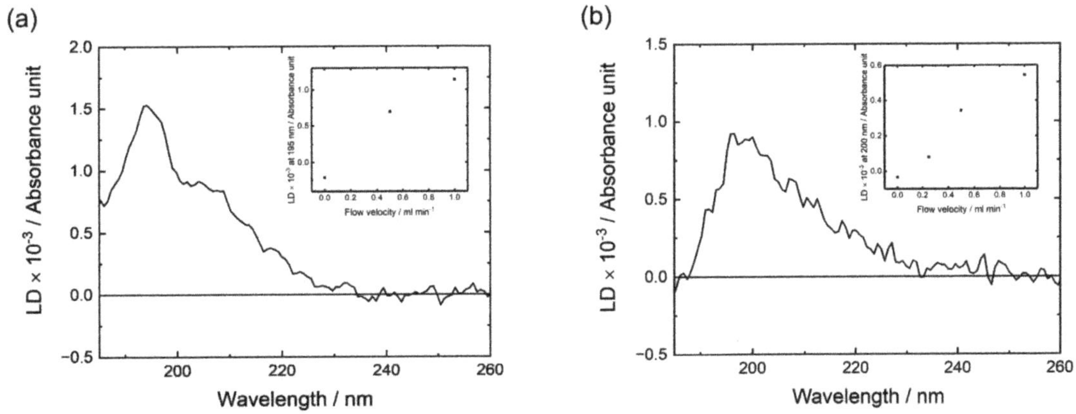

Figure 4. Linear dichroism (LD) spectra of M2 in DPPG liposomes at the lipid-to-peptide ratio (L/P) of (**a**) 25 and (**b**) 4. The insets of (**a**,**b**) show the dependence of flow velocity on the LD at 195 and 200 nm, respectively. All spectra were recorded at 25 °C, pH 7.0, and M2 concentration of 200 μM. Experimental conditions are described in detail in Section 2.4.

The electric or magnetic dipole moments of helical structures are approximately vertical to the helix axis at 190 nm, parallel at 208 nm, and vertical at 222 nm [51]. Furthermore, the absorption spectrum of M2 at an L/P ratio of 25 was successfully reproduced by the three component spectra of Gaussian functions, which have peak positions around 222, 208, and 190 nm, respectively, as shown in Figure S6a. These results allowed us to analyze the LD spectra under two different assumptions: first, all M2 peptides form the same helical orientation with a single helical angle against the membrane surface; second, M2 peptides form perpendicular (transmembrane) and parallel helical structures against the membrane surface. The results under the first assumption showed that the LD spectrum could be interpreted as the angle between the M2 helix axis and the membrane normal (48°) (Figure S6b) while those under the second assumption showed that the ratio of M2 helix axes perpendicular and parallel to the membrane surface was 1:1.2 (Figure S6c). Both fitting analyses reproduced the experimental LD well, as shown in Figure S6d. As for the LD spectrum at an L/P of 4, the electric or magnetic dipole moments of the β-strand structure were approximately vertical to the strand axis at 195 nm, parallel at 219 nm, and vertical at 221 nm [51,52]; however, the fitting analysis, as described in Figure S6, was difficult because the two dipole moments (219 and 221 nm) were very close and had inverse directions, which induced cancelation. However, a positive peak around 200 nm was clearly detected in the LD spectrum and showed that the axis of the β-strands of M2 was perpendicular to the membrane surface on average.

3.3. Fluorescence Anisotropy

To investigate the contribution of the α-helix formation in the monomeric state and the β-strand formation in the oligomeric state to the stability of lipid membranes, we measured the fluorescence anisotropy of DPPG liposomes in the presence of M2 at L/P = 25 and 3. Fluorescence anisotropy can monitor the degree of dynamics or rotational mobility of DPH in liposomes, and the parameters are then used to investigate the lipid packing (ordering) or the membrane fluidity of lipid bilayers [35,53]. The DPH probe is present in the hydrophobic core of liposomes and the motion of the DPH molecules is normally restricted to the gel-phase liposomes, resulting in high fluorescence anisotropy values. However, when the temperature is increased or the phase of the lipid bilayer shifts to the liquid-crystalline phase, the degree of rotational mobility of the DPH molecules increased, leading to a decrease in the anisotropy values. Figure 5 shows the temperature dependence of the fluorescence anisotropy values *r* of DPPG liposomes in the absence of M2 and at L/P = 25 and 3.

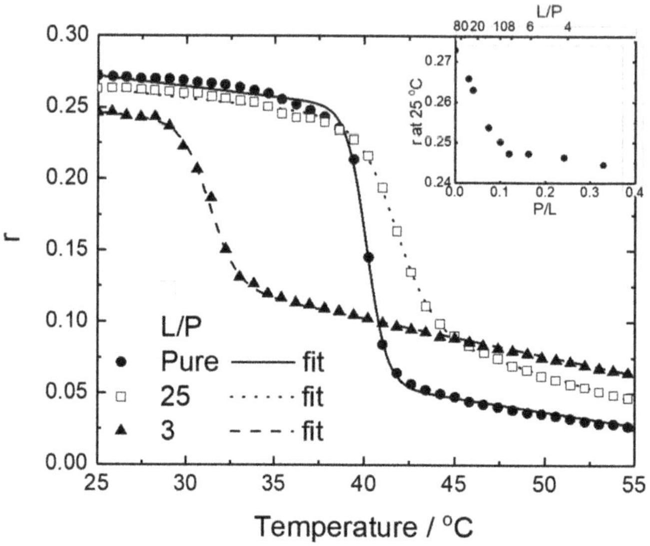

Figure 5. Fluorescence anisotropy values *r* and fitting curves of DPPG liposomes in the absence of M2 and in the presence of M2 at L/P = 25, and 3. The inset shows P/L (L/P) dependence of *r* at 25 °C. All data were recorded at pH 7.0 and DPPG concentration of 400 μM. Experimental conditions are described in detail in Section 2.5.

As shown in the inset of Figure 5, the *r* at 25 °C linearly decreased as L/P decreased, up to 10, whereas *r* did not change below a L/P of 10, implying that the effect of M2 binding on the DPPG membrane changes around L/P = 10. We analyzed the midpoint temperature T_m and the molar enthalpy change ΔH of the phase transition of DPPG liposomes in the presence of M2 using a sigmoid function (Equation (2)). The fitting curves were well fitted to the experimental data (Figure 5). The T_m and ΔH values are listed in Table 1. From this table, pure DPPG liposomes showed a phase-transition temperature (T_m) of approximately 40.1 °C, which is consistent with the main phase-transition temperature of the DPPG membrane (T_m = 41 °C), showing that the DPH fluorescence anisotropy measurement can successfully monitor the change in membrane fluidity of DPPG liposomes. The T_m of DPPG liposomes at a L/P = 25 was 41.7 °C, which was higher than that of pure DPPG liposomes. In contrast, the T_m of DPPG liposomes at L/P = 3 was 31.3 °C, showing a significant decrease. This indicates that M2 in the DPPG membranes at L/P = 3 destabilizes the bilayer structure of the membranes. Since M2 can form a monomeric helical structure and oligomeric strand structure at L/P = 25 and 3, respectively, these findings suggest that the adsorption of monomeric and oligomeric M2 on the DPPG membranes largely contributes to the stabilization and destabilization of the membrane, respectively.

Table 1. Midpoint temperature T_m and molar enthalpy change ΔH obtained by fitting the temperature dependence of the fluorescence anisotropy in the absence of M2 and in the presence of M2 at L/P of 25 and 3 in Figure 5.

L/P	T_m [°C]	ΔH [kcal/mol]
Only liposome	40.12 ± 0.02	351 ± 7
25	41.71 ± 0.02	180 ± 4
3	31.31 ± 0.03	231 ± 7

4. Discussion

In this study, we used SRCD spectroscopy to reveal that the conformation of M2 changed from random coil to α-helix structures via an intermediate state as the L/P ratio increased. The global fitting analysis of the SRCD spectra based on the adsorption model indicated that α-helical M2 monomers assembled and transformed into β-strand M2 oligomers in the intermediate state. LD data also showed that the β-strand M2 oligomers bind to the membrane as the strand axis is perpendicular to the membrane surface. To elucidate the effect of oligomerization on the structure and stability of lipid membranes, further we conducted fluorescence anisotropy measurements of DPPG liposomes in the presence of M2 and revealed that the formation of an oligomeric β-strand structure contributed to the destabilization of the membrane structure. Therefore, our findings demonstrate that the oligomeric β-strand structure of M2 in membranes plays a crucial role in the disruption of the cell membrane.

Our SRCD data showed that M2 formed a helical conformation on the membrane at L/P = 25 and in the monomeric state. The LD result at L/P = 25 could be interpreted as the two possible orientations of M2 with respect to the membrane surface: the helix axis of M2 for the membrane normal was uniformly distributed at an angle of 48° or the helix axes of M2 perpendicular and parallel to the membrane surface were mixed at a ratio of 1:1.2. In addition, our fluorescence anisotropy data showed an increase in the membrane stability at L/P = 25. According to previous research [40], pure electrostatic interactions between negatively charged DPPG headgroups and positively charged residues of peptides contributes to the increase in membrane stability, whereas the hydrophobic interaction of peptides with membrane core regions induces perturbation of lipid chain packing, leading to the destabilization of the membrane structure. Furthermore, Fourier transform infrared (FTIR) and solid-state NMR spectroscopy have indicated that α-helical M2 analogs directly bind to the headgroups of DPPG phospholipids [21]. If the helix axis of M2 was uniformly distributed at an angle of 48° against the membrane normal, M2 would be partially inserted into the membrane, resulting in hydrophobic interactions, destabilizing the membrane structure. Wimley [19] suggested four types of models for the molecular mechanism of the antimicrobial activity of M2, one of which is that M2 peptides can form a transmembrane pore. In this case, the polar faces of peptides whose helical orientation is perpendicular to the membrane surface can interact with the polar headgroups of phospholipids [7], stabilizing the membrane structure. Hence, it is possible that helical M2 interacts with DPPG headgroups and exists in a mixture of perpendicular and parallel helices. Furthermore, because the toroidal pore formation and other models, such as carpet and detergent models do not require specific peptide-peptide interactions [19], it is also possible that M2 peptides in the pore formation could be treated as a monomeric state.

To confirm the presence of pore formation at L/P = 25, we tested M2-induced leakage of calcein molecules entrapped in DPPG liposomes and found that approximately 30% of the dye molecules were released from the liposomes (Figure S7). This result indicates that the pores formed in liposomes at L/P = 25, but the M2-induced disruption of DPPG liposomes would not be completed under this condition. In contrast, the extent of liposome-entrapped calcein leakage at L/P = 4 was 2-fold greater (~60%) than that at L/P = 25. Hence, elucidating what occurred around L/P = 4 might be important for understanding the overall molecular mechanism of DPPG liposome disruption by M2.

We found that the α-helix monomers of M2 assembled and transformed into β-strand oligomers as the L/P ratio decreased from 25 to 4. Although our fitting analysis of the SRCD data did not yield the unique thermodynamic parameters of the M2 adsorption to DPPG membranes due to strong cross-correlation between each parameter and the limited number of data points (Figure S8), the fitting analysis suggested that M2 in the DPPG membranes formed the β-strand z-mer ($z > 2$). We also estimated that the average axis of β-strands was perpendicular to the membrane surface and confirmed that the formation of β-strand oligomers contributed to the destabilization of the membrane structure. Since the hydrophobic interaction of peptides with membrane core regions induces destabilization of the membrane structure [40], the β-strand oligomers of M2 may be buried in the DPPG membrane core or the disruption of the chain packing of DPPG membranes. These results suggest that the oligomeric β-strand M2 has unique conformations, such as the toxic β-sheet channels of protegrin-1 [54] and transmembrane amyloid β (Aβ) oligomers [55]. Ciudad et al. suggested that the β-strand-rich Aβ42 tetramer formed lipid-stabilized pores in a membrane-mimicking environment, in which water permeation occurred [55]. The discussion requires further analyses such as FTIR [21], vibrational sum frequency generation [56], and Raman spectroscopy [22]. However, the β-strand oligomers of M2 may contribute to the formation of stable pores in DPPG membranes, inducing the complete disruption of the membrane structure.

Previous reports [57] suggested that ΔH and T_m of membrane phase transition were basically correlated with each other for several membrane-bound proteins and peptides, meaning that the destabilization of membrane (decrement in T_m) would relate to the decrement in ΔH. In this study, we observed that both ΔH and T_m of the phase transition of DPPG liposome decreased at L/P = 3, compared to those of pure DPPG liposome (Table 1), which identifies with the previous results and suggests that the penetration of oligomeric M2 peptides would contribute the destabilization membrane and the decrement in ΔH. However, the ΔH and T_m at L/P = 25 decreased and increased, respectively. This conflict might be induced by the unique conformations of M2 on the membrane because M2 would form two types of orientations on the membrane (the helix axes of M2 perpendicular and parallel to the membrane surface were mixed at a ratio of 1:1.2). The ΔH and T_m at L/P = 25 might be also interpreted from a balance between the stabilization by the electrostatic interaction and the destabilization by the pore formation.

We investigated the conformation of M2 in dimyristoyl-phosphatidylglycerol liposomes and observed that M2 in the liposomes did not show the presence of an oligomeric state with a β-strand structure (data not shown). It is likely that the formation of oligomeric β-strand structures largely depends on the inherent characteristics of lipid constituents, such as the phase-transition temperature and tail region length. Given that bacteria change the lipid composition, including the acyl chain, of their cell membranes in response to their environment [58–60], to understand the antimicrobial mechanism of M2, it is important to characterize the detailed dependence of lipid constituents. In addition, it is important to investigate the interactions between M2 and bacterial cells [61] to clarify whether M2 can form β-strands when interacting with bacterial cells under physiological conditions.

5. Conclusions

In this study, SRCD, LD, and fluorescence spectroscopy were used to characterize the relationships between the conformations and activities of M2 on the DPPG membrane. We found that M2 formed β-strand oligomers in DPPG liposomes, which induced the destabilization of the membrane structure and the leakage of calcein molecules entrapped in the membrane. It has been commonly recognized that the formation of the helical structure of M2 on the membrane could be the driving force of the destabilization of the membrane or the antimicrobial activity; however, our findings demonstrate that the oligomeric β-strand structure of M2 in membranes also plays a crucial role in the disruption of the cell membrane, and future studies will be necessary to characterize the molecular mechanisms underlying the antimicrobial action of M2, which would be helpful for gaining new insights into the antimicrobial mechanism of AMPs.

Supplementary Materials: The following are available online at https://www.mdpi.com/article/10.3390/membranes12020131/s1, Figure S1: SRCD spectra of M2 in DPPE/DPPG (molar ratio: 3/1) liposomes at 25 °C, Figure S2: SVD analysis results of SRCD spectra of M2 in DPPG liposomes, Figure S3: Error in reproducing SRCD data set of M2 in DPPG liposomes using basis CD spectra from SVD analysis, Figure S4: Optical density at 400 nm of M2-DPPG liposome system, Figure S5: Bar graphs of secondary structures of M2 obtained by CDSSTR and BeStSel programs, Figure S6: (a) absorption spectrum of M2 in DPPG liposomes at the L/P of 25 and deconvoluted absorption spectra, (b) angle dependence of theoretical LD spectra, (c) [Transmembrane]: [On surface] ratio dependence of the theoretical LD spectra, and (d) experimental and calculated LD spectra of M2 in DPPG liposomes at the L/P of 25, Figure S7: results of calcein dye leakage assay, Figure S8: Dependence of the fitting error on *z*.

Author Contributions: SRCD measurements, M.K. and S.S.; LD measurements, M.K.; fluorescence anisotropy measurements, M.K.; calcein leakage experiments, M.K. and R.T.; data analyses, M.K.; writing—original draft preparation, M.K.; writing—review and editing, M.K. and K.M.; supervision, K.M.; project administration, M.K. and K.M.; funding acquisition, K.M. All authors have read and agreed to the published version of the manuscript.

Funding: This research was financially funded by Grants-in-Aid for Scientific Research from the Ministry of Education, Culture, Sports, Science, and Technology of Japan (grant number: 19K06587).

Institutional Review Board Statement: Not applicable.

Informed Consent Statement: Not applicable.

Acknowledgments: The measurements of synchrotron-radiation circular dichroism and linear dichroism spectra were performed with the approval of the Hiroshima Synchrotron Radiation Center (proposal numbers: 21AU011, 20AU014, and 19BG047).

Conflicts of Interest: The authors declare no conflict of interest.

References

1. Hughes, J.M. Preserving the Lifesaving Power of Antimicrobial Agents. *JAMA* **2011**, *305*, 1027–1028. [CrossRef] [PubMed]
2. Zasloff, M. Antimicrobial Peptides of Multicellularorganisms. *Nature* **2002**, *415*, 389–395. [CrossRef]
3. Hancock, R.E.W.; Sahl, H.G. Antimicrobial and Host-Defense Peptides as New Anti-Infective Therapeutic Strategies. *Nat. Biotechnol.* **2006**, *24*, 1551–1557. [CrossRef]
4. Melo, M.N.; Ferre, R.; Castanho, M.A.R.B. Antimicrobial Peptides: Linking Partition, Activity and High Membrane-Bound Concentrations. *Nat. Rev. Microbiol.* **2009**, *7*, 245–250. [CrossRef] [PubMed]
5. Matsuzaki, K. Control of Cell Selectivity of Antimicrobial Peptides. *Biochim. Biophys. Acta* **2009**, *1788*, 1687–1692. [CrossRef]
6. Lei, J.; Sun, L.C.; Huang, S.; Zhu, C.; Li, P.; He, J.; Mackey, V.; Coy, D.H.; He, Q.Y. The Antimicrobial Peptides and Their Potential Clinical Applications. *Am. J. Transl. Res.* **2019**, *11*, 3919–3931. [PubMed]
7. Matsuzaki, K. (Ed.) *Antimicrobial Peptides Basics for Clinical Application*; Springer: Singapore, 2019; ISBN 978-981-13-3587-7.
8. Benfield, A.H.; Henriques, S.T. Mode-of-Action of Antimicrobial Peptides: Membrane Disruption vs. Intracellular Mechanisms. *Front. Med. Technol.* **2020**, *2*, 610997. [CrossRef] [PubMed]
9. Zasloff, M. Magainins, a Class of Antimicrobial Peptides from Xenopus Skin: Isolation, Characterization of Two Active Forms, and Partial cDNA Sequence of a Precursor. *Proc. Natl. Acad. Sci. USA* **1987**, *84*, 5449–5453. [CrossRef]
10. Matsuzaki, K.; Sugishita, K.I.; Ishibe, N.; Ueha, M.; Nakata, S.; Miyajima, K.; Epand, R.M. Relationship of Membrane Curvature to the Formation of Pores by Magainin 2. *Biochemistry* **1998**, *37*, 11856–11863. [CrossRef]
11. Bechinger, B.; Zasloff, M.; Opella, S.J. Structure and Orientation of the Antibiotic Peptide Magainin in Membranes by Solid-State Nuclear Magnetic Resonance Spectroscopy. *Protein Sci.* **1993**, *2*, 2077–2084. [CrossRef]
12. Tamba, Y.; Yamazaki, M. Single Giant Unilamellar Vesicle Method Reveals Effect of Antimicrobial Peptide Magainin 2 on Membrane Permeability. *Biochemistry* **2005**, *44*, 15823–15833. [CrossRef]
13. Wenk, M.R.; Seelig, J. Magainin 2 Amide Interaction with Lipid Membranes: Calorimetric Detection of Peptide Binding and Pore Formation. *Biochemistry* **1998**, *37*, 3909–3916. [CrossRef] [PubMed]
14. Matsuzaki, K.; Murase, O.; Tokuda, H.; Funakoshi, S.; Fujii, N.; Miyajima, K. Orientational and Aggregational States of Magainin 2 in Phospholipid Bilayers. *Biochemistry* **1994**, *33*, 3342–3349. [CrossRef]
15. Matsuzaki, K.; Murase, O.; Miyajima, K. Kinetics of Pore Formation by an Antimicrobial Peptide, Magainin 2, in Phospholipid Bilayers. *Biochemistry* **1995**, *34*, 12553–12559. [CrossRef]
16. Schümann, M.; Dathe, M.; Wieprecht, T.; Beyermann, M.; Bienert, M. The Tendency of Magainin to Associate upon Binding to Phospholipid Bilayers. *Biochemistry* **1997**, *36*, 4345–4351. [CrossRef] [PubMed]
17. Gregory, S.M.; Pokorny, A.; Almeida, P.F.F. Magainin 2 Revisited: A Test of the Quantitative Model for the All-or-None Permeabilization of Phospholipid Vesicles. *Biophys. J.* **2009**, *96*, 116–131. [CrossRef] [PubMed]
18. Travkova, O.G.; Moehwald, H.; Brezesinski, G. The Interaction of Antimicrobial Peptides with Membranes. *Adv. Colloid Interface Sci.* **2017**, *247*, 521–532. [CrossRef]
19. Wimley, W.C. Describing the Mechanism of Antimicrobial Peptide Action with the Interfacial Activity Model. *ACS Chem. Biol.* **2010**, *5*, 905–917. [CrossRef]
20. Jackson, M.; Mantsch, H.H.; Spencer, J.H. Conformation of Magainin-2 and Related Peptides in Aqueous Solution and Membrane Environments Probed by Fourier Transform Infrared Spectroscopy. *Biochemistry* **1992**, *31*, 7289–7293. [CrossRef]
21. Hirsh, D.J.; Hammer, J.; Maloy, W.L.; Blazyk, J.; Schaefer, J. Secondary Structure and Location of a Magainin Analogue in Synthetic Phospholipid Bilayers. *Biochemistry* **1996**, *35*, 12733–12741. [CrossRef]

22. Williams, R.W.; Starman, R.; Taylor, K.M.P.; Gable, K.; Beeler, T.; Zasloff, M.; Covell, D. Raman Spectroscopy of Synthetic Antimicrobial Frog Peptides Magainin 2a and PGLa. *Biochemistry* **1990**, *29*, 4490–4496. [CrossRef] [PubMed]
23. Nguyen, L.T.; Haney, E.F.; Vogel, H.J. The Expanding Scope of Antimicrobial Peptide Structures and Their Modes of Action. *Trends Biotechnol.* **2011**, *29*, 464–472. [CrossRef]
24. Matsuo, K.; Maki, Y.; Namatame, H.; Taniguchi, M.; Gekko, K. Conformation of Membrane-Bound Proteins Revealed by Vacuum-Ultraviolet Circular-Dichroism and Linear-Dichroism Spectroscopy. *Proteins* **2016**, *84*, 349–359. [CrossRef]
25. Matsuo, K.; Kumashiro, M.; Gekko, K. Characterization of the Mechanism of Interaction Between α1-Acid Glycoprotein and Lipid Membranes by Vacuum-Ultraviolet Circular-Dichroism Spectroscopy. *Chirality* **2020**, *32*, 594–604. [CrossRef]
26. Kumashiro, M.; Izumi, Y.; Matsuo, K. Conformation of Myelin Basic Protein Bound to Phosphatidylinositol Membrane Characterized by Vacuum-Ultraviolet Circular-Dichroism Spectroscopy and Molecular-Dynamics Simulations. *Proteins* **2021**, *89*, 1251–1261. [CrossRef] [PubMed]
27. Hope, M.J.; Bally, M.B.; Webb, G.; Cullis, P.R. Production of Large Unilamellar Vesicles by a Rapid Extrusion Procedure. Characterization of Size Distribution, Trapped Volume and Ability to Maintain a Membrane Potential. *Biochim. Biophys. Acta* **1985**, *812*, 55–65. [CrossRef]
28. Ojima, N.; Sakai, K.; Matsuo, K.; Matsui, T.; Fukazawa, T.; Namatame, H.; Taniguchi, M.; Gekko, K. Vacuum-Ultraviolet Circular Dichroism Spectrophotometer Using Synchrotron Radiation: Optical System and On-Line Performance. *Chem. Lett.* **2001**, 522–523. [CrossRef]
29. Matsuo, K.; Sakai, K.; Matsushima, Y.; Fukuyama, T.; Gekko, K. Optical Cell with a Temperature-Control Unit for a Vacuum-Ultraviolet Circular Dichroism Spectrophotometer. *Anal. Sci.* **2003**, *19*, 129–132. [CrossRef] [PubMed]
30. Sreerama, N.; Woody, R.W. Estimation of Protein Secondary Structure from Circular Dichroism Spectra: Comparison of CONTIN, SELCON, and CDSSTR Methods with an Expanded Reference Set. *Anal. Biochem.* **2000**, *287*, 252–260. [CrossRef]
31. Sreerama, N.; Venyaminov, S.Y.U.; Woody, R.W. Estimation of the Number of α-Helical and β-Strand Segments in Proteins Using Circular Dichroism Spectroscopy. *Protein Sci.* **1999**, *8*, 370–380. [CrossRef]
32. Matsuo, K.; Yonehara, R.; Gekko, K. Secondary-Structure Analysis of Proteins by Vacuum-Ultraviolet Circular Dichroism Spectroscopy. *J. Biochem.* **2004**, *135*, 405–411. [CrossRef] [PubMed]
33. Matsuo, K.; Yonehara, R.; Gekko, K. Improved Estimation of the Secondary Structures of Proteins by Vacuum-Ultraviolet Circular Dichroism Spectroscopy. *J. Biochem.* **2005**, *138*, 79–88. [CrossRef]
34. Rodger, A.; Marrington, R.; Geeves, M.A.; Hicks, M.; De Alwis, L.; Halsall, D.J.; Dafforn, T.R. Looking at Long Molecules in Solution: What Happens When They Are Subjected to Couette Flow? *Phys. Chem. Chem. Phys.* **2006**, *8*, 3161–3171. [CrossRef]
35. Lentz, B.R. Use of Fluorescent Probes to Monitor Molecular Order and Motions Within Liposome Bilayers. *Chem. Phys. Lipids* **1993**, *64*, 99–116. [CrossRef]
36. Lakowicz, J.R. *Principles of Fluorescence Spectroscopy*, 3rd ed.; Springer: Boston, MA, USA, 2006; ISBN 978-0387312781.
37. Bartlett, G.R. Phosphorus Assay in Column Chromatography. *J. Biol. Chem.* **1959**, *234*, 466–468. [CrossRef]
38. Chatelier, R.C.; Minton, A.P. Adsorption of Globular Proteins on Locally Planar Surfaces: Models for the Effect of Excluded Surface Area and Aggregation of Adsorbed Protein on Adsorption Equilibria. *Biophys. J.* **1996**, *71*, 2367–2374. [CrossRef]
39. Shashilov, V.A.; Lednev, I.K. Advanced Statistical and Numerical Methods for Spectroscopic Characterization of Protein Structural Evolution. *Chem. Rev.* **2010**, *110*, 5692–5713. [CrossRef]
40. Arouri, A.; Dathe, M.; Blume, A. Peptide Induced Demixing in PG/PE Lipid Mixtures: A Mechanism for the Specificity of Antimicrobial Peptides Towards Bacterial Membranes? *Biochim. Biophys. Acta* **2009**, *1788*, 650–659. [CrossRef]
41. Edwards-Gayle, C.J.C.; Barrett, G.; Roy, S.; Castelletto, V.; Seitsonen, J.; Ruokolainen, J.; Hamley, I.W. Selective Antibacterial Activity and Lipid Membrane Interactions of Arginine-Rich Amphiphilic Peptides. *ACS Appl. Bio Mater.* **2020**, *3*, 1165–1175. [CrossRef]
42. Persson, D.; Thorén, P.E.G.; Herner, M.; Lincoln, P.; Nordén, B. Application of a Novel Analysis to Measure the Binding of the Membrane-Translocating Peptide Penetratin to Negatively Charged Liposomes. *Biochemistry* **2003**, *42*, 421–429. [CrossRef] [PubMed]
43. Manavalan, P.; Johnson, W.C. Protein Secondary Structure from Circular Dichroism Spectra. *J. Biosci.* **1985**, *8*, 141–149. [CrossRef]
44. Gorbenko, G.P.; Ioffe, V.M.; Kinnunen, P.K.J. Binding of Lysozyme to Phospholipid Bilayers: Evidence for Protein Aggregation upon Membrane Association. *Biophys. J.* **2007**, *93*, 140–153. [CrossRef]
45. Zuckermann, M.J.; Heimburg, T. Insertion and Pore Formation Driven by Adsorption of Proteins onto Lipid Bilayer Membrane-Water Interfaces. *Biophys. J.* **2001**, *81*, 2458–2472. [CrossRef]
46. Greenfield, N.; Fasman, G.D. Computed Circular Dichroism Spectra for the Evaluation of Protein Conformation. *Biochemistry* **1969**, *8*, 4108–4116. [CrossRef]
47. Gesell, J.; Zasloff, M.; Opella, S.J. Two-Dimensional 1H NMR Experiments Show That the 23-Residue Magainin Antibiotic Peptide Is an α-Helix in Dodecylphosphocholine Micelles, Sodium Dodecylsulfate Micelles, and Trifluoroethanol/Water Solution. *J. Biomol. NMR* **1997**, *9*, 127–135. [CrossRef]
48. Vermeer, L.S.; Marquette, A.; Schoup, M.; Fenard, D.; Galy, A.; Bechinger, B. Simultaneous Analysis of Secondary Structure and Light Scattering from Circular Dichroism Titrations: Application to Vectofusin-1. *Sci. Rep.* **2016**, *6*, 39450. [CrossRef] [PubMed]
49. Miles, A.J.; Wallace, B.A. Circular Dichroism Spectroscopy of Membrane Proteins. *Chem. Soc. Rev.* **2016**, *45*, 4859–4872. [CrossRef]

50. Micsonai, A.; Wien, F.; Kernya, L.; Lee, Y.H.; Goto, Y.; Réfrégiers, M.; Kardos, J. Accurate Secondary Structure Prediction and Fold Recognition for Circular Dichroism Spectroscopy. *Proc. Natl. Acad. Sci. USA* **2015**, *112*, E3095–E3103. [CrossRef]
51. Rodger, A.; Nordén, B.; Dafforn, T. *Linear Dichroism and Circular Dichroism: A Textbook on Polarized-Light Spectroscopy*; Royal Society of Chemistry: Cambridge, UK, 2010; ISBN 978-1847559029.
52. Rodger, A.; Rajendra, J.; Marrington, R.; Ardhammar, M.; Nordén, B.; Hirst, J.D.; Gilbert, A.T.B.; Dafforn, T.R.; Halsall, D.J.; Woolhead, C.A.; et al. Flow Oriented Linear Dichroism to Probe Protein Orientation in Membrane Environments. *Phys. Chem. Chem. Phys.* **2002**, *4*, 4051–4057. [CrossRef]
53. Lentz, B.R. Membrane "Fluidity" as Detected by Diphenylhexatriene Probes. *Chem. Phys. Lipids* **1989**, *50*, 171–190. [CrossRef]
54. Jang, H.; Ma, B.; Lal, R.; Nussinov, R. Models of Toxic β-Sheet Channels of protegrin-1 Suggest a Common Subunit Organization Motif Shared with Toxic Alzheimer β-Amyloid Ion Channels. *Biophys. J.* **2008**, *95*, 4631–4642. [CrossRef]
55. Ciudad, S.; Puig, E.; Botzanowski, T.; Meigooni, M.; Arango, A.S.; Do, J.; Mayzel, M.; Bayoumi, M.; Chaignepain, S.; Maglia, G.; et al. Aβ(1–42) Tetramer and Octamer Structures Reveal Edge Conductivity Pores as a Mechanism for Membrane Damage. *Nat. Commun.* **2020**, *11*, 3014. [CrossRef] [PubMed]
56. Rzeźnicka, I.I.; Pandey, R.; Schleeger, M.; Bonn, M.; Weidner, T. Formation of Lysozyme Oligomers at Model Cell Membranes Monitored with Sum Frequency Generation Spectroscopy. *Langmuir* **2014**, *30*, 7736–7744. [CrossRef] [PubMed]
57. Papahadjopoulos, D.; Moscarello, M.; Eylar, E.H.; Isac, T. Effects of proteins on the thermotropic phase transitions of phospholipid membranes. *Biochim. Biophys. Acta Biomembr.* **1975**, *401*, 317–335. [CrossRef]
58. Quinn, P.J. The Fluidity of Cell Membranes and Its Regulation. *Prog. Biophys. Mol. Biol.* **1981**, *38*, 1–104. [CrossRef]
59. Veld, G.I.; Driessen, A.J.M.; Konings, W.N. Bacterial Solute Transport Proteins in Their Lipid Environment. *FEMS Microbiol. Rev.* **1993**, *12*, 293–314. [CrossRef] [PubMed]
60. Hazel, J.R.; Williams, E.E. The Role of Alterations in Membrane Lipid Composition in Enabling Physiological Adaptation of Organisms to Their Physical Environment. *Prog. Lipid Res.* **1990**, *29*, 167–227. [CrossRef]
61. Avitabile, C.; D'Andrea, L.D.; Romanelli, A. Circular Dichroism Studies on the Interactions of Antimicrobial Peptides with Bacterial Cells. *Sci. Rep.* **2014**, *4*, 4293. [CrossRef]

membranes

MDPI

Review

Heavy Atom Detergent/Lipid Combined X-ray Crystallography for Elucidating the Structure-Function Relationships of Membrane Proteins

Shinya Hanashima [1], Takanori Nakane [2,†] and Eiichi Mizohata [3,4,*]

1 Department of Chemistry, Graduate School of Science, Osaka University, 1-1 Machikaneyama, Toyonaka 563-0043, Japan; hanashimas13@chem.sci.osaka-u.ac.jp
2 Department of Biological Sciences, Graduate School of Science, The University of Tokyo, Tokyo 113-0033, Japan; tnakane@mrc-lmb.cam.ac.uk
3 Department of Applied Chemistry, Graduate School of Engineering, Osaka University, 2-1 Yamadaoka, Suita 565-0871, Japan
4 Japan Science and Technology Agency, PRESTO, Kawaguchi 332-0012, Japan
* Correspondence: mizohata@chem.eng.osaka-u.ac.jp; Tel.: +81-6-6879-7410; Fax: +81-6-6879-7409
† Present address: MRC Laboratory of Molecular Biology, Cambridge CB2 0QH, UK.

Abstract: Membrane proteins reside in the lipid bilayer of biomembranes and the structure and function of these proteins are closely related to their interactions with lipid molecules. Structural analyses of interactions between membrane proteins and lipids or detergents that constitute biological or artificial model membranes are important for understanding the functions and physicochemical properties of membrane proteins and biomembranes. Determination of membrane protein structures is much more difficult when compared with that of soluble proteins, but the development of various new technologies has accelerated the elucidation of the structure-function relationship of membrane proteins. This review summarizes the development of heavy atom derivative detergents and lipids that can be used for structural analysis of membrane proteins and their interactions with detergents/lipids, including their application with X-ray free-electron laser crystallography.

Keywords: membrane protein; lipid; detergent; chemical synthesis; X-ray free-electron laser (XFEL); serial femtosecond crystallography (SFX); structure-function relationship

1. Introduction

All cellular organisms have biomembranes composed of lipids, which form a boundary between the cell cytoplasm and the surrounding extracellular environment and organize complex structures inside the cell to construct cellular organelles [1]. Membrane proteins localize in the lipid bilayer of biomembranes and play a variety of important functions such as transport of various substances between the inside and outside of the cell, signal transduction, energy synthesis, and cell adhesion. In particular, membrane proteins associated with human diseases are attracting considerable attention as drug targets and are actively studied. Compounds that bind to membrane proteins and inhibit or promote their functions are being explored as drug candidates, and membrane proteins overexpressed on the surface of certain cancer cells are targets for the development of antibody drugs [2–5]. Thus, elucidation of the structure and function of membrane proteins including their interactions with lipids at the molecular level is essential both from the perspective of basic science to understand the mechanisms of life and from the perspective of medical applications. However, the progress of membrane protein research has lagged behind that of water-soluble proteins because of the difficulties associated with isolating hydrophobic membrane proteins in sufficient quantity and quality for analysis. In the field of structural biology, only about 6000 of the approximately 180,000 structural coordinates registered in the Protein Data Bank (PDB) are those of membrane proteins.

Nonetheless, advances in research technology have accelerated the study of membrane protein structures. With the continuous development of genetic engineering, expressing high levels of stable membrane proteins as recombinant proteins and purifying them with a sufficient yield has gradually become possible. High-speed atomic force microscopy has made it possible to image motion during functioning with nanometer-order spatial resolution without destroying the structure of biological samples [6]. Cryo-electron microscopy has undergone remarkable technological improvements over the last decade, enabling structural analysis of membrane proteins and their complexes with near-atomic resolution without requiring crystal preparation [7]. Most attempts to visualize the three-dimensional structure of membrane proteins and their interactions with lipids/detergents at near-atomic resolution or better have been made using X-ray crystallography. Advancing synchrotron radiation facilities and beamline technologies have facilitated the determination of crystal structures of membrane proteins. This trend will continue with the development of fourth-generation synchrotron radiation sources, including X-ray free-electron laser (XFEL) facilities [8].

Solving new structures by X-ray crystallography requires both intensities and phases of structure factors of diffracted X-ray waves; however, measured diffraction patterns only give intensities (so-called "phase problem"). Therefore, experimental (de novo) phase determination methods including single or multiple heavy-atom isomorphous replacement (SIR or MIR), single- or multi-wavelength anomalous diffraction (SAD or MAD) or the combination of SIR or MIR with anomalous scattering (SIRAS or MIRAS) must be applied to solve the phase problem. Isomorphous replacement methods use the difference in reflection intensities between the native crystal and the heavy atom-labeled derivative crystal to determine the phases, whereas the anomalous diffraction method uses anomalous differences between Bijvoet pairs [9]. Both methods require accurate measurement of small intensity differences. However, expressing selenomethionine-labeled recombinant proteins is time-consuming and costly, and even if the native crystals are soaked in a heavy-atom solution, the heavy atoms often do not bind to the protein or the crystallinity easily collapses, resulting in loss of diffraction ability. Thus, the development of heavy atom labeled protein ligands is needed that bind various membrane proteins and facilitate drawing electron density maps efficiently from X-ray diffraction data to reveal three-dimensional structures.

In this review, we summarize the development and use of heavy atom derivative detergents/lipids that can be used to analyze interactions between membrane proteins and detergents/lipids and to determine the structure of membrane proteins. We also discuss the potential of combining recently developed XFEL crystallography with heavy atom detergents/lipids to understand the structure-function relationship of membrane proteins and their interactions with detergents/lipids.

2. Heavy Atom Labeled Protein Ligands

Typical de novo methods to determine crystallographic phases involve heavy-atom derivatization of protein crystals. These crystals can be used for isomorphous replacement methods and anomalous diffraction methods. The period 4 to 6 elements, including lanthanides, are often used as electron-rich heavy atoms for solving phase problems when determining protein structures by X-ray crystallography. Soaking manipulation introduces heavy metal ions or heavy atom-labeled compounds into protein crystals. The successfully soaked ions or compounds in protein crystals are immobilized through interactions with functional groups of proteins mostly by ionic and dipole linkages, H-bonding, and coordination bonding [10,11]. Numerous conditions are typically screened as a rule of thumb to ensure successful crystallization of the target protein and referring to the chemistry of ionic interactions and covalent modifications is often useful for the introduction of heavy atoms into protein crystals.

Protein ligands containing heavy atom moieties have been developed to achieve two purposes. First, these ligands unveil the molecular interactions between proteins and ligands. The strong and characteristic diffraction of heavy atoms, due to their high electron density and anomalous scattering, reveals the ligand position on the protein. Second, the ligand can be used for de novo phasing to determine the protein structure. However, careful design of the heavy atom ligand is essentially required to avoid disturbance for the protein-ligand interaction. Heavy atom-containing detergents/lipids are potentially useful for solving structures of membrane proteins because they have affinity for the hydrophobic surface of membrane proteins and can be used to determine phases.

2.1. Halogens

Protein-ligand analogs carrying a Br or I atom have been used successfully to determine X-ray protein crystal structures with de novo phasing; however, careful design of the analog structure is important because of the properties of halogen atoms. The carbon(sp^3)–halogen bond length (C–Br: 191 pm; C–I: 216 pm) is significantly longer than the corresponding carbon–carbon bond length (151 pm) [12]. Thus, careful consideration may be required when selecting protein crystallization conditions because of the bulkiness and lower bond-dissociation energy of the carbon(sp^3)–halogen bond (C–Br: 72.4 kcal/mol; C–I: 56.9 kcal/mol) [13]. Possible instability can be overcome by installing halogens on an aromatic ring because the bond-dissociation energy between a benzene carbon (Ph) and the halogen atom is higher (Ph–Br: 84 kcal/mol; Ph–I: 67 kcal/mol) than that of carbon(sp^3)–halogen bonds [13].

The high electron density and anomalous scattering from halogen atoms are used to disclose the exact atom position in crystal structures. A powerful application is to elucidate the binding of drugs to membrane proteins. For example, the anomalous scattering of Br installed on Fluoxetine (Br–Fluoxetine), the antidepressant drug named Prozac®®, clearly reported the ligand position at the fenestration of the K2P channel TREC-2, which is regulated by arachidonic acid [14]. In a similar approach, Br-labeled Memantine, a drug used for treating Alzheimer's disease, was used to accurately unveil the binding mode to the prokaryotic pentameric ligand-gated ion channel [15]. The crystal structure indicated that the Br analog blocks the extracellular entryway of the channel pore to inhibit ion passage through the channel.

Halogen labeling has also been used to determine the crystal structure of RNA molecules. RNA is a biopolymer that plays a direct role in the regulation of cellular activities through molecular interactions. Some structures of RNA have been successfully determined through de novo phasing using halogen atoms such as 5-bromouridine [16] and 5-iodouridine [17]. Nucleotides are crucial protein substrates for maintaining cellular homeostasis. ATP mimicking analogs with a halogen atom are good inhibitors of protein kinases. The affinity of protein-inhibitor binding is dependent on the number and position of halogen substitutions, which contribute to "halogen bonding" [18]. Therefore, halogen bonding is another important function of halogen atoms introduced on the protein ligand, which may enhance ligand-protein interactions and possibly modulate the orientation of the ligand [19].

Cell membrane lipids are indispensable for regulating the structure and function of membrane-related proteins. Phospholipids carrying brominated hydrocarbon chains **1** and **2** have been used to examine phase separation and domain distributions (Figure 1) [20–22]. Namely, an electron-rich Br atom selectively positioned in the middle or terminal position of a phospholipid hydrocarbon chain efficiently acts as a collisional quencher for a fluorescence group close to the Br atom. These brominated fatty acids of phospholipids are readily prepared by substituting the terminal alcohol with Br [23] or adding Br_2 to an unsaturated bond [24]. The electron-rich Br atom can be a good reporter in X-ray diffraction studies for detecting particular atomic positions in lipid tails consisting of continuous methylene and terminal methyl groups. X-ray diffraction data of a bilayer membrane composed of mono-bromo dipalmitoylphosphatidylcholine (DPPC) **1** disclosed the terminal position of the hydrocarbon chain in the membrane, where the terminus of lipid chains was substituted with a Br instead of the ω-methyl group [25].

Br mimics the size of the ω-methyl group of the hydrocarbon chain because both have similar van der Waals radii (Br: 185 pm; CH_3: 200 pm). Bromination did not affect the transition enthalpy and cooperativity significantly; although, mono-bromo DPPC showed a significant decrease by 9 °C in the phase transition temperature. Brominated lipids were also used to determine lipid binding sites on the surface of membrane proteins in single-crystal X-ray structural studies of proteins [26]. This was achieved by discriminating between the hydrocarbon chains of lipids and those of detergents that were used to reconstitute membrane proteins into membranes. When labeling phospholipids with Br atoms, the anomalous signals from dibromo phospholipid **2** in Figure 1 and its analogs with different lipid headgroups make it easier to discriminate lipid electron densities of surrounding lipids from those of detergents and other molecules.

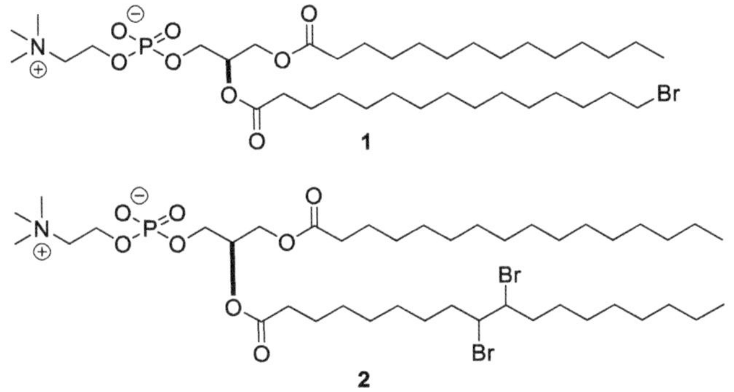

Figure 1. Structures of the bromo-phospholipids.

Brominated lipids can be used to determine lipidic ligand–protein interactions (Figure 2). Chemosensory proteins transport chemicals from air to the receptor and are associated with olfaction or taste processes. 12-Bromododecanol was used in a crystallization study as a surrogate for the hydrophobic ligand of a chemosensory protein [27]. The clear observation of three peaks in the anomalous difference electron density map revealed that three chain positions of the bromohydrocarbon occupied the large ligand cavity of the protein after major structural changes. Pheromone binding proteins are another class of odorant proteins. Similarly, iodohexadecane was used instead of the intrinsic pheromone, bombykol, to determine the conformation of the binding cavity of the insect pheromone binding protein from *Bombyx mori* [28].

We recently developed the synthetic route for ω-doubly brominated 1,2-dimyristoyl-*sn*-glycero-3-phosphocholine (Br_2-DMPC) via the tosyl intermediate (Scheme 1). Esterification of the ω-tosyl fatty acid to the two hydroxy groups on glycerophosphocholine, instead of using the ω-bromo fatty acid, successfully produced an intermediate **3** in moderate yield, and the following substitution reaction using LiBr afforded Br_2-DMPC. These brominated lipids will be used as ligands and as annular lipids in membrane protein crystallization studies.

Large membrane proteins usually consist of multiple transmembrane helices and sizable ectodomains. Therefore, labeling multiple sites of the membrane protein with heavy atoms is an approach to solve phase problems in X-ray diffraction studies. The use of appropriate detergents is often required to manipulate and crystallize membrane proteins. Detergents used for crystallization can be conveniently labeled with heavy atoms, which is suitable for de novo phasing. Accordingly, heavy atom-bearing tripod amphiphiles shown in Figure 3 were developed previously [29]; the iododetergent **4**, which solubilizes membrane proteins, is suitable for manipulating membrane proteins.

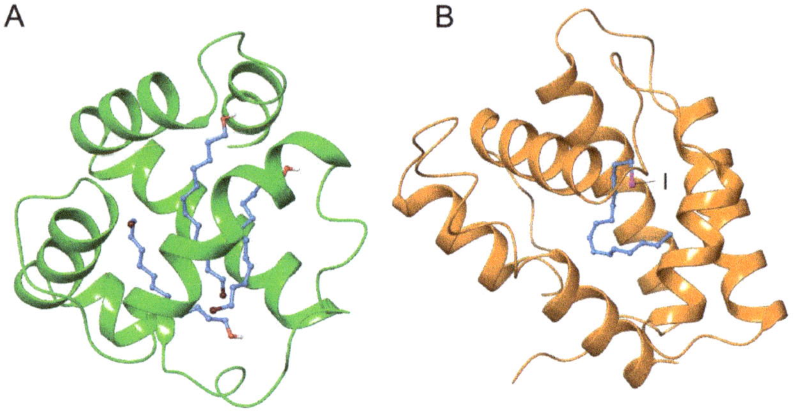

Figure 2. Crystal structures of the complex with lipidic ligands with halogen moiety. Moth chemosensory protein CSPMbraA6 with bromohydrocarbon chain (PDB ID: 1N8U) (**A**). Moth pheromone binding protein BmorPBP with iodohexadecane (PDB ID: 2P71) (**B**). Hydrocarbon chain of the ligands were shown in sky-blue, and Br and I were shown in purple and magenta, respectively.

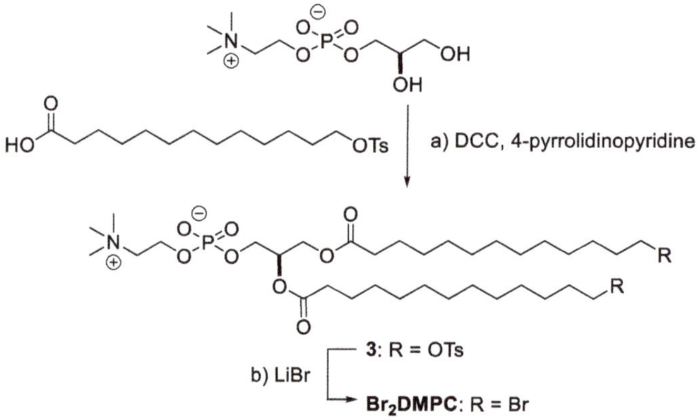

Scheme 1. Synthesis of ω-Br_2-DMPC. Synthetic procedures; (**a**) *N,N′*-dicyclohexylcarbodiimide (DCC), 4-pyrrolidinopyridine, CH_2Cl_2, 40%. (**b**) LiBr, acetone 41%. Ts; *p*-toluenesulfonyl group. Detail of this synthesis was described in Supplementary Materials.

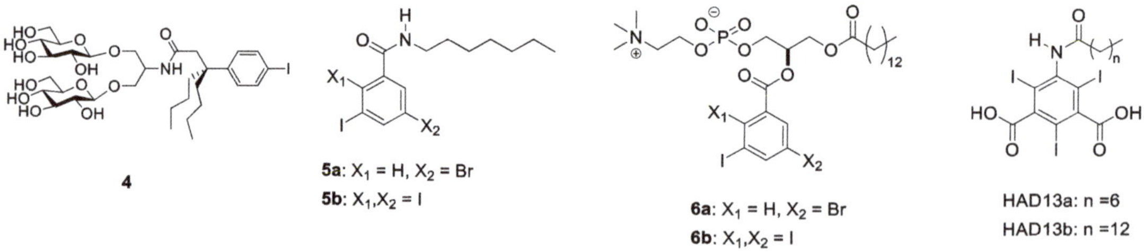

Figure 3. Lipophilic compounds and detergents bearing multiple halogen substitutions that can be used to solve the phasing problem for membrane protein structures.

We developed amphipathic compounds, phospholipid mimics and detergents bearing multiple halogen substitutions on aromatic rings for de novo phasing and revealing lipid-protein interactions. The amphipathic properties manifested by the hydrophilic headgroups linked to the lipophilic hydrocarbon tails are essential for these molecules to mix well with bilayer membranes and interact with membrane proteins in a similar manner to intrinsic lipids (Figure 3). Furthermore, the multiple halogens arrayed on a plane of the aromatic ring assisted with the easy detection of the moiety. Commercially available benzoic acid or phthalic acid derivatives carrying multiple halogen substitutions, some of which have been used as a phasing reagent in protein crystallization [30,31], were coupled with lipophilic moieties through amidation or esterification with alkyl amine, lysophosphatidylcholine or fatty acids to give compounds **5a**, **5b**, **6a** (HAD16), **6b**, HAD13a [32], and HAD13b (see Supplementary Materials for synthetic details). Importantly, HAD13 was developed based on a successful phasing reagent 5-amino-2,4,6-triiodoisophthalic acid (I3C, or magic triangle) [30,31] for application in X-ray imaging. I3C is a common synthetic intermediate for several X-ray contrasting reagents safely used to improve the visibility of vascular structures and organs during radiographic procedures in clinical diagnostics [33,34]. I3C has been used for solving the crystal structure of more than 30 protein structures (mainly soluble proteins) in the PDB. Recently, 5-amino-2,4,6-tribromobenzene-1,3-dicarboxylic acid and tetrabromoterephthalic acid, which are bromo analogs of I3C, were successfully used as phasing reagents in the crystallization of model proteins [35,36]. Further application of I3C and other halogenated benzoates for membrane protein crystallization was achieved by improving the amphiphilicity of the halogenated benzoate by introducing an acyl chain at an amino group or coupling with a lysophospholipid. The amphiphilic nature of these compounds ensures good miscibility with phospholipids and for acting as detergents by surrounding membrane proteins. Phospholipid mimics **6a** (HAD16) and **6b** and detergents HAD13a and HAD13b were miscible in DMPC/CHAPSO bicelles (q = 2.8), which were used for crystallization of membrane proteins [37], whereas hydrophobic compounds **5a** and **5b** were shown to readily form precipitates. In practice, HAD13a was used successfully to perform de novo phasing of XFEL diffraction data obtained from membrane protein microcrystals (see details in Section 3).

2.2. Selenium

Selenium is a major heavy atom that is used to achieve de novo phasing of protein crystallographic data. Oxygen and sulfur belong to the second and third periods of the group 16 elements in the periodic table, thus being potentially substituted with a selenium element in the 4th period. Therefore, recombinant expression using selenocysteine and selenomethionine readily furnishes target proteins with heavy atom labels, which are suitable for the SAD/MAD phasing method [38]. However, care must be taken because the toxicity of selenomethionine sometimes hampers the growth of recombinant organisms used to overexpress target proteins. In addition, selenium derivatives of nucleic acids have been incorporated into DNA and RNA to solve the phase problem in MAD phasing [39].

Selenium-labeled derivatives of ligands that bind to proteins have been synthesized for protein crystallography. The selenium atom on the protein ligand is used to phase X-ray diffraction data and determine the exact atomic position of the bound ligand by using anomalous scattering, as is the case with halogen ligands. In carbohydrate chemistry, selenium-containing sugars and their derivatives have been developed as antioxidants and glycosidase inhibitors [40,41]. Although selenium is a highly toxic element, the potential of these seleno-sugars as lead compounds for therapeutic reagents has been examined [42].

Selenomethyl-*N*-acetylglucosamine (βMeSe-GlcNAc), a selenium-labeled sugar, was first used for protein crystallization to solve the phasing of data obtained on bacterial F17-G adhesin (Scheme 2) [43] because GlcNAc is a native ligand of this adhesin. The selenomethyl group was smoothly installed at the anomeric position of 1-chloro-GlcNAc. βMeSe-GlcNAc successfully mimics the native ligand because the binding constants of GlcNAc and βMeSe-GlcNAc to the adhesin are similar. Imberty and co-workers developed a selenomethyl derivative of fucose (β-MeSe-Fuc) (Scheme 2). βMeSe-Fuc was first used as a seleno-ligand for phasing when solving the crystal structures of fucose-binding lectins derived from a plant pathogenic bacteria [44]. βMeSe-Fuc was also used to determine the co-crystal structure of other lectins AFL [45] and BC2L-C [46], which originate from the opportunistic infection-related fungus *Aspergillus fumigatus* and the bacterium *Burkholderia cenocepacia*, respectively. αMeSe-Fuc was readily synthesized by anomerization of βMeSe-Fuc with a Lewis acid [47] or glycosylation between fucosyl imidate and a selenoacetal acceptor [48]. Recently, Shimabukuro et al. successfully introduced the MeSe group at the 2–, 3–, or 4–OH position of fucose (Figure 4) and these compounds were used as ligands for a fucose binding lectin that originated from *Aspergillus oryzae* [49]. The MeSe substitution disclosed the essential hydroxy group for interaction with the lectin. Currently, the selenium atom can be incorporated into an oligosaccharide structure, which facilitates applications of seleno probes to elucidate oligosaccharide binding to proteins by not only crystallography but NMR because ^{77}Se (I = 1/2, 7.6% natural abundance) is an NMR active nucleus that gives rise to a sharp signal [50].

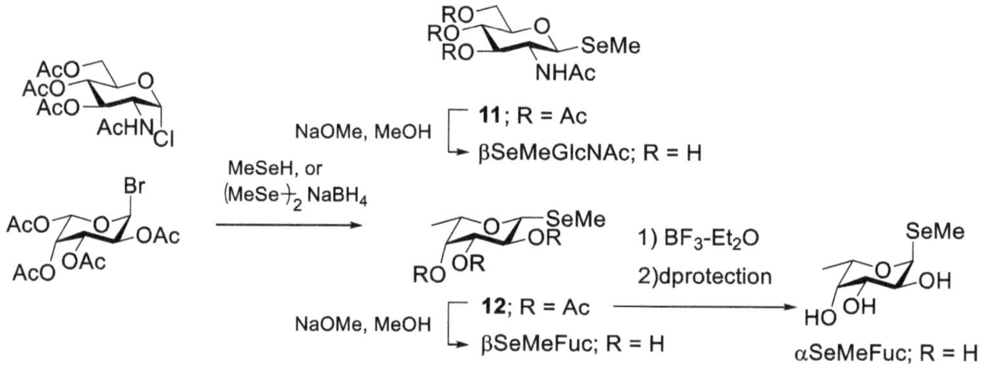

Scheme 2. Synthetic scheme of the selenoglycosides where selenium was substituted at the glycoside oxygen.

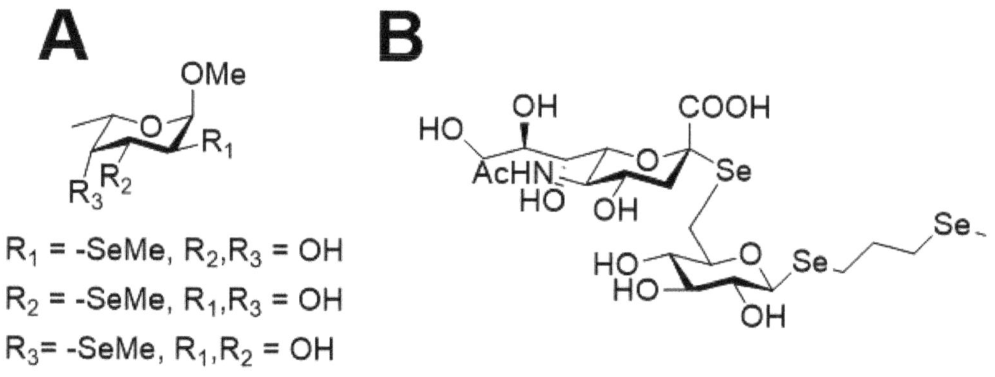

Figure 4. Fucose derivatives with selenomethyl substitutions at the 2, 3, or 4 hydroxy group (**A**) and sialylglucose derivative with three selenomethyl substitutions (**B**).

The hydrophilic sugar moieties are also used as headgroups of detergents to manipulate and crystallize membrane proteins. Detergents carrying a heavy atom can be used in membrane protein studies. Dodecyl-β-D-selenomaltoside (SeDDM), a heavy atom analog of dodecyl-β-D-maltoside (DDM), was used to achieve MAD phasing when solving the structure of leukotriene C4 synthase by X-ray crystallography (Figure 5A) [51]. Furthermore, SeDDM was used successfully as a phasing detergent to solve the structure of the prokaryotic pentameric ligand-gated ion channel [52]; however, the sugar headgroups were poorly resolved (Figure 5B). Detergent bundles of SeDDM were clearly observed inside the pore.

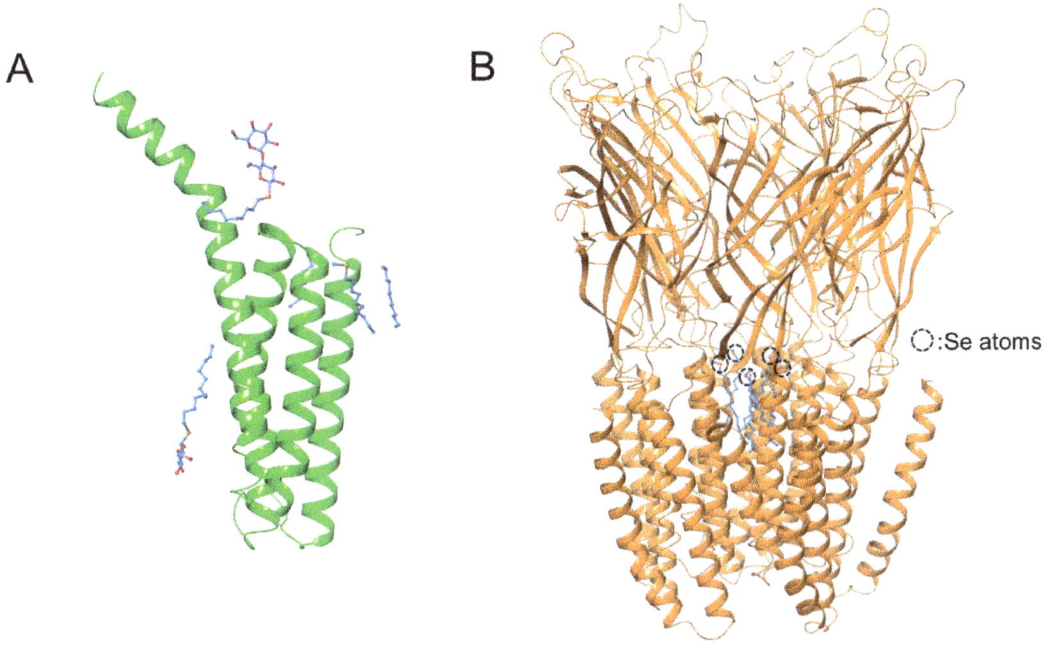

Figure 5. Crystal structures of transmembrane proteins Leukotriene C4 synthase (PDB ID: 3B29) (**A**) and pLGIC pentamer (PDB ID: 4IL4) (**B**) using SeDDM as the phasing detergent. The hydrocarbon chains of Se-DDM are shown in sky-blue.

Fatty acids are amphiphilic compounds ubiquitously present in our body. A selenofatty acid whose methylene group was substituted with a selenium atom was developed because their van der Waals radii are very similar (Se: 190 pm; CH_2: 200 pm). Carbon belongs to group 14 and is thus different from selenium in the periodic table. This difference results in a slightly longer bond length (C–Se: 194.5 pm; C–C: 154.0 pm) and a smaller bond angle (C–Se–C: 96.3°; C–C–C 112.6°). Thus, the distance between the two carbon atoms of a C–Se–C moiety is 289.8 pm, which is 34 pm longer than the distance of 1,3-carbon atoms in an alkane. However, this difference does not affect the conformation of the whole acyl chain significantly, except for the selenoether and its neighboring portions. Therefore, selenium can be used as a bioisostere for a methylene unit, which is particularly useful for introducing a heavy atom into the hydrophobic part of detergents and ligands. For example, Fredga and Lindgren first reported the synthesis of 4-seleno-hexadecanoic acid and 12-seleno-hexadecanoic acid, the seleno analogs of palmitic acid (Scheme 3) [53]. Two or more selenium atoms were also incorporated into a fatty acid using a similar synthetic pathway [54]. Sadek and Basmdjian incorporated radioactive ^{75}Se into fatty acid chains for imaging [55].

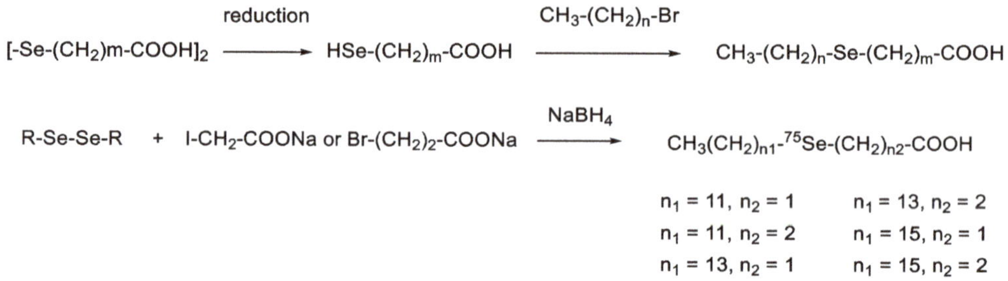

Scheme 3. Conventional synthesis of seleno-fatty acids.

Selenium is an essential trace element and a source of the antioxidant selenoproteins. Ether-type lipids bearing a selenomethyl moiety at the terminus of a hydrocarbon chain were synthesized to gain antioxidant activity [56,57]. Interestingly, seleno fatty acids were also developed as antimicrobial agents [58]. These seleno-lipid derivatives can be used to solve the phase problem in appropriate protein crystallization studies.

Lipidic ligands carrying a heavy atom moiety are also useful for revealing lipid–protein interactions, not only for phase determination but by making the heavy atom position conspicuous with the anomalous difference map. Recently, we developed a heavy atom derivative of α-galactosylceramide (α-GalCer) known as KRN7000, which activates immune responses by inducing cytokine production upon binding to the protein CD1d (Figure 6) [59]. Selenium was incorporated into the fatty acid chain of α-GalCer by substitution of a methylene group through chemical synthesis. The selenium and ω-halo derivatives were potent inducers of IFN-γ and IL-4 production in murine splenocytes. The heavy atom-modified lipid derivatives of α-GalCer were accommodated deeply in the lipid-binding cavity of CD1d.

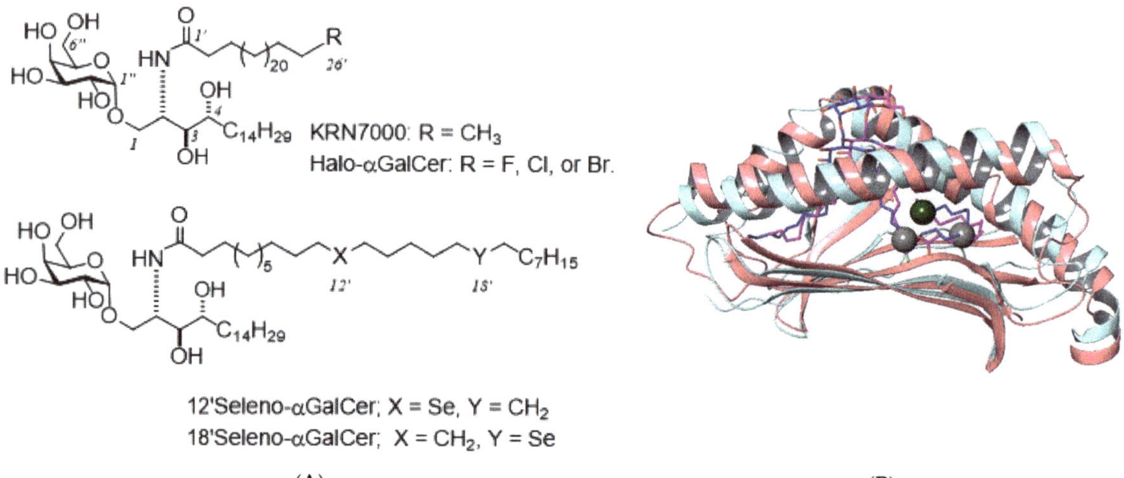

Figure 6. Structures of heavy atom derivatives of α-galactosylceramide and binding model with CD1d. The chemical structures of α-GalCer derivatives with terminal halogens or selenium (**A**), and the superposed structures of murine CD1d complexed with Th1-selective ligand C8PhF (PDB ID: 3GMO; pink ribbon and purple wire) and that with KRN7000 (PDB ID: 3HE6; sky-blue ribbon and blue wire) (**B**). C26′ of KRN7000 is shown as a dark green ball and C12′/C18′ are shown with gray balls, which corresponded to halogen and selenium positions of the synthetic analogues, respectively.

3. Serial Femtosecond Crystallography (SFX)

An important recent innovation in quantum beam technology for structural biology is the emergence of XFEL. The world's first XFEL facility, LCLS [60], was built at Stanford University in the United States in 2009. In 2011, SACLA [61] was established in Japan, followed by the European XFEL [62] in Germany, PAL-XFEL [63] in Korea and SwissFEL [64] in Switzerland. Serial femtosecond crystallography (SFX) is a new data measurement technique of X-ray crystallography, which takes advantage of the ultra-high beam brilliance, femtosecond pulse duration and high spatial coherence of XFEL [65]. At SACLA, an operation mode with a wavelength range of 0.62–2.76 Å, a pulse duration of 2–10 fs, a pulse energy ~400 uJ at 10 keV, and a frequency of 30–60 Hz is available for SFX [66].

In SFX measurements, many microcrystals are ejected at random orientations from an injector and introduced into the orbit of the XFEL beam, and the diffraction image produced when a single pulse of XFEL hits a microcrystal within a femtosecond exposure time is recorded at room temperature. Because the brightness of the XFEL is typically a billion times brighter than synchrotron radiation such as SPring-8, the diffraction intensity obtained with a single XFEL pulse is comparable to the diffraction intensity of a 1-s exposure with synchrotron radiation. Diffraction images of thousands to a half-million microcrystals are collected for structural analysis. In the case of synchrotron radiation crystallography (SRX), at least milliseconds of exposure time are required for data collection, which results in radiation damage and photoreduction, a phenomenon where hydrated electrons that are generated on the order of picoseconds by the interaction of X-rays with water molecules in the crystal react with protein molecules in the crystal to break chemical bonds and reduce metal active centers [67]. Because femtosecond XFEL diffraction in SFX is completed on a shorter timescale than the process of radiation damage and photoreduction, this technique captures the damage-free structure of samples [68].

3.1. De Novo Phasing in SFX

Since the first construction of the XFEL facility in 2009, all SFX structures reported in international journals had been solved by the molecular replacement method using known structures as search models. This is because phase determination by the anomalous scattering method requires accurate measurement of small intensity differences between the reflections of the Bijvoet pairs, which is much more difficult with SFX than with the conventional SRX oscillation method. In SFX, data are collected by irradiating a large number of microcrystals of different sizes and orientations with XFEL pulses that have fluctuations in intensities and wavelength spectra. This leads to large errors in the observed diffraction intensities, and all measurements are partial reflections, making it difficult to determine the phases. In 2014, the first de novo structure determined with SFX was reported by the Gd-SAD method using a model protein lysozyme [69]. Subsequently, we reported de novo phasing of the luciferin-regenerating enzyme by the Hg-SIRAS method [70] and lysozyme by the S/Cl-SAD method [71] in 2015 and successfully determined the structure of copper-containing nitrite reductase by the Cu-SAD method in 2016 [67]. In the same year, the phase determination of BinAB by the Hg/Gd/I-MIRAS method [72] and streptavidin by the Se-SAD method [73] were reported. Thus, successful cases of de novo phasing by SFX were reported. However, these data analyses are all from water-soluble proteins and required tens or hundreds of thousands of high-resolution diffraction images of 1.7–2.3-Å resolution. Therefore, the authors took on the challenge of developing an efficient method for the phase determination of membrane proteins by SFX.

3.2. De Novo Phasing with the HAD13a Detergent

In general, the diffraction quality of membrane protein crystals is lower when compared with that of water-soluble proteins. This is also true for SFX using the most brilliant XFEL light source available today. In the experimental phasing method of SRX, heavy atoms such as Se, Hg, Au, and Pt, which have absorption edges around 1.000 Å, have been selected frequently and used to derivatize relatively large crystals (50–200 μm). This is because, in most cases, the beamlines used for X-ray crystallography of biological macromolecules are optimized for wavelengths around this range. In contrast, because SFX uses microcrystals (1–50 μm), X-ray beams with longer wavelengths are often selected to obtain larger diffraction signals. We selected iodine, which has an excellent anomalous scattering effect at longer wavelengths (f'' = 8.6e at 1.771 Å). HAD13a was synthesized by attaching a hydrophobic alkyl chain (caprylic acid) to I3C to give HAD13a affinity toward hydrophobic surfaces of membrane proteins [32]. HAD13a has detergent properties (CMC: 4.6 mM) and was used to label membrane proteins with heavy atoms by simply mixing it with microcrystals of bacteriorhodopsin crystallized by the bicelle method or G protein-coupled A2a adenosine receptor (A2A GPCR) obtained by the LCP method (Figure 7). The HAD13a-labeled bacteriorhodopsin was successfully phase-determined by SAD, SIR, and SIRAS methods using iodine atoms. For phase determination by the SAD method, 23,000 indexed diffraction images and a resolution of 2.1 Å were required. In contrast, in the SIRAS method with the addition of native crystal data, reflections up to a resolution of 3.3 Å were sufficient for phasing, and when the resolution was extended, only 7000 (4000 derivative and 3000 native) indexed images were required to determine the phase. This indicates that the SIRAS method is more powerful than the SAD method for efficient de novo phasing in SFX. At a similar time as our study, Batyuk et al. succeeded in determining A2A GPCR by the S-SAD method, which required 500,000 images and obtained a resolution of 2.5 Å [74].

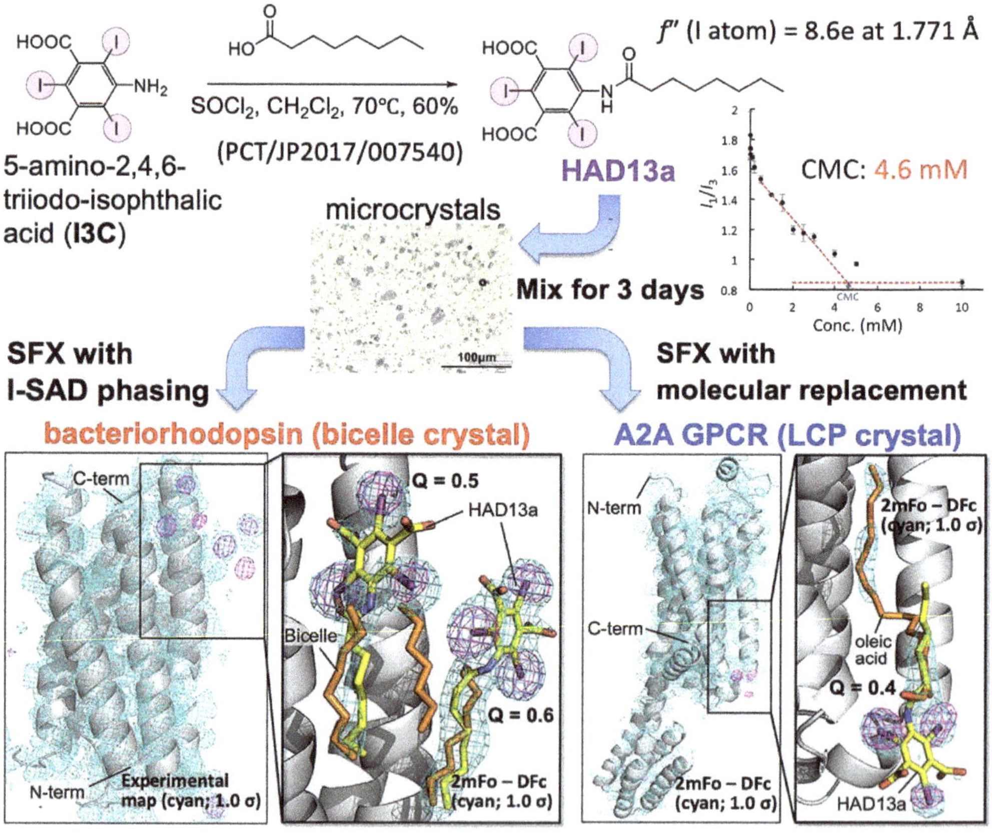

Figure 7. Serial femtosecond crystallography (SFX) of membrane proteins that have interacted with iododetergent, HAD13a. Figure reproduced with modification from Ref. [32].

3.3. *Binding of the HAD16 Lipid to a Membrane Protein*

HAD16 was synthesized by modifying the aromatic head group of HAD16H with multiple heavy-atom halogen groups to furnish a hydrophobic alkyl tail (Scheme 4, Supplementary Materials Section S6). The structural properties of HAD16 mimic a phosphatidylcholine structure. HAD16 can be used as a tool for analyzing lipid-membrane protein interactions. Microcrystals of bacteriorhodopsin obtained by the bicelle method were mixed with HAD16 and then SFX data were collected at SACLA (Supplementary Materials Section S7). We tried de novo phasing the HAD16 dataset. Although two heavy atom sites were located by the SIR, SIRAS or SAD method, auto-tracing was not possible even with all 18,069 lattices obtained. This is most likely because of the smaller isomorphous and anomalous signal from HAD16, which contains only one I atom and one Br atom per molecule instead of three I atoms in HAD13a, as well as only one HAD16 molecule in the asymmetric unit when compared with two HAD13a molecules.

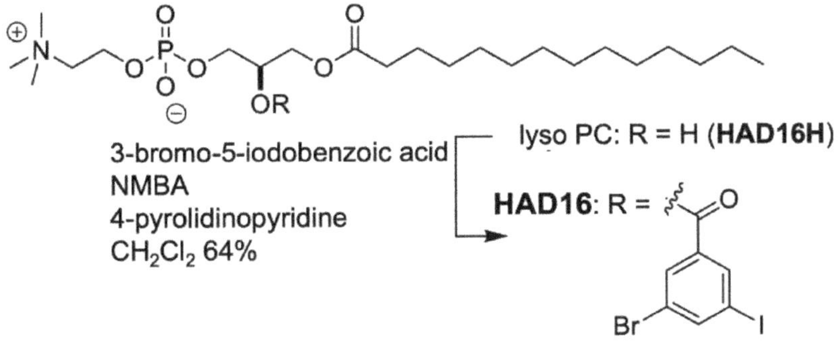

Scheme 4. Synthetic reaction scheme for HAD16. HAD16 was readily produced via a single step synthesis by the esterification. The easy preparation will expand its application range. Details of this synthesis were described in Supplementary Materials.

The HAD16 molecule adopting alternative conformations binds to an interface region among three symmetry-related bacteriorhodopsin molecules (Figure 8). The long alkyl tail of HAD16 interacts with the hydrophobic transmembrane surface. The choline and phosphate group face the bulk solvent. The aromatic ring in the head group ring is stabilized by π–π stacking with Y26 (Figure 8B) and hydrophobic interactions of I/Br atoms with L22, G23, V127, and L221 from a bacteriorhodopsin molecule and W80 from a neighboring molecule. Because a lipid or detergent molecule is located in the corresponding site of the native bacteriorhodopsin structure, we concluded that HAD16 binds to the same site by partially replacing the lipid or detergent. Thus, we modeled lipid/detergent-derived alkyl chains and HAD16 together using partial occupancies. We also collected a dataset from microcrystals soaked with HAD16H lacking the alkyl tail to study the importance of the tail region for bacteriorhodopsin binding (Scheme 4). The anomalous difference map did not show any significant peaks assignable to the compound (data not shown). This establishes that the alkyl chain was indispensable for HAD16 binding to bacteriorhodopsin.

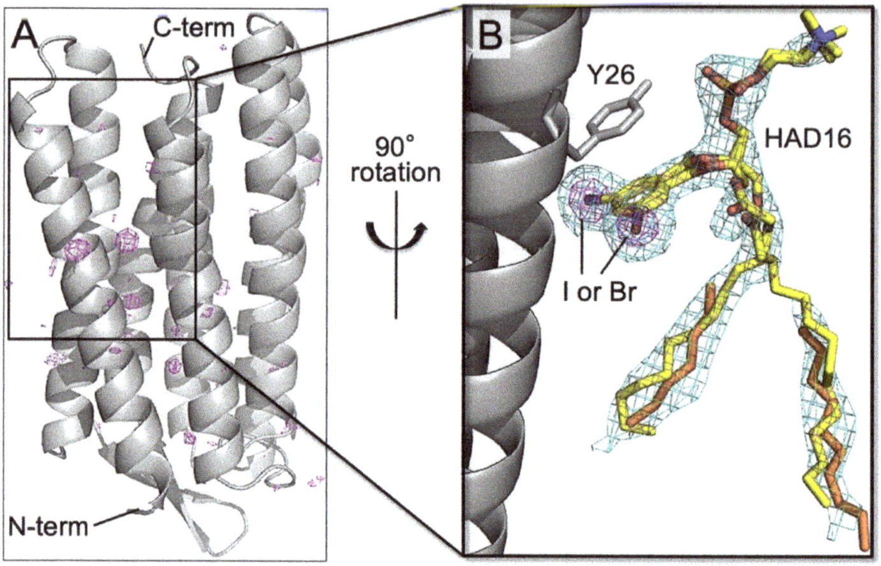

Figure 8. Structure of bacteriorhodopsin in complex with HAD16. (**A**) Final refined structure (gray ribbon model) after structure determination by molecular replacement. Anomalous difference map (purple mesh; contoured at 4.0 σ) is also shown. (**B**) Close-up view of the HAD16 binding site in the final refined structure with $2mF_o$–DF_c map (cyan mesh; 1.0 σ). The C, N, O, I, and Br atoms of HAD16 (stick models), which is assigned as alternative conformations, are colored yellow, blue, red, purple, and brown, respectively. Alkyl chains derived from native lipid or DMPC/CHAPSO bicelles are depicted as orange stick models.

4. Conclusions

Biomembranes are the last frontier in life science and the most challenging subject to study. Biomembranes are composed of a lipid bilayer consisting of phospholipids, glycolipids, sterols, and membrane proteins. We presented here artificial lipids and detergents containing heavy atoms, which can be used as components of model biomembranes for analyzing interactions with membrane proteins. In addition to the examples presented above, there is also a study of a detergent labeled with Hg [75]. In combination with X-ray crystallography, these artificial lipids and detergents can be used to determine novel structures of membrane proteins or to identify the orientation of lipid/detergent molecules by visualizing the position of the heavy atoms using anomalous X-ray scattering. In particular, SFX can observe damage-free structures at physiological temperatures. Future applications of these lipids/detergents for elucidating structure–function relationships of membrane proteins and biomembranes include: (i) labeling of various model lipids to distinguish between outer and inner leaflet regions that bind specifically to membrane proteins; (ii) observation of peripheral lipids surrounding membrane proteins with weak affinity at high resolution; and (iii) visualization of the dynamics of lipid localization on the surface of membrane proteins by time-resolved SFX analysis [76].

Supplementary Materials: The following are available online at https://www.mdpi.com/2077-0375/11/11/823/s1.

Author Contributions: Conceptualization, S.H. and E.M.; chemical synthesis, S.H.; SFX data analysis, T.N. and E.M.; writing manuscript, S.H. and E.M.; and funding acquisition, S.H. and E.M. All authors have read and agreed to the published version of the manuscript.

Funding: This work was supported by the Japan Science and Technology Agency (JST), PRESTO (JPMJPR17GB), a Grant-in-Aid for Scientific Research on Innovative Areas from the Ministry of Education, Culture, Sports, Science, and Technology (MEXT) (19H05780), a Kakenhi Grant

(16H06315) from Japan Society for the Promotion of Science, the ERATO active lipid structure project from JST, and the X-ray Free Electron Laser Priority Strategy Program of MEXT.

Institutional Review Board Statement: Not applicable.

Informed Consent Statement: Not applicable.

Data Availability Statement: Not applicable.

Acknowledgments: The authors gratefully acknowledge Yo Yano and Haruka Saiki for synthetic efforts and Michio Murata for helpful discussion. We also thank the beamline staff at SACLA for technical assistance and the SACLA High Performance Computing system for computational resource.

Conflicts of Interest: The authors declare no conflict of interest.

Abbreviations

Abbreviations	
A2A GPCR	G protein-coupled A2a adenosine receptor
α-GalCer	α-galactosylceramide
β-MeSe-Fuc	selenomethyl derivative of fucose
βMeSe-GlcNAc	Selenomethyl-*N*-acetylglucosamine
CHAPSO	3-[(3-cholamidopropyl)dimethylammonio]-2-hydroxypropanesulfonate
CMC	critical micelle concentration
DDM	dodecyl-β-D-maltoside
DMPC	1,2-dimyristoyl-*sn*-glycero-3-phosphocholine
DPPC	dipalmitoylphosphatidylcholine
ERATO	Exploratory Research for Advanced Technology
I3C	5-amino-2,4,6-triiodoisophthalic acid
JST	Japan Science and Technology Agency
LCLS	Linac Coherent Light Source
LCP	lipidic cubic phase
MAD	multi-wavelength anomalous diffraction
MEXT	Ministry of Education, Culture, Sports, Science and Technology
MIR	multiple heavy-atom isomorphous replacement
MIRAS	multiple heavy-atom isomorphous replacement with anomalous scattering
PAL-XFEL	Pohang Accelerator Laboratory X-ray Free Electron Laser
PDB	Protein Data Bank
PRESTO	Precursory Research for Embryonic Science and Technology
SACLA	SPring-8 Angstrom Compact Free Electron Laser
SAD	single-wavelength anomalous diffraction
SeDDM	dodecyl-β-D-selenomaltoside
SFX	serial femtosecond crystallography
SIR	single heavy-atom isomorphous replacement
SIRAS	single heavy-atom isomorphous replacement with anomalous scattering
SRX	synchrotron radiation crystallography
SwissFEL	Swiss X-ray Free Electron Laser
XFEL	X-ray free-electron laser

References

1. Murata, M.; Sugiyama, S.; Matsuoka, S.; Matsumori, N. Bioactive Structure of Membrane Lipids and Natural Products Elucidated by a Chemistry-Based Approach. *Chem. Rec.* **2015**, *15*, 675–690. [CrossRef] [PubMed]
2. Yamashita, T.; Mizohata, E.; Nagatoishi, S.; Watanabe, T.; Nakakido, M.; Iwanari, H.; Mochizuki, Y.; Nakayama, T.; Kado, Y.; Yokota, Y.; et al. Affinity Improvement of a Cancer-Targeted Antibody through Alanine-Induced Adjustment of Antigen-Antibody Interface. *Structure* **2019**, *27*, 519–527.e515. [CrossRef] [PubMed]

3. Nakayama, T.; Mizohata, E.; Yamashita, T.; Nagatoishi, S.; Nakakido, M.; Iwanari, H.; Mochizuki, Y.; Kado, Y.; Yokota, Y.; Satoh, R.; et al. Structural features of interfacial tyrosine residue in ROBO1 fibronectin domain-antibody complex: Crystallographic, thermodynamic, and molecular dynamic analyses. *Protein Sci. A Publ. Protein Soc.* **2015**, *24*, 328–340. [CrossRef] [PubMed]
4. Sugiyama, A.; Kawamura, T.; Tanaka, T.; Doi, H.; Yamashita, T.; Shinoda, K.; Fujitani, H.; Yamatsugu, K.; Shimizu, Y.; Tatsumi, T.; et al. Cupid and Psyche system for the diagnosis and treatment of advanced cancer. *Proc. Jpn. Acad. Ser. B Phys. Biol. Sci.* **2019**, *95*, 602–611. [CrossRef]
5. Kado, Y.; Mizohata, E.; Nagatoishi, S.; Iijima, M.; Shinoda, K.; Miyafusa, T.; Nakayama, T.; Yoshizumi, T.; Sugiyama, A.; Kawamura, T.; et al. Epiregulin recognition mechanisms by anti-epiregulin antibody 9E5: Structural, functional, and molecular dynamics simulation analyses. *J. Biol. Chem.* **2016**, *291*, 2319–2330. [CrossRef]
6. Uchihashi, T.; Ganser, C. Recent advances in bioimaging with high-speed atomic force microscopy. *Biophys. Rev.* **2020**, *12*, 363–369. [CrossRef] [PubMed]
7. Renard, K.; Byrne, B. Insights into the Role of Membrane Lipids in the Structure, Function and Regulation of Integral Membrane Proteins. *Int. J. Mol. Sci.* **2021**, *22*, 9026. [CrossRef] [PubMed]
8. Mizohata, E.; Nakane, T.; Fukuda, Y.; Nango, E.; Iwata, S. Serial femtosecond crystallography at the SACLA: Breakthrough to dynamic structural biology. *Biophys. Rev.* **2018**, *10*, 209–218. [CrossRef] [PubMed]
9. Hendrickson, W.A.; Teeter, M.M. Structure of the Hydrophobic Protein Crambin Determined Directly from the Anomalous Scattering of Sulfur. *Nature* **1981**, *290*, 107–113. [CrossRef]
10. Agniswamy, J.; Joyce, M.G.; Hammer, C.H.; Sun, P.D. Towards a rational approach for heavy-atom derivative screening in protein crystallography. *Acta Crystallogr. Sect. D Biol. Crystallogr.* **2008**, *64*, 354–367. [CrossRef]
11. Lu, J.; Sun, P.D. A rapid and rational approach to generating isomorphous heavy-atom phasing derivatives. *FEBS J.* **2014**, *281*, 4021–4028. [CrossRef]
12. Allen, F.H.K.O.; Watson, D.G.; Brammer, L.; Orpen, A.G.; Taylor, R. Tables of bond lengths determined by x-ray and neutron diffraction. Part 1. Bond lengths in organic compounds. *J. Chem. Soc. Perkin Trans.* **1987**, *2*, 1–19. [CrossRef]
13. Blanksby, S.J.; Ellison, G.B. Bond dissociation energies of organic molecules. *Acc. Chem. Res.* **2003**, *36*, 255–263. [CrossRef] [PubMed]
14. Dong, Y.Y.; Pike, A.C.; Mackenzie, A.; McClenaghan, C.; Aryal, P.; Dong, L.; Quigley, A.; Grieben, M.; Goubin, S.; Mukhopadhyay, S.; et al. K2P channel gating mechanisms revealed by structures of TREK-2 and a complex with Prozac. *Science* **2015**, *347*, 1256–1259. [CrossRef]
15. Ulens, C.; Spurny, R.; Thompson, A.J.; Alqazzaz, M.; Debaveye, S.; Han, L.; Price, K.; Villalgordo, J.M.; Tresadern, G.; Lynch, J.W.; et al. The prokaryote ligand-gated ion channel ELIC captured in a pore blocker-bound conformation by the Alzheimer's disease drug memantine. *Structure* **2014**, *22*, 1399–1407. [CrossRef] [PubMed]
16. Martick, M.; Scott, W.G. Tertiary contacts distant from the active site prime a ribozyme for catalysis. *Cell* **2006**, *126*, 309–320. [CrossRef]
17. Stagno, J.R.; Yu, P.; Dyba, M.A.; Wang, Y.X.; Liu, Y. Heavy-atom labeling of RNA by PLOR for de novo crystallographic phasing. *PLoS ONE* **2019**, *14*, e0215555. [CrossRef]
18. Czapinska, H.; Winiewska-Szajewska, M.; Szymaniec-Rutkowska, A.; Piasecka, A.; Bochtler, M.; Poznanski, J. Halogen Atoms in the Protein-Ligand System. Structural and Thermodynamic Studies of the Binding of Bromobenzotriazoles by the Catalytic Subunit of Human Protein Kinase CK2. *J. Phys. Chem. B* **2021**, *125*, 2491–2503. [CrossRef]
19. Scholfield, M.R.; Zanden, C.M.; Carter, M.; Ho, P.S. Halogen bonding (X-bonding): A biological perspective. *Protein Sci. A Publ. Protein Soc.* **2013**, *22*, 139–152. [CrossRef]
20. Silvius, J.R. Cholesterol modulation of lipid intermixing in phospholipid and glycosphingolipid mixtures. Evaluation using fluorescent lipid probes and brominated lipid quenchers. *Biochemistry* **1992**, *31*, 3398–3408. [CrossRef]
21. Abrams, F.S.; London, E. Calibration of the parallax fluorescence quenching method for determination of membrane penetration depth: Refinement and comparison of quenching by spin-labeled and brominated lipids. *Biochemistry* **1992**, *31*, 5312–5322. [CrossRef] [PubMed]
22. Bolen, E.J.; Holloway, P.W. Quenching of tryptophan fluorescence by brominated phospholipid. *Biochemistry* **1990**, *29*, 9638–9643. [CrossRef]
23. Halter, M.; Nogata, Y.; Dannenberger, O.; Sasaki, T.; Vogel, V. Engineered lipids that cross-link the inner and outer leaflets of lipid bilayers. *Langmuir* **2004**, *20*, 2416–2423. [CrossRef] [PubMed]
24. Markello, T.; Zlotnick, A.; Everett, J.; Tennyson, J.; Holloway, P.W. Determination of the topography of cytochrome b5 in lipid vesicles by fluorescence quenching. *Biochemistry* **1985**, *24*, 2895–2901. [CrossRef]
25. Cudmore, A.J.; Bradshaw, J.P.; Alecio, M.R. X-Ray-Diffraction Studies Using a Novel Synthetic Phospholipid. *Biophys. Chem.* **1994**, *49*, 71–76. [CrossRef]
26. Roszak, A.W.; Gardiner, A.T.; Isaacs, N.W.; Cogdell, R.J. Brominated lipids identify lipid binding sites on the surface of the reaction center from Rhodobacter sphaeroides. *Biochemistry* **2007**, *46*, 2909–2916. [CrossRef] [PubMed]
27. Campanacci, V.; Lartigue, A.; Hallberg, B.M.; Jones, T.A.; Giudici-Orticoni, M.T.; Tegoni, M.; Cambillau, C. Moth chemosensory protein exhibits drastic conformational changes and cooperativity on ligand binding. *Proc. Natl. Acad. Sci. USA* **2003**, *100*, 5069–5074. [CrossRef]

28. Lautenschlager, C.; Leal, W.S.; Clardy, J. Bombyx mori pheromone-binding protein binding nonpheromone ligands: Implications for pheromone recognition. *Structure* **2007**, *15*, 1148–1154. [CrossRef] [PubMed]
29. Chae, P.S.C.; Cho, K.H.; Bae, H.E. Heavy atom-bearing tripod amphiphiles for the membrane protein study. *New J. Chem.* **2014**, *38*, 2354–2361. [CrossRef]
30. Beck, T.; Krasauskas, A.; Gruene, T.; Sheldrick, G.M. A magic triangle for experimental phasing of macromolecules. *Acta Crystallogr. Sect. D Biol. Crystallogr.* **2008**, *64*, 1179–1182. [CrossRef]
31. Sippel, K.H.; Robbins, A.H.; Reutzel, R.; Domsic, J.; Boehlein, S.K.; Govindasamy, L.; Agbandje-McKenna, M.; Rosser, C.J.; McKenna, R. Structure determination of the cancer-associated Mycoplasma hyorhinis protein Mh-p37. *Acta Crystallogr. Sect. D Biol. Crystallogr.* **2008**, *64*, 1172–1178. [CrossRef] [PubMed]
32. Nakane, T.; Hanashima, S.; Suzuki, M.; Saiki, H.; Hayashi, T.; Kakinouchi, K.; Sugiyama, S.; Kawatake, S.; Matsuoka, S.; Matsumori, N.; et al. Membrane protein structure determination by SAD, SIR, or SIRAS phasing in serial femtosecond crystallography using an iododetergent. *Proc. Natl. Acad. Sci. USA* **2016**, *113*, 13039–13044. [CrossRef]
33. Katzberg, R.W. Urography into the 21st century: New contrast media, renal handling, imaging characteristics, and nephrotoxicity. *Radiology* **1997**, *204*, 297–312. [CrossRef] [PubMed]
34. Yu, S.B.; Watson, A.D. Metal-Based X-ray Contrast Media. *Chem. Rev.* **1999**, *99*, 2353–2378. [CrossRef]
35. Beck, T.; Gruene, T.; Sheldrick, G.M. The magic triangle goes MAD: Experimental phasing with a bromine derivative. *Acta Crystallogr. Sect. D Biol. Crystallogr.* **2010**, *66*, 374–380. [CrossRef] [PubMed]
36. Truong, J.Q.; Nguyen, S.; Bruning, J.B.; Shearwin, K.E. Simplified heavy-atom derivatization of protein structures via co-crystallization with the MAD tetragon tetrabromoterephthalic acid. *Acta Cryst. F Struct. Biol. Commun.* **2021**, *77*, 156–162. [CrossRef]
37. Faham, S.; Boulting, G.L.; Massey, E.A.; Yohannan, S.; Yang, D.; Bowie, J.U. Crystallization of bacteriorhodopsin from bicelle formulations at room temperature. *Protein Sci. A Publ. Protein Soc.* **2005**, *14*, 836–840. [CrossRef] [PubMed]
38. Hendrickson, W.A.; Horton, J.R.; LeMaster, D.M. Selenomethionyl proteins produced for analysis by multiwavelength anomalous diffraction (MAD): A vehicle for direct determination of three-dimensional structure. *EMBO J.* **1990**, *9*, 1665–1672. [CrossRef]
39. Lin, L.; Sheng, J.; Huang, Z. Nucleic acid X-ray crystallography via direct selenium derivatization. *Chem. Soc. Rev.* **2011**, *40*, 4591–4602. [CrossRef] [PubMed]
40. Merino-Montiel, P.L.; López, Ó.; Fernández-Bolaños, J.G. L-Isofucoselenofagomine and derivatives: Dual activities as antioxidants and as glycosidase inhibitors. *Tetrahedron* **2012**, *68*, 3591–3595. [CrossRef]
41. Mangiavacchi, F.; Coelho Dias, I.F.; Di Lorenzo, I.; Grzes, P.; Palomba, M.; Rosati, O.; Bagnoli, L.; Marini, F.; Santi, C.; Lenardao, E.J.; et al. Sweet Selenium: Synthesis and Properties of Selenium-Containing Sugars and Derivatives. *Pharmaceuticals* **2020**, *13*. [CrossRef]
42. Davies, M.J.S.; Schiesser, C.H. 1,4-Anhydro-4-seleno-d-talitol (SeTal): A remarkable selenium-containing therapeutic molecule. *New J. Chem.* **2019**, *43*, 9759–9765. [CrossRef]
43. Buts, L.; Loris, R.; De Genst, E.; Oscarson, S.; Lahmann, M.; Messens, J.; Brosens, E.; Wyns, L.; De Greve, H.; Bouckaert, J. Solving the phase problem for carbohydrate-binding proteins using selenium derivatives of their ligands: A case study involving the bacterial F17-G adhesin. *Acta Crystallogr. Sect. D Biol. Crystallogr.* **2003**, *59*, 1012–1015. [CrossRef]
44. Kostlanova, N.; Mitchell, E.P.; Lortat-Jacob, H.; Oscarson, S.; Lahmann, M.; Gilboa-Garber, N.; Chambat, G.; Wimmerova, M.; Imberty, A. The fucose-binding lectin from Ralstonia solanacearum. A new type of beta-propeller architecture formed by oligomerization and interacting with fucoside, fucosyllactose, and plant xyloglucan. *J. Biol. Chem.* **2005**, *280*, 27839–27849. [CrossRef] [PubMed]
45. Houser, J.; Komarek, J.; Kostlanova, N.; Cioci, G.; Varrot, A.; Kerr, S.C.; Lahmann, M.; Balloy, V.; Fahy, J.V.; Chignard, M.; et al. A soluble fucose-specific lectin from Aspergillus fumigatus conidia–structure, specificity and possible role in fungal pathogenicity. *PLoS ONE* **2013**, *8*, e83077. [CrossRef] [PubMed]
46. Sulak, O.; Cioci, G.; Delia, M.; Lahmann, M.; Varrot, A.; Imberty, A.; Wimmerova, M. A TNF-like trimeric lectin domain from Burkholderia cenocepacia with specificity for fucosylated human histo-blood group antigens. *Structure* **2010**, *18*, 59–72. [CrossRef] [PubMed]
47. Sommer, R.; Hauck, D.; Varrot, A.; Imberty, A.; Kunzler, M.; Titz, A. O-Alkylated heavy atom carbohydrate probes for protein X-ray crystallography: Studies towards the synthesis of methyl 2-O-methyl-L-selenofucopyranoside. *Beilstein. J. Org. Chem.* **2016**, *12*, 2828–2833. [CrossRef] [PubMed]
48. Shimabukuro, J.; Makyio, H.; Suzuki, T.; Nishikawa, Y.; Kawasaki, M.; Imamura, A.; Ishida, H.; Ando, H.; Kato, R.; Kiso, M. Synthesis of seleno-fucose compounds and their application to the X-ray structural determination of carbohydrate-lectin complexes using single/multi-wavelength anomalous dispersion phasing. *Bioorg. Med. Chem.* **2017**, *25*, 1132–1142. [CrossRef] [PubMed]
49. Makyio, H.; Shimabukuro, J.; Suzuki, T.; Imamura, A.; Ishida, H.; Kiso, M.; Ando, H.; Kato, R. Six independent fucose-binding sites in the crystal structure of Aspergillus oryzae lectin. *Biochem. Biophys. Res. Commun.* **2016**, *477*, 477–482. [CrossRef]
50. Suzuki, T.; Hayashi, C.; Komura, N.; Tamai, R.; Uzawa, J.; Ogawa, J.; Tanaka, H.N.; Imamura, A.; Ishida, H.; Kiso, M.; et al. Synthesis and Glycan-Protein Interaction Studies of Se-Sialosides by (77)Se NMR. *Org. Lett.* **2019**, *21*, 6393–6396. [CrossRef] [PubMed]

51. Saino, H.; Ago, H.; Ukita, Y.; Miyano, M. Seleno-detergent MAD phasing of leukotriene C4 synthase in complex with dodecyl-beta-D-selenomaltoside. *Acta Crystallogr. Sect. F Struct. Biol. Cryst. Commun.* **2011**, *67*, 1666–1673. [CrossRef] [PubMed]
52. Sauguet, L.; Poitevin, F.; Murail, S.; Van Renterghem, C.; Moraga-Cid, G.; Malherbe, L.; Thompson, A.W.; Koehl, P.; Corringer, P.J.; Baaden, M.; et al. Structural basis for ion permeation mechanism in pentameric ligand-gated ion channels. *EMBO J.* **2013**, *32*, 728–741. [CrossRef] [PubMed]
53. Fredga, A.L.; Lindgren, A. Selena-hexanoic and Selena-pentanoic Acids. *Acta Chem. Scand.* **1961**, *15*, 938–939. [CrossRef]
54. Golmohammadi, R. Studies on the methods of preparation of di- and tri-selena-straight-chain fatty acids and corresponding amides, with some remarks on the infrared spectra of the mono-, di- and tri-selena-fatty acids. *Acta Chem. Scand.* **1966**, *20*, 563–571. [CrossRef]
55. Sadek, S.A.B.; Basmadjian, G.P. Selenium labeled fatty acids as potential myocardial imaging agents. *J. Label. Compd. Radiopharm.* **1983**, *20*, 487–494. [CrossRef]
56. Yanishlieva, N.R.; Marinova, E.; Houte, H.; Partali, V.; Sliwka, H.R. 11-Selenadodecylglyceryl-1-ether in lipid autoxidation. *J. Am. Oil Chem. Soc.* **2001**, *78*, 691–696. [CrossRef]
57. Houte, H.; Partali, V.; Sliwka, H.R.; Quartey, E.G. Synthesis of structured lipids and etherlipids with antioxidants: Combination of a selena fatty acid and a selena fatty alcohol with a carotenoic acid in glyceride molecules. *Chem. Phys. Lipids* **2000**, *105*, 105–113. [CrossRef]
58. Kabara, J.J.; Vrable, R. Antimicrobial lipids: Natural and synthetic fatty acids and monoglycerides. *Lipids* **1977**, *12*, 753–759. [CrossRef] [PubMed]
59. Hossain, M.I.; Hanashima, S.; Nomura, T.; Lethu, S.; Tsuchikawa, H.; Murata, M.; Kusaka, H.; Kita, S.; Maenaka, K. Synthesis and Th1-immunostimulatory activity of alpha-galactosylceramide analogues bearing a halogen-containing or selenium-containing acyl chain. *Bioorg. Med. Chem.* **2016**, *24*, 3687–3695. [CrossRef] [PubMed]
60. Emma, P.; Akre, R.; Arthur, J.; Bionta, R.; Bostedt, C.; Bozek, J.; Brachmann, A.; Bucksbaum, P.; Coffee, R.; Decker, F.J.; et al. First lasing and operation of an angstrom-wavelength free-electron laser. *Nat. Photonics* **2010**, *4*, 641–647. [CrossRef]
61. Ishikawa, T.; Aoyagi, H.; Asaka, T.; Asano, Y.; Azumi, N.; Bizen, T.; Ego, H.; Fukami, K.; Fukui, T.; Furukawa, Y.; et al. A compact X-ray free-electron laser emitting in the sub-angstrom region. *Nat. Photonics* **2012**, *6*, 540–544. [CrossRef]
62. Decking, W.; Abeghyan, S.; Abramian, P.; Abramsky, A.; Aguirre, A.; Albrecht, C.; Alou, P.; Altarelli, M.; Altmann, P.; Amyan, K.; et al. A MHz-repetition-rate hard X-ray free-electron laser driven by a superconducting linear accelerator. *Nat. Photonics* **2020**, *14*, 391–397. [CrossRef]
63. Kang, H.S.; Min, C.K.; Heo, H.; Kim, C.; Yang, H.; Kim, G.; Nam, I.; Baek, S.Y.; Choi, H.J.; Mun, G.; et al. Hard X-ray free-electron laser with femtosecond-scale timing jitter. *Nat. Photonics* **2017**, *11*, 708–713. [CrossRef]
64. Prat, E.; Abela, R.; Aiba, M.; Alarcon, A.; Alex, J.; Arbelo, Y.; Arrell, C.; Arsov, V.; Bacellar, C.; Beard, C.; et al. A compact and cost-effective hard X-ray free-electron laser driven by a high-brightness and low-energy electron beam. *Nat. Photonics* **2020**, *14*, 748–754. [CrossRef]
65. Chapman, H.N.; Fromme, P.; Barty, A.; White, T.A.; Kirian, R.A.; Aquila, A.; Hunter, M.S.; Schulz, J.; DePonte, D.P.; Weierstall, U.; et al. Femtosecond X-ray protein nanocrystallography. *Nature* **2011**, *470*, 73–77. [CrossRef] [PubMed]
66. Tono, K.; Nango, E.; Sugahara, M.; Song, C.; Park, J.; Tanaka, T.; Tanaka, R.; Joti, Y.; Kameshima, T.; Ono, S.; et al. Diverse application platform for hard X-ray diffraction in SACLA (DAPHNIS): Application to serial protein crystallography using an X-ray free-electron laser. *J. Synchrotron Radiat.* **2015**, *22*, 532–537. [CrossRef]
67. Fukuda, Y.; Tse, K.M.; Nakane, T.; Nakatsu, T.; Suzuki, M.; Sugahara, M.; Inoue, S.; Masuda, T.; Yumoto, F.; Matsugaki, N.; et al. Redox-coupled proton transfer mechanism in nitrite reductase revealed by femtosecond crystallography. *Proc. Natl. Acad. Sci. USA* **2016**, *113*, 2928–2933. [CrossRef]
68. Barty, A.; Caleman, C.; Aquila, A.; Timneanu, N.; Lomb, L.; White, T.A.; Andreasson, J.; Arnlund, D.; Bajt, S.; Barends, T.R.; et al. Self-terminating diffraction gates femtosecond X-ray nanocrystallography measurements. *Nat. Photonics* **2012**, *6*, 35–40. [CrossRef] [PubMed]
69. Barends, T.R.; Foucar, L.; Botha, S.; Doak, R.B.; Shoeman, R.L.; Nass, K.; Koglin, J.E.; Williams, G.J.; Boutet, S.; Messerschmidt, M.; et al. De novo protein crystal structure determination from X-ray free-electron laser data. *Nature* **2014**, *505*, 244–247. [CrossRef] [PubMed]
70. Yamashita, K.; Pan, D.; Okuda, T.; Sugahara, M.; Kodan, A.; Yamaguchi, T.; Murai, T.; Gomi, K.; Kajiyama, N.; Mizohata, E.; et al. An isomorphous replacement method for efficient de novo phasing for serial femtosecond crystallography. *Sci. Rep.* **2015**, *5*, 14017. [CrossRef] [PubMed]
71. Nakane, T.; Song, C.; Suzuki, M.; Nango, E.; Kobayashi, J.; Masuda, T.; Inoue, S.; Mizohata, E.; Nakatsu, T.; Tanaka, T.; et al. Native sulfur/chlorine SAD phasing for serial femtosecond crystallography. *Acta Crystallogr. Sect. D Biol. Crystallogr.* **2015**, *71*, 2519–2525. [CrossRef] [PubMed]
72. Colletier, J.P.; Sawaya, M.R.; Gingery, M.; Rodriguez, J.A.; Cascio, D.; Brewster, A.S.; Michels-Clark, T.; Hice, R.H.; Coquelle, N.; Boutet, S.; et al. De novo phasing with X-ray laser reveals mosquito larvicide BinAB structure. *Nature* **2016**, *539*, 43–47. [CrossRef] [PubMed]
73. Hunter, M.S.; Yoon, C.H.; DeMirci, H.; Sierra, R.G.; Dao, E.H.; Ahmadi, R.; Aksit, F.; Aquila, A.L.; Ciftci, H.; Guillet, S.; et al. Selenium single-wavelength anomalous diffraction de novo phasing using an X-ray-free electron laser. *Nat. Commun.* **2016**, *7*. [CrossRef]

74. Batyuk, A.; Galli, L.; Ishchenko, A.; Han, G.W.; Gati, C.; Popov, P.A.; Lee, M.Y.; Stauch, B.; White, T.A.; Barty, A.; et al. Native phasing of x-ray free-electron laser data for a G protein-coupled receptor. *Sci. Adv.* **2016**, *2*, e1600292. [CrossRef] [PubMed]
75. Cogdell, R.J.; Gardiner, A.T.; Roszak, A.W.; Stoncius, S.; Kocovsky, P.; Isaacs, N.W. Mapping lipid and detergent molecules at the surface of membrane proteins. *Biochem. Soc. Trans.* **2011**, *39*, 775–779. [CrossRef] [PubMed]
76. Nango, E.; Royant, A.; Kubo, M.; Nakane, T.; Wickstrand, C.; Kimura, T.; Tanaka, T.; Tono, K.; Song, C.; Tanaka, R.; et al. A three-dimensional movie of structural changes in bacteriorhodopsin. *Science* **2016**, *354*, 1552–1557. [CrossRef]

membranes

MDPI

Article

Cholesterol Modulates the Interaction between HIV-1 Viral Protein R and Membrane

Chun-Hao Liu [1,2,3], Shing-Jong Huang [4,*] and Tsyr-Yan Yu [3,5,*]

1 Chemical Biology and Molecular Biophysics, Taiwan International Graduate Program, Academia Sinica, Taipei 11529, Taiwan; gerard1057@gmail.com
2 Institute of Bioinformatics and Structural Biology, National Tsing Hua University, Hsin Chu 300044, Taiwan
3 Institute of Atomic and Molecular Sciences, Academia Sinica, Taipei 10617, Taiwan
4 Instrumentation Center, National Taiwan University, Taipei 10617, Taiwan
5 International Graduate Program of Molecular Science and Technology, National Taiwan University, Taipei 10617, Taiwan
* Correspondence: shingjonghuang@ntu.edu.tw (S.-J.H.); tyyu@pub.iams.sinica.edu.tw (T.-Y.Y.); Tel.: +886-2-3366-1152 (S.-J.H.); +886-2-2362-8210 (T.Y.Y.)

Abstract: Being a major metabolite for maintaining cellular homeostasis, as well as an important structural component in lipid membrane, cholesterol also plays critical roles in the life cycles of some viruses, including human immunodeficiency virus-1 (HIV-1). The involvement of cholesterol in HIV-1 infectivity, assembly and budding has made it an important research target. Viral protein R (Vpr) is an accessory protein of HIV-1, which is involved in many major events in the life cycle of HIV-1. In addition to its multi-functional roles in the HIV-1 life cycle, it is shown to interact with lipid membrane and form a cation-selective channel. In this work, we examined the effect of cholesterol on the interaction of Vpr and lipid membrane. Using calcein release assay, we found that the membrane permeability induced by the membrane binding of Vpr was significantly reduced in the presence of cholesterol in membrane. In addition, using solid-state NMR (ssNMR) spectroscopy, Vpr was shown to experience multiple chemical environments in lipid membrane, as indicated by the broad line shape of carbonyl ^{13}C resonance of Cys-76 residue ranging from 165–178 ppm, which can be attributed to the existence of complex Vpr-membrane environments. We further showed that the presence of cholesterol in membrane will alter the distribution of Vpr in the complex membrane environments, which may explain the change of the Vpr induced membrane permeability in the presence of cholesterol.

Keywords: HIV-1; Vpr; cholesterol; lipid composition; membrane; calcein release; REDOR; NMR

1. Introduction

Cholesterol is generally viewed as a structural component in cell membranes, which plays an important role in membrane stability and fluidity. It is also an essential precursor for the biosynthesis of several steroid hormones and bile acids [1]. Recently, the roles of cholesterol in the life cycles of various viruses, in the aspects of viral infectivity, assembly, and structural stability, have attracted significant attention [2]. High cholesterol content of influenza virus particles was reported to be a critical factor in the viral fusion process [3]. Lipid rafts, containing high cholesterol levels, were shown to be essential in the assembly of viral particles of parainfluenza virus type 3 [4]. For HIV-1, the viral particle contains high levels of cholesterol [5,6] and the involvement of cholesterol in HIV-1 infectivity, assembly, budding, etc., has been reported [7–13].

HIV-1 viral protein R (Vpr) is a functional accessory protein, which is involved in many aspects of the viral life cycle, including the nuclear import of the pre-integration complex [14–16], cell cycle arrest at G2 phase [17], cell apoptosis [18–20], transactivation of the long-terminal repeat (LTR) genes [21], etc. [22–25]. In addition to its functional roles in HIV-1 viral cycle, the interactions between Vpr and lipid membrane also draw significant attention. Vpr could form a cation-selective channel on lipid membrane [23]. The peptides derived from Vpr sequence were demonstrated to behave as cell-penetrating peptides (CPPs), showing the potential applications such as DNA transfection and gene delivery [26–28]. Previously, we reported a novel strategy for producing Vpr protein in large quantity using an *E. coli* expression system and further demonstrated that Vpr has a stronger binding affinity to DOPG lipids compared to DOPC lipids, suggesting a lipid-composition dependent of Vpr-membrane interaction [29]. During the HIV-1 virus particle assembly at the plasma membrane, lipid rafts containing high amount of cholesterol are targeted by Gag proteins [9,10]. In addition, depletion of cellular cholesterol significantly reduces HIV-1 particle production [30]. Once the viral particles are released out from the lipid raft, a higher cholesterol content in the HIV-1 virus envelope may benefit to its stability [31]. Moreover, significant amount of Vpr proteins are packaged into the virus particle by Vpr/Gag ratio to be around 1:7 [32], which contribute to the early infection of the virus. Its association with the p6 protein is essential for the translocation from the nuclear membrane to the plasma membrane as well as packaging into the virus particle [33,34]. Thus, it is reasonable to investigate if cholesterol, being an important structural component in the plasma membrane, may play a role in the interaction of Vpr and membrane. We first used calcein release assay to probe the effect of cholesterol on Vpr induced membrane permeability and showed that the Vpr induced membrane permeability was reduced as the content of cholesterol in membrane was increased. This result is similar to what were observed from the studies of a pore-forming peptide [35] and CPPs [36]. The analysis using circular dichroism (CD) spectroscopy showed that no significant change of the secondary structure of Vpr was observed with the addition of cholesterol in Vpr proteoliposomes. NMR spectroscopy is a straightforward tool for acquiring residue specific information of the binding event. In addition, the chemical shift of a NMR resonance signal is a result of local electron shielding, and thus, it reflects the local chemical environment. Solution NMR was used here to examine if Vpr protein has specific interaction with cholesterol. As a result, we did not observe specific binding of cholesterol to Vpr in DPC micelles. Even though cholesterol does not bind to Vpr specifically, nor affect the secondary structure directly, cholesterol may still modulate the Vpr-membrane interaction by changing the protein-lipid complex arrangement. REDOR NMR is a powerful solid-state NMR method for providing the distance information between two hetero nuclei. Through our ssNMR study of [1-^{13}C, 99%] Cysteine selective labeled Vpr embedded in liposome, we showed the existence of complex Vpr-membrane environments and the distribution of Vpr among these environments was altered with the presence of cholesterol in membrane. We believe that more insights into cholesterol modulating Vpr-membrane interaction can be revealed by other methods, such as computational simulations. In summary, we show that cholesterol, without interacting with Vpr directly, can affect the Vpr-membrane interaction.

2. Materials and Methods

2.1. Materials

1,2-dioleoyl-sn-glycero-3-phospho-(1′-rac-glycerol) (DOPG) and 1,2-dioleoyl-sn-glycero-3-phosphocholine (DOPC) were purchased from Avanti Polar Lipids (Alabaster, AL, USA). [^{15}N, 99%] ammonium chloride, [U-^{13}C, 99%] D-glucose, and [1,1′-$^{13}C_2$, 99%] L-Cystine were purchased from Cambridge Isotope Laboratories (Andover, MA, USA). [$^2H_{25}$, 98%] dodecylphosphocholine (DPC) was purchased from Frankenstein Bio Reagents, (Cambridge, MA, USA). Chloroform was purchased from Avantor, Radnor, PA, USA. 2,2,2-Trifluoroethanol (TFE) was purchased from Alfa Aesar, Hersham, UK. 2,2,2-Trifluoroacetic acids (TFA) was purchased from ACROS Organics, Morris Plains, NJ, USA. Deuterium oxide (D_2O), Triton X-100 (TX-100), calcein, cholesterol, 3-(trimethylsilyl)-1-propanesulfonic acid sodium salt (DSS) were purchased from Sigma Aldrich, St. Louis, MO, USA. ISOGRO was purchased from Silantes, Munich, DE or Sigma Aldrich, St. Louis, MO, USA.

2.2. Vpr Protein Production and Purification

We followed the established protocol [29] to produce, purify and characterize HIV-1 Vpr protein. Briefly, an overnight culture of BL21(DE3) *E. coli* (RBC Bioscience, Birmingham, UK), transformed with His-tagged B1 domain of streptococcal protein g (GB1)-fused Vpr construct, was prepared at 37 °C in LB medium, and then sub-cultured at 20 °C in a defined growth medium described in the supporting information of reference number 29. To prepare [U-^{2}H,^{13}C,^{15}N] labeled Vpr protein, 1 g of [^{15}N, 99%] ammonium chloride, 2 g of [U-^{13}C, 99%] D-glucose and 1 L of D_2O (98%) per liter culture were used to replace the natural abundance chemicals in the growth medium. This culture was also supplemented with the 1 g of [U-^{13}C, ^{15}N] ISOGRO per liter after induction with 0.1 mM isopropyl β-D-thiogalactoside (IPTG). To prepare [1-^{13}C, 99%] Cysteine selective labeled Vpr protein, a mixture of amino acids and nucleobases was used to replace the ammonium chloride in the defined growth medium. The mixture of amino acids and nucleobases contains 125 mg of [1,1′-$^{13}C_2$, 99%] L-Cystine, 200 mg of each of the rest of nineteen non-labeled amino acids, and the five nucleobases, including 125 mg of adenine, 125 mg of guanine, 125 mg of thymine, 125 mg of cytosine, and 50 mg thymine per liter.

The culture was induced with 0.1 mM IPTG, while the OD_{600} reached the range of 0.7–0.9, and the culture was continued to grow at 18 °C for 24 h to allow for the expression of the soluble His-tagged GB1-fused Vpr protein. Pure Vpr protein precipitates can be obtained following the established protocol [29]. The Vpr protein was examined by sodium dodecyl sulfate polyacrylamide gel electrophoresis (SDS-PAGE) and matrix-assisted laser desorption ionization-time of flight (MALDI-TOF) mass spectra (microflex LRF, Bruker, MA, USA) to ensure the purity as shown in an example in Figure S1.

2.3. DPC Solubilized Vpr Sample for Sequence-Specific Resonance Assignments

The Vpr protein precipitates were obtained by treating [U-^{2}H,^{13}C,^{15}N] labeled His-tagged GB1-fused Vpr with tobacco etch virus (TEV) protease to remove His-tagged GB1 tag. The [U-^{2}H,^{13}C,^{15}N] labeled Vpr protein precipitates dissolved in a denaturing buffer, consisting of 50 mM sodium phosphate at pH 6.5, 50 mM sodium chloride (NaCl), 0.5 mM tris(2-carboxyethyl)phosphine (TCEP), 1 mM ethylenediaminetetraacetic acid (EDTA), and 6 M guanidine hydrochloride (GdnHCl), to the concentration of 1 mg/mL. The protein solution was then added in a dropwise manner at a flow rate of 50 µL/min to a refolding buffer, consisting of 50 mM sodium phosphate at pH 6.5, 50 mM NaCl, 0.5 mM TCEP, 1 mM EDTA, and 9 mM of the tail deuterated [$^{2}H_{25}$, 98%] DPC. The protein solution was further dialyzed against a buffer containing 25 mM sodium phosphate at pH 6.5, 25 mM NaCl, 1 mM dithiothreitol (DTT), and 1 mM EDTA at 4 °C using a 6–8 kDa molecular weight cut-off (MWCO) dialysis membrane (Orange Scientific, Braine-l'Alleud, Belgium). The protein solution was concentrated and further purified by size exclusion chromatography using a Superdex increase s200 (10/300) GL column (Cytiva, Marlborough, MA, USA) equilibrated with a buffer containing 25 mM sodium phosphate at pH 6.5, 25 mM NaCl, 1 mM TCEP, 1 mM EDTA, and 1.8 mM [$^{2}H_{25}$, 98%] DPC. The NMR samples were further concentrated with 5 kDa molecular weight cut-off (MWCO) centricon (Vivaspin, Göttingen, Germany). The samples for the 3D triple resonance experiments and for the cholesterol titration experiment were of the concentrations of 552 µM and 280 µM, respectively. D_2O, containing DSS at the concentration of 20 mg/mL, was included in each NMR sample to 6% (wt %).

2.4. Vpr Proteoliposomes Preparation

To produce Vpr incorporated in DOPG liposomes, DOPG lipid was dissolved in chloroform (50 mg/mL) and dried on the side of a round bottom flask using a gentle stream of nitrogen gas. To produce Vpr incorporated in DOPG containing 5%, 10%, and 30% of cholesterol liposomes, the lipid mixes were first solubilized in chloroform with the molar ratio of DOPG lipid to cholesterol as 95 to 5, 90 to 10 and 70 to 30, respectively. Vpr protein precipitates were dissolved in a TFE solution containing 0.1% TFA to the concentration of 1 mg/mL, and then added to the dried lipid film containing flask. The lipid and protein mixture were prepared with the molar ratio of lipid to Vpr as 50 to 1. A gentle stream of nitrogen gas was used to remove the organic solvent, followed by placing the flask under vacuum overnight. The protein-lipid film formed on the glass wall was re-suspended in a buffer containing 5 mM sodium phosphate at pH 7.5, 5 mM potassium sulfate, and 1 mM TCEP. The Vpr proteoliposomes were further homogenized using an extruder with polycarbonate membrane with a defined pore size of 0.4 μm in diameter. For secondary structure analyses using circular dichroism spectroscopy, the Vpr proteoliposomes were purified by size exclusion chromatography using an ENrich 70 (10/300) column (Bio-Rad, Hercules, CA, USA) in a buffer containing 5 mM sodium phosphate at pH 7.5, 5 mM potassium sulfate, and 1 mM TCEP. To prepare the samples for the ssNMR experiment, [1-^{13}C] cysteine selective labeled Vpr protein was used in the proteoliposomes production, and a DOPG liposome without Vpr protein incorporated was also prepared as a reference ssNMR sample. The size distribution of Vpr proteoliposomes were characterized by dynamic light scattering to ensure the size homogeneity (DelsaNano C particle size and zeta potential analyzer, Beckman Coulter, Indianapolis, IN, USA), as indicated in Figure S2A–F.

2.5. Calcein Release Assay

We used calcein release assay to study the Vpr induced membrane permeabilization of liposomes [29]. When the membrane-impermeable fluorescent probe calcein is encapsulated in liposome, the fluorescence signal is low due to the self-quenching effect of calcein at high concentration. The fluorescence signal will increase accompanied by the release of calcein from liposomes. To examine the effect of cholesterol on the Vpr induced membrane permeabilization, we prepared liposomes containing various amount of cholesterol. The four different liposomes were prepared with the four different molar ratios of cholesterol to lipid, including 20 to 80, 30 to 70, 40 to 60 and 50 to 50, while an equal amount of DOPC lipid and DOPG lipid were used in each liposome preparation. The liposomes were prepared in a buffer containing 25 mM Tris at pH 7.5, 25 mM NaCl, 1 mM TCEP, and 0.5 mM EDTA. The size homogeneity of calcein encapsulated liposomes with increasing cholesterol content were confirmed using dynamic light scattering as shown in Figure S2G–J.

Time-resolved fluorescence spectra were recorded using a luminescence spectrometer (LS55, PerkinElmer, Waltham, MA, USA), to monitor the increase of the fluorescence signal due to the release of calcein from liposomes. The fluorescence spectra at each time point after the addition of His-tagged GB1-fused Vpr to the liposomes solution were recorded from 505 to 535 nm with the excitation light at the wavelength of 495 nm. The interval between each scan was 30 s and the total monitoring time for each experiment was 40 min. The positive control experiments were carried out by recording the time-resolved fluorescence spectra for 40 min after mixing the liposomes solution with a buffer containing 25 mM Tris at pH 7.5, 25 mM NaCl, 1 mM TCEP, 0.5 mM EDTA, and 0.3% TX-100, where the negative control experiments were carried out by monitoring the time-resolved fluorescence spectra of liposomes solution without the addition of His-tagged GB1-fused Vpr or the buffer containing TX-100. Each experimental condition was performed with three replicates.

The calcein release kinetics, quantified by the fluorescence intensity at 517 nm, were quantified using Equation (1).

$$calcein\ release(t)\% = \frac{I_{Vpr}(t)\ -\ I_{buffer}(t)}{I_{TX-100}(t)\ -\ I_{buffer}(t)} \times 100\% \quad (1)$$

where $I_{Vpr}(t)$ is the fluorescence intensity at the time t after the addition of His-tagged GB1-fused Vpr to the liposome solution, $I_{buffer}(t)$ the reference fluorescence intensity at the time t, and $I_{TX\text{-}100}(t)$ the fluorescence intensity at the time t after the addition of TX-100 containing buffer.

2.6. Spectroscopic Characterizations

The CD spectra between 190–260 nm were recorded using Jasco CD spectrometer (J-815, Tokyo, Japan) with a cuvette of 1 mM path length. For each CD spectrum, 8 scans were accumulated and averaged, with the bandwidth of 2 nm, the scanning speed at 50 nm/min and the step resolution of 0.2 nm. The secondary structure content was analyzed by DichroWeb using CDSSTR program and SP175 as reference data set, which was optimized for the range of 190–240 nm. The lowest data point used in the analysis was set to 190 nm, and the scaling factor was 0.95.

For sequence-specific resonance assignment, a set of standard NMR spectra, including 2D [^{1}H,^{15}N] TROSY-HSQC and 3D TROSY-HNCA, 3D TROSY-HNCOCA, 3D TROSY-HNCACB spectra of [U-^{2}H,^{13}C,^{15}N] Vpr solubilized in [$^{2}H_{25}$, 98%] DPC detergent micelles were recorded at 37 °C using Bruker AVANCE III 850 MHz NMR spectrometer (Bruker BioSpin, Billerica, MA, USA) equipped with a TCI (^{1}H/^{13}C/^{15}N) 5mm CryoProbe with Z gradient. The assignment was performed with computer aided resonance assignment (CARA) program [37]. A Vpr sample with the addition of cholesterol at the ratio of cholesterol to Vpr as 1 to 2.6 was prepared and then a 2D [^{1}H,^{15}N] TROSY-HSQC spectrum was recorded to examine if cholesterol binds specifically to Vpr. To further increase the amount of cholesterol in the sample, additional cholesterol was first dissolved in ethanol to the concentration of 6 mg/mL due to the low solubility of cholesterol in DPC buffer. Then, 10 µL of the cholesterol in ethanol was added to the Vpr sample reaching the final ratio of cholesterol to Vpr as 2.4 to 1. An additional 2D [^{1}H,^{15}N] TROSY-HSQC spectrum was recorded for the comparison with cholesterol-free spectrum.

To prepare the samples for ssNMR studies, proteoliposome pellets were first collected by ultracentrifugation (Optima XE-90, Beckman Coulter, Indianapolis, IN, USA) at 200,000 g for 4 h. Each pellet was then transferred into a 3.2 mM rotor with the aid of a homemade centrifuge adapter using ultracentrifugation at 200,000 g for 4 h.

All ssNMR experiments were carried out on a Bruker wide-bore 11.7-T Avance III 500 MHz spectrometer (Bruker BioSpin, Billerica, MA, USA) equipped with a 3.2 mM triple-resonance magic angle spinning (MAS) probehead. The samples were cooled down with a nitrogen gas stream under heat exchanging in a 25 L liquid nitrogen dewar and the sample temperature was ca. 233 K. The sample spinning rate was 8 kHz. The ^{13}C {^{31}P} rotational-echo double-resonance (REDOR) sequence comprises cross polarization (CP) of the ^{13}C spin from proton, followed by a series of ^{13}C π pulses applied at the end of each rotor cycle for rotational echoes, and a series of ^{31}P π pulses at the middle of each rotor cycle for dipolar dephasing. XY-8 phase cycling was applied to both π-pulse chains. The rf field strengths for ^{13}C and ^{31}P π pulse was 35 and 50 kHz, respectively. ^{1}H continuous wave (CW) and two-pulse phase –modulated (TPPM) decoupling schemes of 70 kHz were applied during the π-pulse chains and signal acquisition, respectively. The recycle delay was 2 s and the CP contact time was 1.5 ms.

3. Results and Discussions

3.1. Secondary Structure Analysis

In our previous work [29], we showed that Vpr solubilized in DPC detergent micelles exhibited a significant fraction of α-helix features. Here, we examined the secondary structure content of Vpr incorporated into lipid bilayer environment. As mentioned in the experimental section, Vpr proteoliposomes were produced by dissolving the protein-lipid film with a buffer and then homogenized with an extruder. The protein concentration in the liposome solution was determined using the extinction coefficient at 280 nm as 19,480 $M^{-1}cm^{-1}$, calculated using ExPasy server [38]. We found the incorporation yield of Vpr in proteoliposomes made with DOPG lipid was in a range from 30% to 50%. No significant difference in terms of the Vpr incorporation yield was observed in cholesterol containing proteoliposomes. In contrast, the incorporation yield of Vpr in proteoliposomes made with DOPC alone was not detectable. This observation demonstrated the selective membrane incorporation of Vpr, which was consistent with our previous observation.

The CD spectrum of Vpr incorporated in DOPG proteoliposomes is shown in Figure 1 and the calculated secondary structure content analyzed by DichroWeb using the CDSSTR program and SP175 as a reference set [39] was shown in Table 1. The secondary structure of Vpr incorporated in the DOPG proteoliposomes was found to have similar α-helical structure content as in DPC detergent micelle reported previously [29]. We further included 5% and 10% cholesterol in the DOPG proteoliposomes and examine the secondary structure of the incorporated Vpr. As shown in the CD spectra, Figure 2, we found that similar α-helical feature was still preserved. Therefore, the secondary structure of Vpr in lipid bilayer environment do not change significantly in the presence of cholesterol.

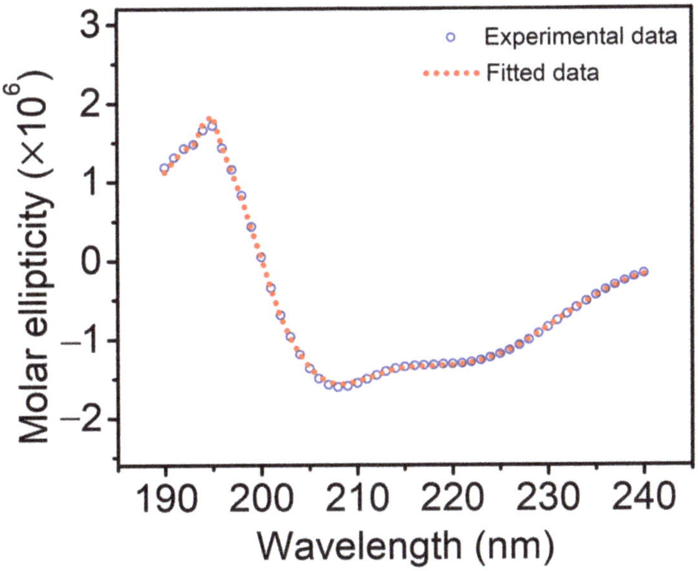

Figure 1. The experimental (open circles) and fitted (dotted line) CD spectra of Vpr incorporated in DOPG proteoliposomes.

Table 1. Calculated secondary structure fractions of Vpr incorporated in DOPG proteoliposomes.

Helix 1	Helix 2	Strand 1	Strand 2	Turns	Unordered
0.31	0.20	0.03	0.04	0.11	0.32

The secondary structure content was analyzed by DichroWeb using the CDSSTR program and SP175 as a reference set which was optimized for 190–240 nm.

3.2. Calcein Release Assay

In our previous work, we showed that membrane permeabilization caused by the Vpr binding to membrane was lipid composition dependent. We observed that the calcein release kinetics was increased when the ratio of DOPG in membrane was increased. In the present work, we examined if the concentration of cholesterol in lipid bilayer membrane affects Vpr induced membrane permeabilization. Calcein release assay in Figure 3 reveals that the calcein release kinetics was decreased as the cholesterol concentration in membrane was increased, suggesting that the cholesterol may play a role to modulate the Vpr-membrane interaction. Our result is similar to the previous studies, where the presence of cholesterol in membrane was shown to reduce the membrane permeability induced by a pore-forming peptide [35] and CPPs [36].

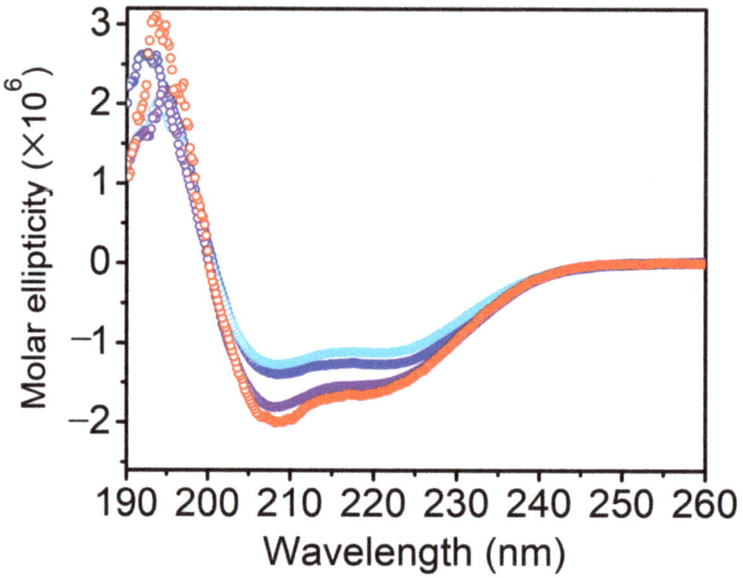

Figure 2. The CD spectra of Vpr solubilized in DPC detergent micelles (red), in DOPG proteoliposomes (purple), and DOPG proteoliposomes in the presence of 5% (blue) and 10% cholesterol (light blue). The ellipticity corresponds to the α-helical content of Vpr is similar in DPC and in DOPG liposome. Vpr maintains similar α-helical structure content in different environments.

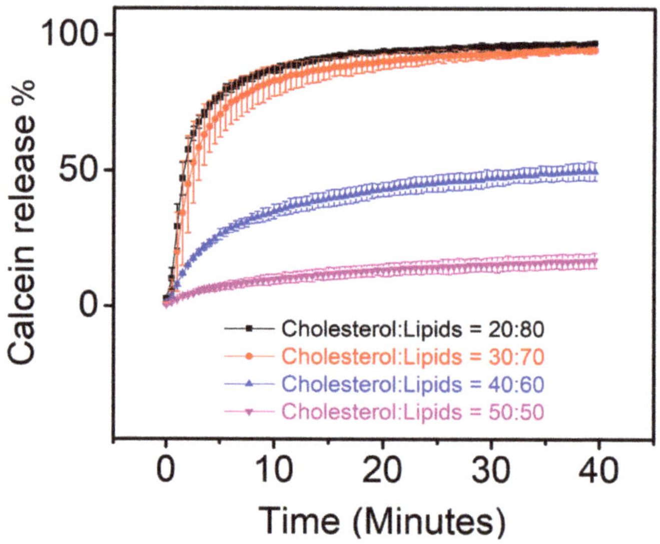

Figure 3. Cholesterol concentration dependent permeabilization of liposomes induced by the membrane binding of His-tagged GB1-fused Vpr. The molar ratio of cholesterol to lipids (DOPC:DOPG = 1:1) in liposomes was increased from 20% to 50%. The reported values are the averages of the triplicate experiments and the standard deviation of each data point is represented as the error bars.

3.3. NMR Characterization

As indicated in calcein release assay, membrane permeability was induced by the binding of Vpr to membrane. The fact that cholesterol plays a role in modulating membrane permeability indicates cholesterol alter the interaction between Vpr and membrane. Here, we attempted to employ solution NMR spectroscopy to examine if cholesterol binds to Vpr specifically. We prepared [U-^{2}H,^{13}C,^{15}N] labeled Vpr protein solubilized in [$^{2}H_{25}$, 98%] DPC detergent micelles for solution NMR characterizations. The 2D [^{1}H,^{15}N] TROSY-HSQC spectrum of Vpr in [$^{2}H_{25}$, 98%] DPC micelles in Figure 4a shows that Vpr exhibited significant fractions of α-helix content consistent with the result of CD characterization. It is worth mentioning that the 2D [^{1}H,^{15}N] TROSY-HSQC spectrum of Vpr stayed unchanged for several weeks, suggesting the sample was of very good stability. The strip plot of TROSY-HNCA and TROSY-HNCACB spectra with the residues assigned from W54 to A59 is shown in Figure 4b to demonstrate the feasibility of protein backbone assignments. By using Talos+ [40], the secondary structure for the segment from W54 to A59 could be predicted based on their chemical shifts. The predictions obtained were well fitted as left-handed α-helical structures in the Ramachandran plot (Figure S3), which is consistent with the structure solved in the protein data bank [41]. No significant chemical shift perturbation was observed when the sample was titrated with cholesterol, as shown in Figure 4a, indicating no evidence of specific cholesterol binding to Vpr solubilized in DPC detergent micelles. The saturation transfer difference (STD) type of experiment [42,43] was not performed due to the overlapping signals of cholesterol and DPC micelles and the low concentration of cholesterol in sample due to the poor solubility of cholesterol.

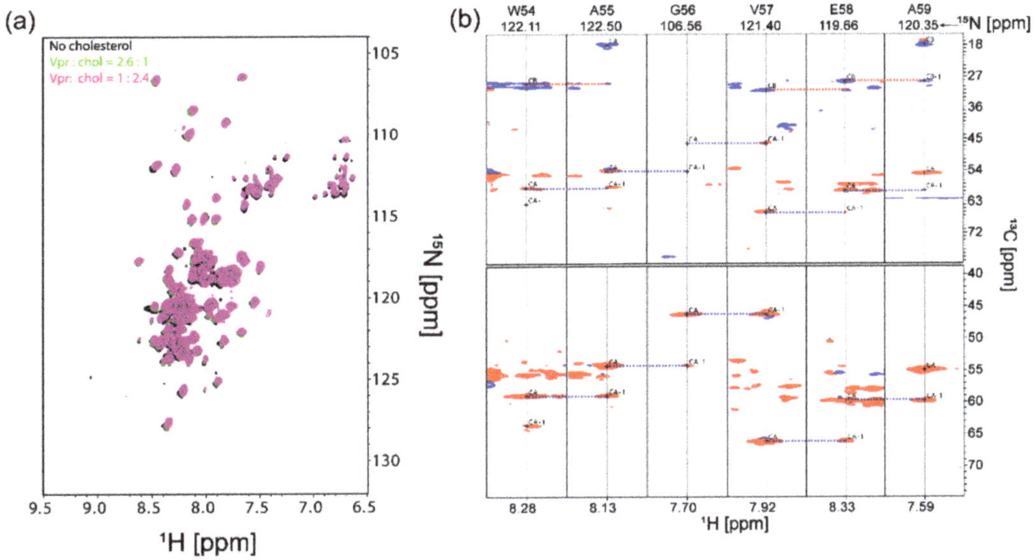

Figure 4. Solution NMR characterizations of [U-^{2}H,^{13}C,^{15}N]-Vpr in [$^{2}H_{25}$, 98%] DPC micelles. (**a**) 2D TROSY-HSQC spectra of [U-^{2}H,^{13}C,^{15}N]-Vpr solubilized in [$^{2}H_{25}$, 98%] DPC detergent micelles with and without the addition of cholesterol. (**b**) The strips for residues 54–59 from a 3D TROSY-HNCACB (top) and 3D TROSY-HNCA experiments of [U-^{2}H, ^{13}C,^{15}N]-Vpr in [$^{2}H_{25}$, 98%] DPC. Red and blue peaks arise from C_{α} and C_{β} nuclei, respectively. The blue and red dashed lines indicate the sequential connection between the strips.

To further examine Vpr in lipid environment, the samples of [1-^{13}C] cysteine selective labeled Vpr embedded in liposomes with and without cholesterol were prepared and characterized using ^{13}C{^{31}P} REDOR NMR spectroscopy. This strategy has been used to study the insertion of HIV fusion peptides in membranes [44–46]. Here, we applied REDOR NMR to probe the interaction of Vpr and membrane. A significant ^{13}C{^{31}P} REDOR dephasing would be observed with 20 ms of the dephasing time if the distance between a ^{13}C and ^{31}P spin pair is less than 10 Å. In other words, ^{13}C{^{31}P} REDOR NMR allows us to observe the carbonyl groups of the Cys residue of Vpr that are in proximity of the lipid phosphate groups (<10 Å). The ^{13}C chemical shift of carbonyl carbons is around 170 ppm using methylene carbon of adamantane at 38.48 ppm as the external chemical shift reference [47]. It is important to mention that the ^{13}C signal observed around 170 ppm is mainly contributed from [1-^{13}C] cysteine selective labeled Vpr since no significant contribution of the carbonyl carbons of lipids was observed as shown in Figure S4. The ^{13}C signal observed around 170 ppm reflect the chemical environments of the cysteine residue, Cys-76, of Vpr since it is the only one cysteine residue in Vpr sequence (Figure S5). We observed a sharp resonance peak at 181 ppm and a broad resonance peak ranging from 165 to 178 ppm. Typically, the line width of a ^{13}C signal from a single-site amino acid residue would be less than 5 ppm [44]. Since different chemical environments will result in different ^{13}C chemical shift, the observed broad peak ranging from 165 to 178 ppm indicated the multiple chemical environments which the Vpr Cys-76 residue experiences. Lipid membranes have been shown to play active roles in modulating the functions of membrane associated proteins or membrane proteins [48–52]. Thus, the membrane cannot be viewed as a passive supporting component for the membrane protein due to the couplings between protein and membrane. Rather, we should consider membrane and membrane associated protein as a complex. The broad signal around 170 ppm can be interpreted as the existence of complex Vpr-membrane environments, which may be a result of the existence of multiple Vpr conformations in lipid membrane. While studying the details of the multiple Vpr conformations is interesting, it is out of the scope of the present work. The existence of complex protein-membrane environments has also been observed previously in HIV fusion peptides [44]. As suggested by the REDOR results in Figure 5a,b, some Vpr proteins were in close proximity of the lipid phosphate groups, indicated by the ~40% REDOR dephasing of the broad peak, while some Vpr proteins were not in close contact with the lipids since no dephasing observed for the sharp peak. We further compared the two S_0 spectra, as shown in Figure 5c. The ratio of the broad peak with respect to the sharp peak increased with the presence of cholesterol in membrane. In addition, the line shape of the broad peak was altered with the addition of cholesterol in the membrane. These results indicated that the distribution of Vpr among these environments was changed in response to the addition of cholesterol in membrane, which may account for our observation of cholesterol modulating the Vpr-membrane interaction as described in Section 3.2.

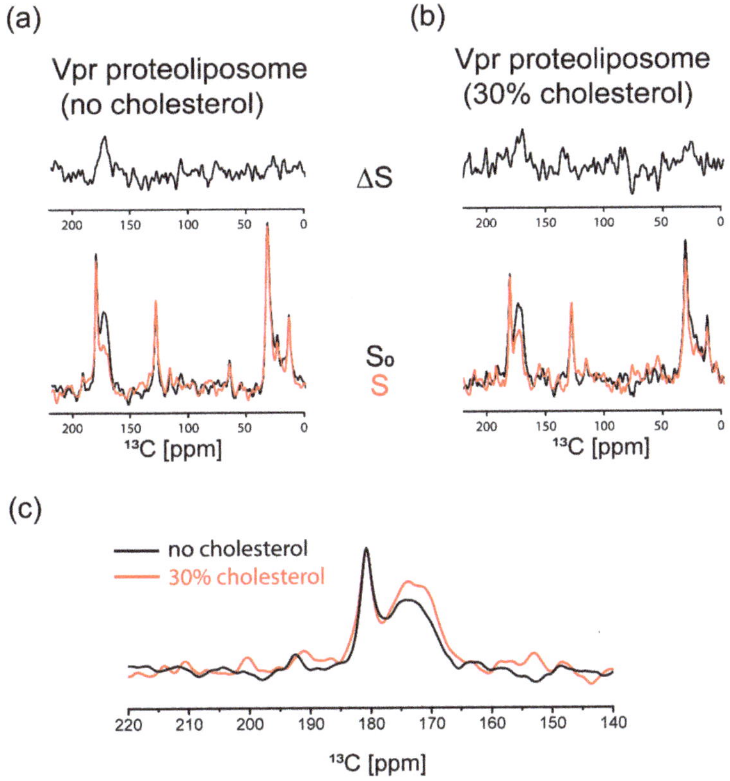

Figure 5. The $^{13}C\{^{31}P\}$ REDOR NMR spectra of [1-^{13}C] cysteine selective labeled Vpr embedded in liposomes. (**a**) $^{13}C\{^{31}P\}$ REDOR NMR spectra of [1-^{13}C] cysteine selective labeled Vpr embedded in DOPG liposomes. (**b**) $^{13}C\{^{31}P\}$ REDOR NMR spectra of [1-^{13}C] cysteine selective labeled Vpr embedded in liposomes (cholesterol: DOPG lipids = 30:70). (**c**) The comparison of the S_0 spectra of these two samples.

4. Conclusions

Based on the results of calcein release assay, the cholesterol concentration in membrane affects the Vpr induced membrane permeability, suggesting a role of cholesterol in the Vpr-membrane interaction. No obvious evidence of specific cholesterol binding to Vpr was observed in DPC detergent micelles based on the result of the NMR titration experiments. Nonetheless, ssNMR spectroscopy was further used to show that the presence of cholesterol in membrane would affect the distribution of Vpr in the various membrane environments, which may partially account for cholesterol playing a role in the Vpr-membrane interaction. We believe that lipid membrane is as an active component regulating membrane protein functions, rather than just a passive component to support membrane protein. Cholesterol, being an active player in the membrane, can affect the function of membrane protein directly or indirectly.

Supplementary Materials: The following are available online at https://www.mdpi.com/2077-0375/11/10/784/s1, Figure S1. Examples of the characterizations of Vpr proteins. Figure S2. Intensity distribution of liposomes measured by dynamic light scattering. Figure S3 Ramachandran plot of the assigned protein segments from A55 to E58 shows well defined left-handed α-helical structure predicted by Talos^{+}. Figure S4. The ^{13}C CPMAS spectrum and $^{13}C\{^{31}P\}$REDOR S_0 spectrum of

DOPG liposomes. Figure S5. The sequence of Vpr protein and its structure (PDB: 1M8L) with 3 α-helical segments (green) and Cys-76 (magenta) specified.

Author Contributions: Conceptualization, T.-Y.Y.; methodology, S.-J.H. and T.-Y.Y.; validation, C.-H.L., T.-Y.Y. and S.-J.H.; formal analysis, C.-H.L. and T.-Y.Y.; investigation, C.-H.L. and T.-Y.Y.; resources, T.-Y.Y. and S.-J.H.; data curation, C.-H.L. and T.-Y.Y.; writing—original draft preparation, C.-H.L., T.-Y.Y. and S.-J.H.; writing—review and editing, C.-H.L. and T.-Y.Y.; visualization, C.-H.L. and T.-Y.Y.; supervision, T.-Y.Y.; project administration, T.-Y.Y.; funding acquisition, T.-Y.Y. All authors have read and agreed to the published version of the manuscript.

Funding: This research was funded by MOST 105-2113-M-001-021-MY2 and AS-iMATE 107-31.

Institutional Review Board Statement: Not applicable.

Informed Consent Statement: Not applicable.

Data Availability Statement: Not applicable.

Acknowledgments: The authors thank MOST and AS-iMATE for providing financial support.

Conflicts of Interest: The authors declare no conflict of interest.

References

1. Berg, J.M.; Tymoczko, J.L.; Stryer, L. Important Derivatives of Cholesterol Include Bile Salts and Steroid Hormones. In *Biochemistry*, 5th ed.; Freeman, W.H.: New York, NY, USA, 2002; Section 26.4.
2. Takahashi, T.; Suzuki, T. Function of membrane rafts in viral lifecycles and host cellular response. *Biochem. Res. Int.* **2011**, *2011*, 245090. [CrossRef]
3. Sun, X.; Whittaker, G.R. Role for influenza virus envelope cholesterol in virus entry and infection. *J. Virol.* **2003**, *77*, 12543–12551. [CrossRef]
4. Tang, Q.; Liu, P.; Chen, M.; Qin, Y. Virion-Associated Cholesterol Regulates the Infection of Human Parainfluenza Virus Type 3. *Viruses* **2019**, *11*, 438. [CrossRef] [PubMed]
5. Aloia, R.C.; Tian, H.; Jensen, F.C. Lipid composition and fluidity of the human immunodeficiency virus envelope and host cell plasma membranes. *Proc. Natl. Acad. Sci. USA* **1993**, *90*, 5181–5185. [CrossRef] [PubMed]
6. Brugger, B.; Glass, B.; Haberkant, P.; Leibrecht, I.; Wieland, F.T.; Krausslich, H.G. The HIV lipidome: A raft with an unusual composition. *Proc. Natl. Acad. Sci. USA* **2006**, *103*, 2641–2646. [CrossRef] [PubMed]
7. Campbell, S.M.; Crowe, S.M.; Mak, J. Lipid rafts and HIV-1: From viral entry to assembly of progeny virions. *J. Clin. Virol.* **2001**, *22*, 217–227. [CrossRef]
8. Campbell, S.M.; Crowe, S.M.; Mak, J. Virion-associated cholesterol is critical for the maintenance of HIV-1 structure and infectivity. *AIDS* **2002**, *16*, 2253–2261. [CrossRef]
9. Ono, A.; Waheed, A.A.; Freed, E.O. Depletion of cellular cholesterol inhibits membrane binding and higher-order multimerization of human immunodeficiency virus type 1 Gag. *Virology* **2007**, *360*, 27–35. [CrossRef]
10. Dick, R.A.; Goh, S.L.; Feigenson, G.W.; Vogt, V.M. HIV-1 Gag protein can sense the cholesterol and acyl chain environment in model membranes. *Proc. Natl. Acad. Sci. USA* **2012**, *109*, 18761–18766. [CrossRef]
11. Yandrapalli, N.; Lubart, Q.; Tanwar, H.S.; Picart, C.; Mak, J.; Muriaux, D.; Favard, C. Self assembly of HIV-1 Gag protein on lipid membranes generates PI(4,5)P2/Cholesterol nanoclusters. *Sci. Rep.* **2016**, *6*, 39332. [CrossRef]
12. Meher, G.; Sinha, S.; Pattnaik, G.P.; Ghosh Dastidar, S.; Chakraborty, H. Cholesterol Modulates Membrane Properties and the Interaction of gp41 Fusion Peptide To Promote Membrane Fusion. *J. Phys. Chem. B* **2019**, *123*, 7113–7122. [CrossRef]
13. Kwon, B.; Mandal, T.; Elkins, M.R.; Oh, Y.; Cui, Q.; Hong, M. Cholesterol Interaction with the Trimeric HIV Fusion Protein gp41 in Lipid Bilayers Investigated by Solid-State NMR Spectroscopy and Molecular Dynamics Simulations. *J. Mol. Biol.* **2020**, *432*, 4705–4721. [CrossRef]
14. Popov, S.; Rexach, M.; Zybarth, G.; Reiling, N.; Lee, M.A.; Ratner, L.; Lane, C.M.; Moore, M.S.; Blobel, G.; Bukrinsky, M. Viral protein R regulates nuclear import of the HIV-1 pre-integration complex. *EMBO J.* **1998**, *17*, 909–917. [CrossRef]
15. Kamata, M.; Nitahara-Kasahara, Y.; Miyamoto, Y.; Yoneda, Y.; Aida, Y. Importin-alpha promotes passage through the nuclear pore complex of human immunodeficiency virus type 1 Vpr. *J. Virol.* **2005**, *79*, 3557–3564. [CrossRef] [PubMed]
16. Jenkins, Y.; McEntee, M.; Weis, K.; Greene, W.C. Characterization of HIV-1 vpr nuclear import: Analysis of signals and pathways. *J. Cell Biol.* **1998**, *143*, 875–885. [CrossRef]
17. He, J.; Choe, S.; Walker, R.; Di Marzio, P.; Morgan, D.O.; Landau, N.R. Human immunodeficiency virus type 1 viral protein R (Vpr) arrests cells in the G2 phase of the cell cycle by inhibiting p34cdc2 activity. *J. Virol.* **1995**, *69*, 6705–6711. [CrossRef] [PubMed]
18. Sabbah, E.N.; Druillennec, S.; Morellet, N.; Bouaziz, S.; Kroemer, G.; Roques, B.P. Interaction between the HIV-1 protein Vpr and the adenine nucleotide translocator. *Chem. Biol. Drug Des.* **2006**, *67*, 145–154. [CrossRef]
19. Qiao, H.; McMillan, J.R. Gelsolin segment 5 inhibits HIV-induced T-cell apoptosis via Vpr-binding to VDAC. *FEBS Lett.* **2007**, *581*, 535–540. [CrossRef]

20. Yasuda, J.; Miyao, T.; Kamata, M.; Aida, Y.; Iwakura, Y. T cell apoptosis causes peripheral T cell depletion in mice transgenic for the HIV-1 vpr gene. *Virology* **2001**, *285*, 181–192. [CrossRef] [PubMed]
21. Kogan, M.; Rappaport, J. HIV-1 accessory protein Vpr: Relevance in the pathogenesis of HIV and potential for therapeutic intervention. *Retrovirology* **2011**, *8*, 25. [CrossRef]
22. Guenzel, C.A.; Herate, C.; Benichou, S. HIV-1 Vpr-a still "enigmatic multitasker". *Front. Microbiol.* **2014**, *5*, 127. [CrossRef]
23. Piller, S.C.; Ewart, G.D.; Premkumar, A.; Cox, G.B.; Gage, P.W. Vpr protein of human immunodeficiency virus type 1 forms cation-selective channels in planar lipid bilayers. *Proc. Natl. Acad. Sci. USA* **1996**, *93*, 111–115. [CrossRef] [PubMed]
24. Gibbs, J.S.; Lackner, A.A.; Lang, S.M.; Simon, M.A.; Sehgal, P.K.; Daniel, M.D.; Desrosiers, R.C. Progression to Aids in the Absence of a Gene for Vpr or Vpx. *J. Virol.* **1995**, *69*, 2378–2383. [CrossRef]
25. Rice, A.P.; Kimata, J.T. Subversion of Cell Cycle Regulatory Mechanisms by HIV. *Cell Host Microbe* **2015**, *17*, 736–740. [CrossRef]
26. Gross, D.A.; Leborgne, C.; Chappert, P.; Masurier, C.; Leboeuf, M.; Monteilhet, V.; Boutin, S.; Lemonnier, F.A.; Davoust, J.; Kichler, A. Induction of tumor-specific CTL responses using the C-terminal fragment of Viral protein R as cell penetrating peptide. *Sci. Rep.* **2019**, *9*, 3937. [CrossRef]
27. Henklein, P.; Bruns, K.; Sherman, M.P.; Tessmer, U.; Licha, K.; Kopp, J.; de Noronha, C.M.; Greene, W.C.; Wray, V.; Schubert, U. Functional and structural characterization of synthetic HIV-1 Vpr that transduces cells, localizes to the nucleus, and induces G2 cell cycle arrest. *J. Biol. Chem.* **2000**, *275*, 32016–32026. [CrossRef]
28. Coeytaux, E.; Coulaud, D.; Le Cam, E.; Danos, O.; Kichler, A. The cationic amphipathic alpha-helix of HIV-1 viral protein R (Vpr) binds to nucleic acids, permeabilizes membranes, and efficiently transfects cells. *J. Biol. Chem.* **2003**, *278*, 18110–18116. [CrossRef] [PubMed]
29. Zhong, P.B.; Liu, C.H.; Chen, Y.T.; Yu, T.Y. The Study of HIV-1 Vpr-Membrane and Vpr-hVDAC-1 Interactions by Graphene Field-Effect Transistor Biosensors. *ACS Appl. Bio Mater.* **2020**, *3*, 6351–6357. [CrossRef]
30. Ono, A.; Freed, E.O. Plasma membrane rafts play a critical role in HIV-1 assembly and release. *Proc. Natl. Acad. Sci. USA* **2001**, *98*, 13925–13930. [CrossRef] [PubMed]
31. Kalyana Sundaram, R.V.; Li, H.; Bailey, L.; Rashad, A.A.; Aneja, R.; Weiss, K.; Huynh, J.; Bastian, A.R.; Papazoglou, E.; Abrams, C.; et al. Impact of HIV-1 Membrane Cholesterol on Cell-Independent Lytic Inactivation and Cellular Infectivity. *Biochemistry* **2016**, *55*, 447–458. [CrossRef] [PubMed]
32. Muller, B.; Tessmer, U.; Schubert, U.; Krausslich, H.G. Human immunodeficiency virus type 1 Vpr protein is incorporated into the virion in significantly smaller amounts than Gag and is phosphorylated in infected cells. *J. Virol.* **2000**, *74*, 9727–9731. [CrossRef]
33. Fritz, J.V.; Dujardin, D.; Godet, J.; Didier, P.; De Mey, J.; Darlix, J.L.; Mely, Y.; de Rocquigny, H. HIV-1 Vpr oligomerization but not that of Gag directs the interaction between Vpr and Gag. *J. Virol.* **2010**, *84*, 1585–1596. [CrossRef]
34. Solbak, S.M.; Reksten, T.R.; Hahn, F.; Wray, V.; Henklein, P.; Henklein, P.; Halskau, O.; Schubert, U.; Fossen, T. HIV-1 p6—A structured to flexible multifunctional membrane-interacting protein. *Biochim. Biophys. Acta* **2013**, *1828*, 816–823. [CrossRef] [PubMed]
35. Nicol, F.; Nir, S.; Szoka, F.C., Jr. Effect of cholesterol and charge on pore formation in bilayer vesicles by a pH-sensitive peptide. *Biophys. J.* **1996**, *71*, 3288–3301. [CrossRef]
36. Pae, J.; Saalik, P.; Liivamagi, L.; Lubenets, D.; Arukuusk, P.; Langel, U.; Pooga, M. Translocation of cell-penetrating peptides across the plasma membrane is controlled by cholesterol and microenvironment created by membranous proteins. *J. Control. Release* **2014**, *192*, 103–113. [CrossRef]
37. Keller, R. *The Computer-Aided Resonance Assignment Tutorial*; Cantina Verlag: Goldau, Switzerland, 2004; CARA can be download from cara.nmr.ch.
38. Gasteiger, E.; Gattiker, A.; Hoogland, C.; Ivanyi, I.; Appel, R.D.; Bairoch, A. ExPASy: The proteomics server for in-depth protein knowledge and analysis. *Nucleic Acids Res.* **2003**, *31*, 3784–3788. [CrossRef]
39. Sreerama, N.; Woody, R.W. Estimation of protein secondary structure from circular dichroism spectra: Comparison of CONTIN, SELCON, and CDSSTR methods with an expanded reference set. *Anal. Biochem.* **2000**, *287*, 252–260. [CrossRef] [PubMed]
40. Shen, Y.; Delaglio, F.; Cornilescu, G.; Bax, A. TALOS+: A hybrid method for predicting protein backbone torsion angles from NMR chemical shifts. *J. Biomol. NMR* **2009**, *44*, 213–223. [CrossRef] [PubMed]
41. Morellet, N.; Bouaziz, S.; Petitjean, P.; Roques, B.P. NMR structure of the HIV-1 regulatory protein VPR. *J. Mol. Biol.* **2003**, *327*, 215–227. [CrossRef]
42. Mayer, M.; Meyer, B. Characterization of Ligand Binding by Saturation Transfer Difference NMR Spectroscopy. *Angew. Chem. Int. Ed. Engl.* **1999**, *38*, 1784–1788. [CrossRef]
43. Takahashi, H.; Nakanishi, T.; Kami, K.; Arata, Y.; Shimada, I. A novel NMR method for determining the interfaces of large protein-protein complexes. *Nat. Struct. Biol.* **2000**, *7*, 220–223. [PubMed]
44. Jia, L.H.; Liang, S.; Sackett, K.; Xie, L.; Ghosh, U.; Weliky, D.P. REDOR solid-state NMR as a probe of the membrane locations of membrane-associated peptides and proteins. *J. Magn. Reson.* **2015**, *253*, 154–165. [CrossRef]
45. Ratnayake, P.U.; Sackett, K.; Nethercott, M.J.; Weliky, D.P. pH-dependent vesicle fusion induced by the ectodomain of the human immunodeficiency virus membrane fusion protein gp41: Two kinetically distinct processes and fully-membrane-associated gp41 with predominant beta sheet fusion peptide conformation. *Biochim. Biophys. Acta* **2015**, *1848*, 289–298. [CrossRef] [PubMed]
46. Schmick, S.D.; Weliky, D.P. Major antiparallel and minor parallel beta sheet populations detected in the membrane-associated human immunodeficiency virus fusion peptide. *Biochemistry* **2010**, *49*, 10623–10635. [CrossRef]

47. Morcombe, C.R.; Zilm, K.W. Chemical shift referencing in MAS solid state NMR. *J. Magn. Reson.* **2003**, *162*, 479–486. [CrossRef]
48. Marsh, D. Protein modulation of lipids, and vice-versa, in membranes. *Biochim. Biophys. Acta* **2008**, *1778*, 1545–1575. [CrossRef] [PubMed]
49. Yeagle, P.L. Modulation of membrane function by cholesterol. *Biochimie* **1991**, *73*, 1303–1310. [CrossRef]
50. Laganowsky, A.; Reading, E.; Allison, T.M.; Ulmschneider, M.B.; Degiacomi, M.T.; Baldwin, A.J.; Robinson, C.V. Membrane proteins bind lipids selectively to modulate their structure and function. *Nature* **2014**, *510*, 172–175. [CrossRef] [PubMed]
51. Ernst, M.; Robertson, J.L. The Role of the Membrane in Transporter Folding and Activity. *J. Mol. Biol.* **2021**, *433*, 167103. [CrossRef]
52. Jodaitis, L.; van Oene, T.; Martens, C. Assessing the Role of Lipids in the Molecular Mechanism of Membrane Proteins. *Int. J. Mol. Sci.* **2021**, *22*, 7267. [CrossRef]

membranes

MDPI

Review

Lipid Membrane Mimetics in Functional and Structural Studies of Integral Membrane Proteins

Saman Majeed [1], Akram Bani Ahmad [1], Ujala Sehar [1] and Elka R. Georgieva [1,2,*]

[1] Department of Chemistry and Biochemistry, Texas Tech University, Lubbock, TX 79409, USA; saman.majeed@ttu.edu (S.M.); abaniahm@ttu.edu (A.B.A.); usehar@ttu.edu (U.S.)
[2] Department of Cell Physiology and Molecular Biophysics, Texas Tech University Health Science Center, Lubbock, TX 79409, USA
* Correspondence: elgeorgi@ttu.edu

Abstract: Integral membrane proteins (IMPs) fulfill important physiological functions by providing cell–environment, cell–cell and virus–host communication; nutrients intake; export of toxic compounds out of cells; and more. However, some IMPs have obliterated functions due to polypeptide mutations, modifications in membrane properties and/or other environmental factors—resulting in damaged binding to ligands and the adoption of non-physiological conformations that prevent the protein from returning to its physiological state. Thus, elucidating IMPs' mechanisms of function and malfunction at the molecular level is important for enhancing our understanding of cell and organism physiology. This understanding also helps pharmaceutical developments for restoring or inhibiting protein activity. To this end, in vitro studies provide invaluable information about IMPs' structure and the relation between structural dynamics and function. Typically, these studies are conducted on transferred from native membranes to membrane-mimicking nano-platforms (membrane mimetics) purified IMPs. Here, we review the most widely used membrane mimetics in structural and functional studies of IMPs. These membrane mimetics are detergents, liposomes, bicelles, nanodiscs/Lipodisqs, amphipols, and lipidic cubic phases. We also discuss the protocols for IMPs reconstitution in membrane mimetics as well as the applicability of these membrane mimetic-IMP complexes in studies via a variety of biochemical, biophysical, and structural biology techniques.

Keywords: integral membrane proteins; lipid membrane mimetics; detergent micelles; bicelles; nanodiscs; liposomes

1. Introduction

Integral membrane proteins (IMPs) (Figure 1) reside and function in the lipid bilayers of plasma or organelle membranes, and some IMPs are located in the envelope of viruses. Thus, these proteins are encoded by organisms from all living kingdoms. In almost all genomes, approximately a quarter of encoded proteins are IMPs [1,2] that play critical roles in maintaining cell physiology as enzymes, transporters, receptors, and more [3–5]. However, when modified via point mutations, deletion, or overexpression, these proteins' function becomes abnormal and often yields difficult- or impossible-to-cure diseases [6,7]. Because of IMPs' important role in physiology and diseases, obtaining their high-resolution three-dimensional (3D) structure in close to native lipid environments; elucidating their conformational dynamics upon interaction with lipids, substrates, and drugs; and ultimately understanding their functional mechanisms is highly important. Such comprehensive knowledge will greatly improve our understanding of physiological processes in cellular membranes, help us develop methodologies and methods to overcome protein malfunction, and improve the likelihood of designing therapeutics for protein inhibition. Notably, it is remarkable that almost 40% of all FDA-approved drugs exploit IMPs as their molecular targets [8,9].

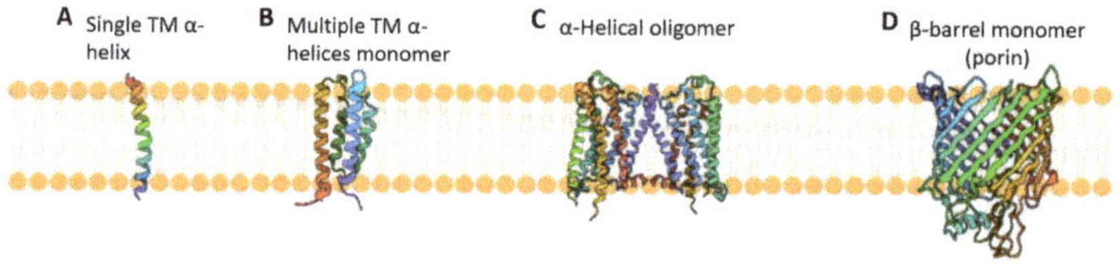

Figure 1. Representative types of IMPs: The α-helical IMPs can have just one helix (**A**) or multiple helices (**B**) that traverse the membrane; they can be multimeric as well (**C**). The β-barrel membrane proteins typically have multiple membrane-traversing strands (**D**) and can be either monomeric or oligomeric. The lipid membrane bilayer is shown in orange. The structures of IMPs with PDB accession codes 5EH6 (**A**), 2KSF (**B**), 5OR1 (**C**), and 4GPO (**D**) are shown in the figure. The membrane orientation was not considered.

Undeniably, functional and structural studies of IMPs have greatly advanced in recent decades by developing diverse in-cell and in-vitro functional assays [10–13]; advancing the X-ray crystallography applications for membrane proteins in detergents [14,15], bicelles, nanodiscs, and lipidic cubic phases [15–20] to determine the structure at a typical 3 Å or even higher resolution; improving data detection and processing for single-particle cryo-electron microscopy (cryoEM) to increase the number of resolved IMPs' structures at ca. 3.5–3 Å resolution [21–23]; the contribution from single-molecule FRET spectroscopy (smFRET) toward understanding IMPs' conformational dynamics in real time under physiological environment conditions [24–26]; the growing number of highly sophisticated studies using EPR spectroscopy via continuous wave (CW) and pulse methods to uncover the short- and long-range conformational dynamics underlying IMPs' functional mechanisms [27–33]; advancing NMR spectroscopy [34–36] and particularly solid-state NMR applied to proteins in lipid-like environments [37–39]; conducting extensive studies using site-directed mutagenesis to identify the roles of certain amino acid residues in the IMPs' function [40–42], molecular dynamics computational studies [43–45]; and more. Despite this substantial progress, IMPs are still understudied and require further research.

The enormous diversity and complexity of IMPs challenges researchers because they must uncover and characterize numerous diverse functional mechanisms. Any step in the workflow, from gene to characterizing IMPs' structure and function can present challenges, such as poor solubilization efficiency from the host cell membrane, limited long-term stability, low protein expression, and more [46–48]. Another serious issue is identifying and developing appropriate membrane protein hosts, i.e., lipid membrane-like mimetics, to which IMPs are transferred from the native membranes where they are expressed, or from inclusion bodies in the case of eukaryotic or viral proteins produced in *E. coli* [49]. This is needed for further purification and in vitro functional and structural studies [50–54]. In general, IMPs are difficult to solubilize away from their native environment in the cell membrane due to their hydrophobic regions [55]. Also, removing these proteins from their native cellular form sometimes results in evident functional and structural implications [54]. Thus, selecting a suitable membrane mimetic for each particular protein is critical for obtaining samples of functional proteins for in vitro studies on active or purposely inhibited protein states. Furthermore, the isolated and purified IMPs often need to be obtained at concentrations and purity, which are satisfactory for the biochemical and biophysical techniques used for these proteins' characterization.

Due to the high importance of membrane mimetics for accommodating and maintain IMPs' native state, special attention must be paid to the current state and further prospective when developing these nano-sized membrane platforms. Therefore, we focus here on reviewing the most widely used and emerging membrane mimetics, which are detergents, multilamellar lipid emulsions, unilamellar liposomes, Lipodisqs®/nanodiscs, bicelles, amphipols, and lipidic cubic phases (LCPs), in IMP purification and structure–function studies. Additionally, we describe applications of these mimetics for particular IMPs and discuss how selecting a membrane mimetic affects these proteins' properties. Of course, due to rapidly increasing contributions in the field and space limitations, this review cannot cover all the developments and applications of membrane mimetic systems and their applications in membrane functional and structural molecular biology studies.

2. An Overview of the Most Widely Used Lipid Membrane Mimetics and Their Applications in Functional and Structural Studies of Integral Membrane Proteins

The development of lipid membrane mimetics to make IMPs amenable for isolation, purification, and in vitro characterization has a long history. Generally, a membrane mimetic should reproduce the lipid bilayer properties, or at least recreate the hydrophobic core environment of a lipid bilayer in its most fundamental form [54,56]. Although detergents have been the most widely used substitute for the membrane environment, in the recent decades a great deal of effort has been invested to expand the diversity of membrane mimetics and to use more lipid bilayer-like structures, which together with the incorporated proteins have high solubility and stability. These novel membrane mimetics provide the following advantages for the incorporated IMPs: (i) convenience to investigate them via research technologies that are impossible or difficult to execute in the presence of detergents, (ii) improved stability, and (iii) provision of an environment with a chemical composition and/or physical characteristics closer to the native lipid membrane bilayer environment [57]. However, all of these membrane mimetics have pros and cons, and not all are compatible with various protein research techniques.

Here, we further describe these membrane mimetics and discuss their applications in studying IMPs.

2.1. Detergents and Detergent Micelles in Studies of Integral Membrane Proteins

2.1.1. General Properties of Detergents and Detergent Micelles

Detergents are the archetypal lipid membrane mimetics and have been extensively utilized for the solubilization and characterization of IMPs. They are amphipathic molecules and, above a certain so-called critical micelle concentration (CMC), self-aggregate to form micelles in aqueous solutions (Figure 2A). Saponins and naturally occurring bile salts were the first detergents used for biochemical studies [58,59]. Currently, ample diverse detergents with variable biochemical and biophysical characteristics are available.

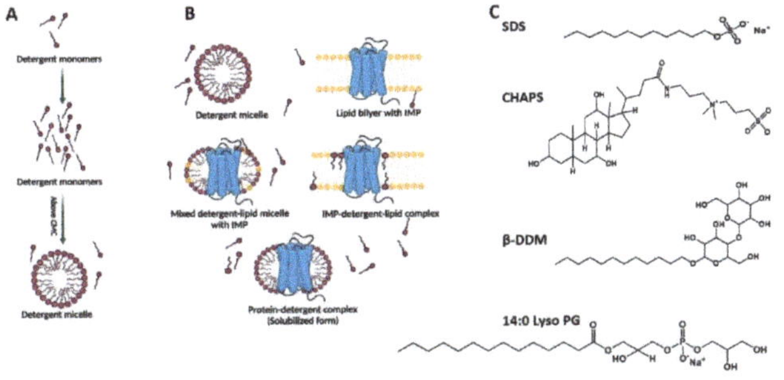

Figure 2. IMPs in detergents: (**A**) In aqueous solution, above a certain concentration (CMC), detergent molecules self-associate to form close to spherical aggregates (micelles) with hydrophilic and hydrophobic portions facing the aqueous environment and the micelle interior, respectively. (**B**) Detergents are used for the extraction of IMPs from the native membrane of expression host—detergent at a high concentration, much above its CMC, is mixed with the native membranes containing the IMP of interest; due to its hydrophobic properties the detergent mixes with the membrane lipids and solubilizes the membrane; as a result, mixed IMP–lipid–detergent, IMP–detergent or detergent–lipid complexes are formed; thereafter, the lipid molecules are removed in the next purification steps unless specific lipids are tidily bound to the IMP. (**C**) The chemical formulas of some of the most widely used in studies of IMPs detergents are shown: SDS is negatively charged, CHAPS is zwitterionic, DDM is non-charged; and 14:0 Lyso PG is negatively charged.

Detergents fit into three major classes (Figure 2C): *ionic detergents* have either positively or negatively charged headgroups and are strong denaturants or harsh membrane mimetics owing to their effect on IMPs' structure, e.g., sodium dodecyl sulfate (SDS) has negatively charged headgroups; *zwitterionic detergents*, e.g., the traditional 3-[(3-cholamidopropyl)dimethyl-ammonio]-1-propane-sulfonate (CHAPS) or Lauryldimethylamine-N-oxide (LDAO), have zero overall molecular charge, exhibit a less pronounced denaturation effect compared to ionic detergents and a stronger solubilization potential compared to non-ionic detergents, and are hence categorized as an intermediate between non-ionic and ionic detergents; and *non-ionic detergents* are comparatively mild, have non-charged hydrophilic groups, tend to shield the inter- and intra-molecular protein–protein interactions and maintain the structural integrity of solubilized proteins, e.g., dodecyl-L-D-maltoside (DDM), lauryl-maltose neopentyl-glycol (LMNG), and octyl-L-D-glucoside (OG) [54,60,61]. Phospholipid-like detergents are either charged, like 14:0 Lyso PG (1-myristoyl-2-hydroxy-sn-glycero-3-phospho-[1′-rac-glycerol]) and 16:0 Lyso PG (1-palmitoyl-2-hydroxy-sn-glycero-3-phospho-[1′-rac-glycerol]), or zwitterionic, like 14:0 Lyso PC (1-myristoyl-2-hydroxy-sn-glycero-3-phosphocholine) and Fos-Choline 12. These have also been extensively used in studies of IMPs [62,63].

2.1.2. Detergent Applications in Integral Membrane Proteins Solubilization, Purification, and Stabilization

Typically, the first step in transmembrane protein purification is extracting it from the host membrane or inclusion body. The protein extraction from the host membrane is carried out by adding an appropriate detergent at a high concentration (several times above the CMC) to the homogenized proteo-lipid membrane, which solubilizes the membrane (Figure 2B). Initially, destabilization and fragmentation of lipid bilayer occur due to inserting the detergent molecules into the membrane. Subsequently, the lipid membrane is dissolved, and then IMP-detergent, lipid-detergent, and lipid-IMP-detergent mixed compositions are formed [64]. Various detergents exhibit different capacities for solubilizing biological membranes. Similarly, the type of detergent used for solubilization can affect the preservation of specifically bound lipid molecules in the IMP's final detergent-solubilized state [65]. Multiple detergents must be screened to identify those that maintain the IMP's structural integrity and functional activity, and suit downstream applications [54]. For instance, detergents with a low CMC can effectively solubilize most membranes but are less appropriate for methods requiring detergent removal because they can be difficult to remove later [66]. Also, using a mild detergent that only binds to the transmembrane region of a given IMP and can retain key lipid interactions is essential for successful studies [67]. Once solubilized, the IMPs' purification follows the same principles as for purifying soluble proteins, utilizing chromatographic methods like affinity, gel filtration, and/or ion-exchange chromatography. Alternatively, when IMPs are deposited into inclusion bodies, such as eukaryotic proteins or prokaryotic outer membrane proteins expressed in *E. coli*, their refolding into detergent micelles is an efficient approach to obtain solubilized membrane proteins in a physiologically-relevant state. Thus, due to their convenience and large variability, detergents are one of the most extensively used membrane mimetics and are almost unavoidably utilized for extracting and solubilizing IMPs from host membranes and for screening for optimal IMP stability [68,69]. In many studies, detergents are also used as intermediate IMP hosts from which the IMP is transferred into more lipid-like and lipid-bilayer-like mimetics, such as nanodiscs, liposomes, and other for additional downstream investigations [54].

On the other hand, the hydrophobic tails of detergent molecules in the micelle, which are shorter and more mobile compared to lipids' alkyl tails, make an inadequate mimic of the lipid bilayer. Due to a mismatch in hydrophobic thicknesses, the isolated IMPs and the detergent micelle can also influence each other's shape, leading to the adoption of non-physiological IMP conformations [70]. In addition, the hydrophobic packing in proteo-micelles is weaker than those for IMPs in a lipid bilayer, allowing increased water penetration into the detergent micelle and leading to IMPs' structural instability [71].

Despite these deficiencies, the detergents and detergent micelles are currently among the most widely used membrane mimetics for in vitro studies of IMPs.

2.1.3. Applications of Detergents in Functional Studies of Integral Membrane Proteins

Although IMPs' activity assays have been conducted mostly in lipid bilayers and predominantly on liposome-reconstituted IMPs, functional studies of detergent-solubilized IMPs have also been carried out. Studies have investigated substrates' binding affinities to characterize a critical stage initiating the substrate translocation via membrane transporters and channels. These studies monitored the binding of a radioactively labeled substrate in the case of the prokaryotic Na/tyrosine transporter (Tyt1) [13], and isothermal titration calorimetry (ITC) studies elucidated the binding of ligands (ions and other substrates) to transporter/channel or receptor IMPs [72–75]. The ATPase activity of ABC transporters in detergents was also examined [76,77]. It was found in such studies that a LmrA transporter in FC-16 detergent has higher ATPase activity and ligand binding compared to LmrA solubilized in DDM [78].

2.1.4. Detergent Applications in Studies of Integral Membrane Proteins Using Biophysical and Structural Biology Methods

Detergent-solubilized IMPs have been extensively studied by almost all available biophysical and structural biology techniques to determine physiologically relevant or disease-linked protein conformations and conformational transitions with and without ligands, e.g., substrates or inhibitors, bound to the protein molecules. Currently, most existing atomic-resolution X-ray crystal structures are of detergent-solubilized IMPs. Importantly, IMPs' proper folding and monodispersity are critical for a successful crystallization. Several approaches have been utilized to assess the IMP homogeneity: size exclusion chromatography (SEC) with light scattering and sedimentation equilibrium centrifugation analyses [79], fluorescence-detection SEC [80], polypeptide thermal stability using a thiol-specific fluorescent reporter to monitor cysteine residue accessibility upon denaturation [81], *nano*DSF with light scattering [82], and thermal or chemical denaturation using circular dichroism (CD) spectroscopy to monitor the stability of IMPs' secondary structure [83,84]. Thus, multiple detergents must be screened, and those that maintain protein homogeneity and integrity are considered for further use [82,85]. Still, other factors appear key to successful IMP crystallization. Given that not just the protein, but the protein–detergent complex must crystallize [86], several analyses searched for a trend in the conditions used for obtaining high-quality IMP crystals [87]. Regarding the detergent used, statistics as of 2015 show that half of IMP crystal structures were obtained in alkyl maltopyranosides, followed by the alkyl glucopyranosides (23%), amine oxides (7%), and polyoxyethylene glycols (7%) [87]. The most successful alkyl maltopyranoside detergent is n-dodecyl-β-D-maltopyranoside (DDM), followed by n-decyl-β-D-maltopyranoside (DM) [87]. Thus, in addition to maintaining protein stability, detergents with shorter chain provide a good environment for IMP crystallization because they form smaller micelles, which facilitate tighter packing in the crystal lattice and higher-quality crystal diffraction [82,88–90]. The IMP structures from diverse families have been solved, and some of these structures capture the same protein in distinct conformations. This information is invaluable for elucidating functional and/or inhibition mechanisms. IMPs crystallized in detergent include glutamate receptor GluA2 [91], neurotransmitter transporter homologue LeuT [92,93], betaine transporter BetP [94], and many more. The protein data bank (PDB) provides detailed information about IMPs' deposited crystal structures in detergents.

In the last decade, EM and single-particle cryoEM in particular have made historic progress in studying detergent-solubilized IMPs by expanding this technique's applications to diverse families of IMPs and by determining these proteins' 3D structure at high resolution down to ca. 3 Å [21,95]. In contrast to X-ray crystallography, EM does not require protein-crystal formation and has much more potential to deal with conformationally heterogeneous proteins and protein complexes. Nevertheless, successful

IMP structure determination via EM requires high stability and proper folding of the detergent-solubilized protein [95]. For this reason, detergents are screened similarly to the crystallization of IMPs. In addition, EM sometimes experiences specific problems with detergents suitable for crystallization, including the detergents DDM or LMNG. It can be difficult to distinguish the protein particle from a detergent via a negative EM stain, as found in the study of citrate transporter CitS in DDM and DM [96]. To reduce the background and facilitate visualizing protein particles, free detergent micelles can be removed prior to the EM experiments [97]. In contrast, other studies found that detergents with low CMC, such as DDM and maltose-neopentyl glycols (MNGs), provide a better platform for a single-particle cryoEM of IMPs [98]. Another detergent used in cryoEM structure determination is digitonin (an amphipathic steroidal saponin) [99]. Fluorinated Fos-Choline-8 detergent was also used to stabilize and determine the structure of a homo-oligomeric serotonin receptor in its apo, serotonin-bound, and drug-bound states [100–102].

Solution NMR spectroscopy has also benefited from detergent-solubilization in studying the high-resolution structure of full-length (FL) IMPs or truncated IMP constructs and in monitoring the conformational transitions in IMPs' monomers and complexes [103]. Specifically for NMR, despite the significant technical and methodological advancements in recent decades, this method is still limited by the protein's size; in the case of IMPs, this includes the size of a membrane mimetic-protein complex. Thus, the slow tumbling of large-protein objects in a solution significantly shortens the traverse relaxation times resulting in NMR line broadening, and ultimately causes a loss of NMR sensitivity [103]. The large size of protein molecules also produces overcrowded NMR spectra, which are difficult to interpret. Therefore, the current size limit for proteins and protein complexes studied by NMR in solution does not exceed 70 kDa even when advantageous pulse sequences are applied [103–105]. Given this, solution NMR studies on IMPs require detergent micelles to be as compact (small) as possible but still adequately mimic the membrane environment [103]. Care must be taken to achieve high monodispersity of the studied IMP. The length of IMP transmembrane segments should also generally match the micelle hydrophobic core to avoid inconsistent NMR data [106]. Historically, "harsh" detergents like dodecylphosphocholine (DPC) and lauryldimethylamine-N-oxide (LDAO) that form small micelles (20–25 kDa) and maintain IMPs functional states have been used to study the human VDAC-1 [107], the human voltage-dependent anion channel [108], the outer membrane protein G [109], and more. Mild detergents, like DM and DDM have been used in NMR solution studies of bacteriorhodopsin [110], G-protein-coupled receptors (GPCRs) [111,112], voltage-dependent K^+ channels [113], and more. IMPs solubilized in micelles of anionic lysolipids (e.g., 14:0 PG and 1-palmitoyl-sn-glycero-3-phospoglycerol [16:0 PG]) and short-chain lipids (e.g., 1,2-dihexanoyl-sn-glycero-3-phosphocholine [DHPC]) have been studied by NMR in solution [114–117].

EPR spectroscopy, continuous wave (CW), and pulse, in combination with spin labeling [27,30,31,118–123], have provided invaluable information about the conformational dynamics and function/inhibition of IMPs. These studies were conducted exclusively or partly on detergent-solubilized IMPs. Large structural rearrangements in DDM–solubilized membrane transporters, which report on protein dynamics along the transport cycle or the assembly into functional units, were uniquely captured by pulse EPR distance measurements [28,32,124–131]. Viral, bacterial, and eukaryotic channels [29,132,133], receptors [134,135], and more were also studied in detergent micelles (DDM, DM, lauryl maltose neopentyl glycol [MNG], etc.) via CW and pulse EPR spectroscopy. Importantly, EPR spectroscopy experiments have no specific requirements for the detergent used insofar as the detergent supports protein stability. Also, there is no restriction on IMP's size, given that the protein can be successfully spin-labeled. Moreover, EPR spectroscopy can investigate IMPs within a broad range of concentrations (e.g., ca. <5 μM to >100 μM), allowing researchers to capture multimeric IMP intermediates in detergent micelles [29].

Another informative technique in studies of detergent-residing IMPs is fluorescence spectroscopy/microscopy, exemplified by Förster resonance energy transfer (FRET) spectroscopy and particularly by the single-molecule FRET (smFRET) version [136]. It captures conformational motions within one protein molecule/complex in real time, although measurements on many molecules/complexes are needed to average the effect of modulating protein conformation by, for example, ligand binding. smFRET has been used in multiple studies on detergent-residing IMPs to monitor their conformational responses to ligands, changes in pH, or other stimuli [137–139].

2.2. Bicelles in Studies of Integral Membrane Proteins

2.2.1. General Properties of Bicelles

Introduced by Prestegard and colleagues in 1988, bicelles (binary/bi-layered mixed micelles) are recognized as the first lipid membrane mimetic system capable of incorporating a substantial amount of lipids to create a bilayer-like environment for membrane proteins [140]. Bicelles are disc-shaped nanoaggregates comprising bilayer-forming long-chain phospholipids mixed with either detergent molecules or short-chain phospholipids in an aqueous environment [69,140] (Figure 3A). They are an attractive membrane mimetic for studying the structure and structural dynamics of membrane proteins. For example, isotropic bicelles can be formed in aqueous solutions by mixing the long-chain lipid 1,2-dimyristoyl-sn-glycero-3-phosphocholine (DMPC) with the detergent 3-[(3-cholamidopropyl)dimethyl-ammonio]-1-propane sulfonate (CHAPS). They can also be formed by mixing the long-chain lipids 1,2-dimyristoil-sn-Glycero-3-[Phospho-rac-(1-glycerol)] (DMPG) and DMPC with the short-chain lipid DHPC [141,142]. Bicellar nanostructures comprising various lipids with incorporated cholesterol, ceramides, cardiolipin, and more have also been developed [143–145].

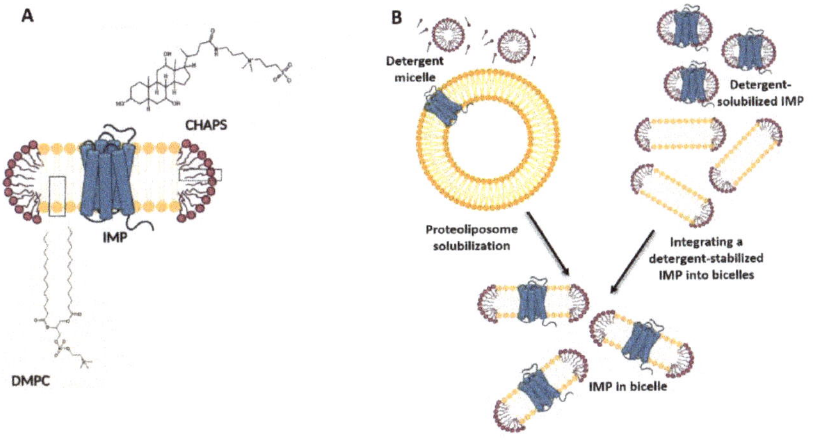

Figure 3. IMPs in bicelles. (**A**) Bicelle-residing IMP containing multiple transmembrane helices is shown; the bicelle is composed of a patch of bilayer-forming lipids (e.g., DMPC) stabilized by short-chain lipid or detergent (e.g., CHAPS). The size of bicelles depends on the molar ratio between long- and short-chain lipids used in their preparation (Equation (1)). In addition, bicelle size is affected also upon dilution of the bicellar solution. (**B**) Two major protocols for incorporation of IMPs into bicelles are outlined: detergent/detergent micelles are mixed with proteoliposomes (**left**) or IMP in detergent micelles are mixed with lipids and bicelle-forming detergent (**right**). The figure shows simplified procedures.

Generally, geometric arguments can help to estimate the bicelle's size using the molar ratio between long- and short-chain lipids (or detergent); this so-called q value

(Equation (1)) to calculate the radius of the bicelle's bilayer region (R) directly, in addition to the bicelle's topology and size [146–148].

$$q = \frac{total\ molarirty\ of\ long-chain\ lipid}{total\ molarity\ of\ detergent\ (short-chain\ lipid) - CMC\ of\ detergent\ (short-chain\ lipid)} \quad (1)$$

In addition, dynamic light scattering and NMR can also be used to experimentally determine bicelles' size and morphology in an aqueous buffer at a constant total lipid/detergent concentration [149,150].

Bicelles with a higher q value are formed from low concentrations of short-chain lipids/detergents in relation to the concentration of long-chain lipids, and they are typically larger than the low q-value bicelles. Bicelles with smaller q values ($q \leq 0.6$) are more "detergent-rich" and "lipid-poor", so the phospholipid environment they provide can perturb the bicelle-incorporated IMP [146]. However, it is difficult to precisely estimate bicelle size. For example, bicelles made of DMPC/DHPC had an estimated average size of 20 nm at q = 2 [143], and those made of DMPC/DMPG/DHPC at q = 2.6 had an estimated average size of 10 nm [149]. This discrepancy can be explained by the limitations of different methods used to determine bicelles' size. IMPs have been reconstituted and studied in both large and small bicelles [146,147].

Due to bicelles' small size, their suspensions are effectively homogeneous and translucent even after incorporating membrane proteins [151,152]. One major advantage of this membrane mimetic system is its resemblance to a small fragment of lipid bilayer. In addition, embedding IMPs in a native-like environment and a simple variation in the q value can help in the system's size scalability [153]. Furthermore, native bicelles made of lysed eukaryotic-cell lipids mixed with DHPC were also prepared to provide diverse lipid types for specific interactions with proteins [154]. Thus, bicelles outperform detergents in maintaining membrane proteins' functional state. In addition, paramagnetic ions can be added to the lipid mixtures, so the resulting bicelles can align in an external magnetic field, aiding magnetic resonance studies on IMPs [155,156].

Notably, the presence of detergent-like short-chain lipids and a bilayer size is insufficient to provide membrane-like lateral pressure and may perturb the structure and dynamics of bicelle-residing IMPs [54,69,157]. Another disadvantage of conventional bicelles is that their size and geometry depend on the total lipid concentration in the solution; therefore, any dilution changes the system properties. At high dilutions, bicelle-to-vesicle transitions can occur [143], so care must be taken to maintain constant lipid concertation throughout the experiment. Attempts were made to overcome this deficiency via kinetically stable bicelles, such as those comprising a mixture of the phospholipid 1,2-dipalmitoyl-sn-glycero-3-phosphatidylcholine (DPPC) and a sodium cholate-derived surfactant (SC-C5) at room temperature. These bicelles' stability results from the high melting temperature of DPPC (41 °C) and a very low SC-C5 CMC (<0.5 mM) [158].

2.2.2. Applications of Bicelles in Solubilizing and Stabilizing Integral Membrane Proteins

Typically, IMPs expressed in host membranes are first extracted and solubilized in detergents and then reconstituted in bicelles. Two basic protocols exist for reconstituting an IMP into bicelles: formulating the bicelles via the addition of detergent to proteoliposomes or integrating a detergent-stabilized IMP into bicelles [159,160] (Figure 3B). In addition, some studies on synthesized and usually truncated IMPs or on other membrane-associated protein constructs have used bicelles for direct solubilization. These hydrophobic proteins and protein constructs are first dissolved in an organic co-solvent, such as chloroform or TFE, and then mixed with the lipids before being lyophilized and dissolved in an appropriate buffer to form bicelles [161].

2.2.3. Applications of Bicelles in Studies on Integral Membrane Proteins Using Biophysical and Structural Biology Methods

Small isotropic bicelles have been a highly preferred membrane mimetic platform in studies of IMP structure and dynamics by solution NMR spectroscopy, since they provide both a close-to-native lipid environment and fast enough tumbling to average out anisotropic effects, yielding good quality NMR spectra [146,160,162]. Still, IMP size is a serious limitation for solution NMR; and the need to produce isotopically labeled IMPs, given that their expression levels are typically small, introduces additional difficulty [36,151]. Nevertheless, the structures of several bicelle-reconstituted relatively large IMPs, such as sensory rhodopsin II [163], EmrE dimer [164], and the transmembrane domain of the receptor tyrosine kinase ephA1 [165], have been solved using solution NMR. Large bicelles have been the choice of solid-state NMR studies because they provide a greater bilayer surface and structural stabilization of the embedded IMPs. Beside the fact that large IMPs can be incorporated, the orientation of large bicelles in the external magnetic field can be controlled. Such bicelles can also be spun at the magic angle, enhancing spectral resolution for the embedded IMPs [151,166,167].

X-ray crystallography has also utilized bicelles to determine the high-resolution structure of IMPs in their native lipid environment, particularly in cases when detergents could not stabilize the IMP structure for crystallization [168]. Bicelle–IMP complexes can be handled similarly to detergent–IMPs and are compatible even with high-throughput robot-aided crystallization [169]. Thus, after the first successful crystallization of bicelle-residing bacteriorhodopsin [170], the crystal structures of several other IMPs, such as β2-adrenergic G-protein coupled receptor-FAB complex [171], rhomboid protease [172], and VDAC-1 [173] were solved.

Studies using EPR spectroscopy, pulse, and CW with spin labeling have also used bicelles as a lipid mimetic to study the conformational dynamics of IMPs. Magnetically aligned bicelles were used to probe the topology and orientation of the second transmembrane domain (M2δ) of the acetylcholine receptor using spin labeling and CW EPR [174]. Further, the immersion depth of the spin-labeled M2δ peptide at different positions in bicelles was determined. Here, CW EPR was used to monitor the decrease in nitroxide spin label spectrum intensity due to nitroxide radical reduction upon the addition of ascorbic acid [175]. Pulse EPR distance measurements on spin-labeled McjD membrane transporter in bicelles revealed functionally relevant conformational transitions [176].

2.3. *Nanodiscs in Studies of Integral Membrane Proteins*

2.3.1. General Properties of Nanodiscs

Sligar and colleagues were first to illustrate nanodisc technology in 1998 in a study focused on liver microsomal NADPH-cytochrome reductase enzyme, the CYP450 reductase [177,178]. The first nanodiscs were proteolipid systems made of lipid bilayer fragments surrounded by high-density lipoprotein (HDL). Thereafter, the diversity of nanodiscs expanded to include lipid nanostructures held intact by a belt of lipoprotein (membrane scaffold protein, MSP) [179,180], saposin [181], peptide [182], or copolymer [183]. All these membrane mimetics are self-assembled, nano-sized, and generally disc-shaped lipid bilayer structures (Figure 4). A major advantage of the nanodisc technology is the absence of detergent molecules and the ability to maintain integrity and shape upon dilution. This overcomes the shortcomings of lipid bicelles and provides a more native-like membrane environment compared to detergents [184,185]. Other advantages of nanodiscs are good accessibility of soluble domains in IMPs, sample homogeneity, and isolation of defined IMP oligomeric states by controlling the size of the nanodisc [186].

Currently, nanodisc systems are classified based mostly on the belt used. The most common type is *MSP nanodiscs* made by using the repeat domain of apolipoprotein A1 (ApoA1), the main component of DHL, which is referred to as membrane scaffold protein (MSP) [177] (Figure 4A). The formation of these nanodiscs requires two copies of the amphipathic α-helical MSP, which wraps up and stabilizes a small disc of lipid bilayer [151,177]. Both copies of MSP are arranged antiparallel to each other [187]. The size of nanodiscs can be controlled by using one or more MSP repeat regions, which are produced by protein engineering. For example, MSP1 consists of one repeat of 10 helices and MSP2 consists of two equivalent repeats each consisting of 10 helices [188,189].

Further modification in just one repeat, e.g., adding identical helices produced longer than the MSP1 constructs MSP1E1, MSP1E2, and MSP1E3, or deletions in MSP1 produced shorter constructs denoted MSLP1D1 and MSP1D2 [189]. Thus, any variation in the number of these amphipathic helical repeats results in different nanodisc diameters/sizes. For an empty nanodisc (one with no IMP incorporated), the type of phospholipid and the MSP construct establish the number of phospholipids in each particle, typically ~20 to 400 [184,188,189]. Sligar and colleagues [188,190] suggested the following correlation between the number of lipid molecules in the nanodiscs (N_L) and amino acids in the scaffold protein (M):

$$N_L \cdot S = (0.423 \cdot M - 9.75)^2 \tag{2}$$

where S represents the mean surface area per lipid used to form the nanodisc, measured in Å^2.

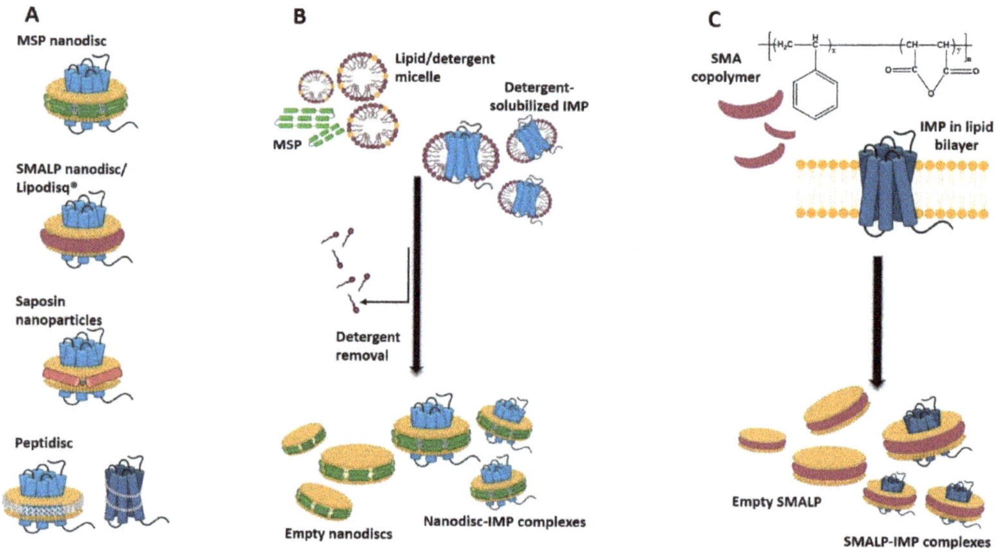

Figure 4. IMPs in nanodiscs. (**A**) IMP-nanodisc complexes of different types are shown. These are discoidal structures containing a segment of lipid bilayer with incorporated IMP surrounded by a belt of different nature that stabilizes the nanoparticle. Depending on the belt used, nanodisc can be IMP–MSP nanodisc, IMP–SMALP/Lipodisq®, IMP–Saposin nanoparticles, and IMP–peptidiscs with and without lipids incorporated. The size of nanodiscs can be controlled by changing the belt length to accommodate just one monomeric IMP or IMP oligomeric complex. (**B**) Typically, the detergent solubilized IMPs are transferred in nanodiscs by mixing IMP in detergent, MSP, detergent-solubilized lipids or mixed detergent–lipid micelles, incubated and the detergents are removed, in most of the cases by using BioBeads. As a result, IMP–nanodisc complexes and empty nanodiscs are formed. The empty nanodiscs can be removed further. (**C**) The IMP–SMALP/Lipodisq® complexes can be formed by mixing CMA copolymer with liposome- or native membrane-residing IMPs. This is an advantage of using CMA copolymers, since they do not require the detergent-solubilization of lipid bilayer prior to IMP reconstitution, and can extract IMPs from the native membranes of expression host.

The prototypical MSP1 construct forms nanodiscs with diameters of about 10 nm and has an overall molecular mass of approximately 150 kDa [188], but the modified MSP1 and MSP2 constructs can form smaller or larger nanodiscs with diameters ranging from about 8.4 nm to 17 nm [184,189]. Recently, nanodiscs with covalently linked N and C termini of newly engineered variants based on ApoA1 were developed, and termed covalently circularized nanodiscs (cNDs) [191].

Copolymer nanodiscs were introduced by Knowles and colleagues [192], who purified an IMP in polymer nanodiscs, i.e., Styrene–maleic acid–lipid particles (SMALPs). These nanodiscs were termed Lipodisq® and are discoidal structures comprising of a segment of lipid bilayer surrounded by a polymer belt [193]. This belt is made of a styrene-maleic acid (SMA) copolymer formed by the hydrolysis of styrene-maleic anhydride (SMAnh) precursor and composed of 1:2 or 1:3 ratios of maleic acid to styrene [192]. The main distinction between MSPs and Lipodisqs is that SMA copolymer can directly cut out patches from the lipid bilayer without the use of detergents [192]. The principle of SMA-bound particles is centered on the interaction of the hydrophobic edge of a planar bilayer membrane with the styrene phenyl rings of the SMA polymer. This interaction stabilizes the disc-shaped SMALPs [69]. Monodisperse lipid discs with 140 lipid molecules and 10–11-nm diameter are formed with the help of SMA for the isolation of target membrane protein [194]. Lipodisqs with different incorporated lipids, e.g., palmitoyl-oleoyl-phosphocholine (POPC) [195] or DMPC [196], have been prepared and used. A major consideration when working with Lipodisqs is their pH-dependent stability, as they precipitate at pH values below 6.5 due to maleic acid moiety protonation, which is a disadvantage when studying IMPs at lower pHs. SMA polymer chelates divalent cations (e.g., Mg^{2+} and Ca^{2+}) that are used for signaling assays, leading to Lipodisqs' insolubility. To overcome these deficiencies, chemical modifications of maleimide carboxylates of SMA polymers with positively charged quaternary ammonium compounds (SMA-QA) or ethanolamine have been employed [197,198]. Another copolymer called DIBMA (di-isobutylene/maleic acid) was also developed—it is less harsh than SMA, stable in the presence of divalent cations owing to the absence of aromatic moiety, and does not interfere with far-UV optical spectroscopy [199].

Synthetic peptide-based nanodiscs (also termed "peptidiscs") are formed by short amphipathic peptides aligned in an antiparallel fashion around the hydrophobic rim of a phospholipid membrane [182,200,201]. Bi-helical peptides displace detergent molecules by wrapping around the hydrophobic parts of detergent-purified membrane proteins [148, 182]. Another example is a peptide derived from the ApoA1, which consists of 18 amino acids that form a single alpha helix of almost the same length as that of the apolipoprotein A1 helix [200,202,203]. Among the major benefits of peptidiscs is that their size can be adjusted by a simple variation in the peptide-to-lipid ratio. Also, peptide nanodiscs encapsulate IMPs irrespective of initial lipid content, so there is no need to consume exogenous lipids to match the diameter of the scaffold membrane as in the case of MSP nanodiscs. Furthermore, peptide stoichiometry is self-determined because the size and shape of the integrated IMP guide the binding of the peptide skeleton [69,204,205]. However, the comparatively high cost of custom peptide synthesis and its low stability due to their noncovalent assembly compared to the stability of other types of nanodisc systems are among the cons of the peptide nanodisc system [69,206].

Saposin nanoparticles are protein-stabilized lipid structures utilizing Saposin lipoprotein variants [207]. Salipro®, a Saposin A (SapA) disc, is the most suitable approach for IMP studies, since it can tolerate a wide range of lipid-to-Saposin ratios [208]. Salipro nanodiscs are composed of two or more SapA proteins that are joined together and assembled in V shapes around a small lipid disc, which makes them relatively flexible/tunable to accommodate different sizes of IMPs [181,209].

2.3.2. Applications of Nanodiscs in Integral Membrane Protein Solubilization and Stabilization

Typically, detergent-solubilized IMPs are reconstituted into nanodiscs of different types, starting either from a whole solubilized membrane or after purification. Currently, the most widely used procedure is to transfer the purified detergent-solubilized IMP into nanodiscs—This is done by mixing the IMP, lipid and scaffold protein or polymer; thereafter, the detergent is removed using BioBeads and the nanodiscs with or without incorporated IMP are formed [190] (Figure 4B). Optimization to determine the optimum scaffold protein, polymer, or peptide, as well as lipid concentration to accommodate each particular IMP in its native oligomeric state, must be performed [186,210]. Procedures for the direct transfer of IMPs from the membrane into nanodiscs with minimal involvement of detergent have been utilized [211]. Lipodisqs have also been used to purify IMPs in native host membranes without any detergent, preserving the IMPs' native state intolerance to detergents and preferences for particular lipids or lipid bilayers [53,212,213]. Furthermore, some advantageous technologies for cell-free expression of IMPs utilize direct incorporation and folding of the synthesized proteins into nanodiscs, which also benefits from the opportunity to tune the nanodiscs' lipid composition [214–216].

2.3.3. Applications of Nanodiscs in Functional Studies of Integral Membrane Proteins

As discussed above, one significant advantage of nanodiscs is that the soluble domains of IMPs reconstituted in them are well accessible. Therefore, binding of ligands, e.g., substrates, inhibitors, etc., and protein partners—all relevant to the IMP function—can easily be studied in a native-like environment. Thus, fluorescence correlation spectroscopy was used to assay fluorescently labeled IMPs' binding interactions via an autocorrelation function, which depends on the diffusion coefficients of the bound vs. unbound species [217,218]. Scintillation proximity assay was used to assess radio–ligand binding to membrane transporters residing in nanodiscs, overcoming the protein activity reduction caused by detergents [219]. An assay measuring ATP hydrolysis by MsbA transporter in nanodiscs demonstrated the importance of MsbA–lipid interactions by varying the nanodisc lipid composition [220]. It was also found that nanodiscs facilitate the identification of monoclonal antibodies targeting multi-pass IMPs, which is important for antibody-based pharmaceutical developments [221].

2.3.4. Applications of Nanodiscs in Studies of Integral Membrane Proteins Using Biophysical and Structural Biology Methods

Since their initial development, nanodiscs have been widely used in studies of IMPs' structure and conformational dynamics due to their suitability to a variety of techniques and methods. As yet, crystallization of IMPs in nanodiscs for X-ray structure determination has proven a difficult task. However, crystallization of IMPs can be assisted by transferring them from nanodiscs/Lipodisqs to lipidic cubic phases (LCPs); high quality crystals of bacteriorhodopsin and rhodopsin crystals were obtained and the structures of these proteins solved at and below 2 Å resolution [17,221].

On the other hand, EM has greatly benefited from nanodiscs, and the first EM studies were on negatively stained nanodisc-IMPs, such as the dimeric bc_1 complex and reaction centers from antenna-free membranes [222,223]. However, the structural resolution achieved was insufficient. Further technical developments in single-particle cryoEM have since made it possible to determine the high-resolution structure of IMPs in native lipid environments, capturing multiple functional protein conformations and oligomeric states [224,225]. Still, only proteins with sufficient molecular weight, typically about or above 150 kDa, can be visualized by the available advanced EM approaches and data processing. Thus, the structure of the ca. 320 kDa trimeric bacterial multidrug efflux transporter AcrB was resolved at a resolution of 3.2 Å in Lipodisqs, uncovering a well-organized lipid-bilayer structure associated with the protein transmembrane domain [226]. Also, the structure of nanodisc-embedded full-length glycine receptor at 3 to 3.5 Å resolution was resolved in the ligand-free, glycine-bound, and allosteric modulator-bound states, providing a comprehensive map of the functionally relevant conformational isomerizations [227]. CryoEM on SthK, a prokaryotic cyclic nucleotide-gated channel, also yielded high-resolution structures of channel apo, cAMP-bound, and cGMP-bound states in nanodiscs [228]. Remarkably, the structures of small IMPs were also resolved by EM in nanodiscs [229]. However, in these studies engineering of fusion protein or antibody/antigen-binding fragment (Fab) was utilized to increase the protein size and stability and succeed in the structure determination. For instance, the structure of 49 kDa *P. falciparum* CQ-resistance transporter PfCRT in complex with Fab was resolved at 3.2 Å resolution [230]. Consequently, nanodisc technology greatly improved the likelihood of understanding the structure of functionally relevant IMP conformations and visualizing essential protein–lipid interactions.

Nanodiscs have been particularly useful in studies of IMPs using NMR spectroscopy as well. Solution NMR has benefited from the fast tumbling of the nanodisc–IMP complex providing correlation times in the nanosecond range [34]. Still, the limitation of IMP size persists. Careful optimization of several parameters must be performed to obtain homogeneous samples with desired size: the scaffold protein/copolymer-to-lipid molar ratio; lipid composition, to provide hydrophobic match to the transmembrane part of IMP and/or specific interactions; and optimizations of nanodisc-to-IMP molar ratios [148,231]. This is true to an extent for all other structural biology techniques utilizing nanodiscs. Also, for solution NMR, reduced-size nanodiscs of 60–120 kDa with faster tumbling are more appropriate to obtain good NMR data quality [38,184]. Solid-state NMR studies have been conducted on complexes oriented in external magnetic field nanodisc/Lipodisq–IMP without magic angle spinning and on isotropic nanodisc/Lipodisq–IMP complexes with magic angle spinning [232]. Such studies open the opportunity to elucidate the high-resolution structure and conformational dynamics of IMPs in native-like environments. Nanodiscs have been useful in NMR applied to GPCRs and other physiologically and biomedically important IMPs [233,234].

EPR spectroscopy studies of spin-labeled IMPs' structure–function relationships and conformational dynamics have also utilized nanodiscs as a membrane-mimetic platform [30,123]. Thus, double electron–electron resonance distance (DEER) measurements were conducted on a nanodisc-incorporated LmrP eukaryotic multidrug transporter [235]. In this study, the lipid makeup of the nanodiscs greatly affected the functional conformational state of the transporter. Lipodisq nanoparticles were used to assess the conformational dynamics of the human KCNQ1 voltage sensing domain [236]: The powerful combination of CW EPR and DEER confirmed the stabilization effect of the Lipodisqs on protein structure. In this study, the superior DEER data quality compared to liposomes highlighted the high potential of these membrane mimetics in studies of IMPs. The Aer primary energy sensor for motility in *E. coli* was also reconstituted in nanodiscs and studied by EPR [237]; although the DEER distances between the protein's native Flavin radicals were very similar in detergent (DDM) and nanodisc environments, the observed protein activity was indeed higher in nanodiscs.

Nanodiscs were used in studies of IMPs by fluorescence-based techniques: internal reflection fluorescence microscopy (TIRFM), fluorescence correlation spectroscopy (FCS), and FRET were all applied to nanodisc-reconstituted cytochrome P450 3A4 and possible mechanisms for protein allosteric regulation were proposed [238,239]. Lipodisq-reconstituted KirBac1.1 potassium channels were studied by using smFRET to probe the structural changes that occur in this multimeric channel upon activation and inhibition [240]. IMPs in native nanodiscs, i.e., copolymer-solubilized native membranes, have also been studied using FRET [241].

2.4. Liposomes in Studies of Integral Membrane Proteins

2.4.1. General Properties of Liposomes

Liposomes were introduced in 1961 by Bangham et al. [242] They are nano- and micro-sized vesicles that can have just one (unilamellar) or multiple (multilamellar) lipid bilayers [243,244] (Figure 5A). Unilamellar vesicles can range in size from 20 nm to more than 1 μM, and depending on their size are classified as small (20–100 nm), large (larger than 100 nm), or giant (larger than 1 μM), with the latter vesicles being closer to the size of a cell. Multilamellar vesicles have multilayer morphology and are greater than 500 nm in diameter. The inside lumen and the space between the lipid bilayers of the unilamellar and multilamellar vesicles are filled with water-based solution, and liposomes present a good artificial mimetic of a cell. Liposomes can be prepared from synthetic bilayer-forming phospholipids, but native membrane-extracted lipids have also been used [245]. Further, the physical and chemical properties of the lipid bilayer in liposomes can be tuned by varying the types and concentrations of lipids, and the amount of cholesterol added [246]. Generally, extrusion through polycarbonate filters can be used to prepare large unilamellar vesicles (LUVs) with a diameter of about 100–500 nm. Low-power bath sonication of lipid suspensions spontaneously forms small unilamellar vesicles (SUVs) with a diameter of about 20–50 nm. Hydrated phospholipids can be used to prepare giant unilamellar vesicles (GUVs) with a diameter greater than 500 nm by applying low-frequency electric fields. Other methods to produce liposomes include freeze-thawing and detergent extraction; hydration of lipid powders or films resulting inthe spontaneous formation of multilamellar vesicles (MLVs) with an overall size between 1 and 10 μm, as well [151,247–249]. Based on their properties that closely mimic biological membranes, liposomes have been extensively used in drug delivery due to their nontoxic nature and ability to encapsulate both hydrophilic and hydrophobic compounds [243,246,250,251].

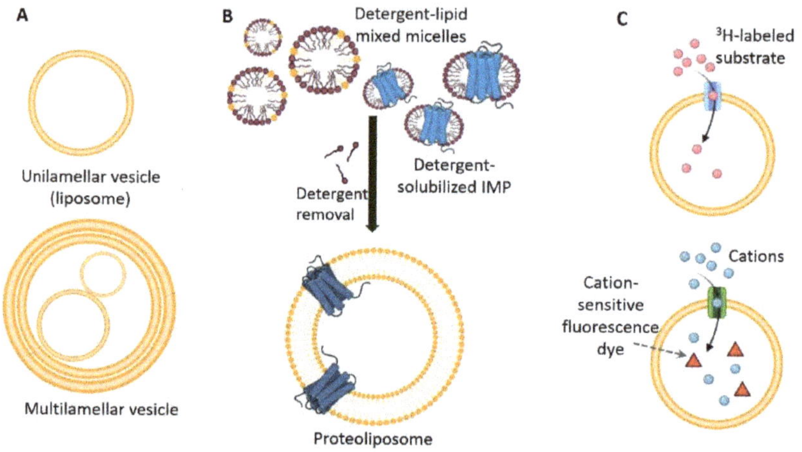

Figure 5. IMPs in liposomes. (**A**) Unilamellar and multilamellar vesicles are shown. These are continuous lipid bilayer structures with incorporated inside water-based solution, so the environment inside and outside of liposomes can be controlled. Typically, the unilamellar liposomes have just one lipid bilayer, whereas multilamellar vesicles have multi-bilayer onion-like structure with solution-filled compartments between the bilayers. (**B**) Typically, the IMPs are reconstituted in liposomes from detergent-solubilized form, which are mixed with detergent-solubilized lipids in the form of mixed detergent-lipid micelles. After some period of incubation to ensure IMP–lipid interactions, the detergent(s) are removed and the proteoliposomes, which usually have close to unilamellar morphology are formed. (**C**) The substrate uptake assay can be carried on liposome-reconstituted membrane transporters or channels: The uptake of radioactively (^{3}H)-labeled substrate in the liposome can be quantified reporting on the IMP activity (upper panel) or the uptake of ions (cations) can be quantified by the changes in the fluorescence intensity of the liposome-incorporated dye, which is sensitive to the presence of substrate (lower panel).

Liposomes are also a great platform to reconstitute and study membrane proteins [248,252, 253]. To this end, liposomes offer several unique advantages compared to other membrane mimetic systems. To begin with, multicomponent systems such as lipid, protein, and substrate complexes can be reconstituted in the liposomes because of the large size of this system [254]. Furthermore, liposomes sustain membrane potential because their hydrophobic bilayer introduces compartments in the aqueous phase, just like the native cells. In addition, liposomes represent a continuous membrane because they are not constrained by a solubilizing scaffold structure. This stands in contrast to other membrane mimetics, which only approximate a membrane bilayer. The diffusion behavior and native lateral pressure of phospholipids and proteins can be studied because of the continuous nature of liposome membranes [255]. All of these properties and the broad range of possible lipid compositions make these membrane mimetics an important tool to study

IMPs' conformational dynamics, substrate relocation across the membrane, folding, etc. at the molecular level [28,29,132,256–258]. In addition to liposomes, vesicles with similar properties termed "polymersomes", which are made of amphiphilic polymers, have also been utilized in studies of biological processes at the membrane, or in drug delivery [259]. However, despite their high potential as membrane mimetics, the current applications of these membrane mimetics in IMPs structure-function studies are fewer compared to phospholipid liposomes, and therefore, their detailed description is beyond the scope of this review.

2.4.2. Reconstitution of Integral Membrane Proteins in Liposomes

Typically, IMPs are transferred in liposomes from a detergent-solubilized state (Figure 5B). First, the desired lipids or lipid mixtures are transferred into a glass vial and dissolved in organic solvent. Then, the solvent is evaporated under a stream of nitrogen or argon gas and then under vacuum to remove the organic solvent completely; the preferred buffer for downstream experiments is added to the dry lipid film, and the lipids are hydrated for approximately 1 h at room temperature or 4 °C. depending on the lipid polycarbon chain saturation and temperature stability, vortexing or sonication can be applied as well. After complete lipid hydration, multilamellar vesicles are formed. Next, aliquots of the lipid suspension are taken in amounts needed to produce the desired final lipid-to-protein molar or w/w ratios and solubilized in mild detergent, e.g., Triton x-100. The detergent-solubilized IMP is mixed with the detergent-solubilized lipids and incubated for approximately 1 h at room temperature or a different temperature, if required. Finally, the detergents are removed to form proteoliposomes [28,29,132,249]. In the last step, the detergent can be removed by either dialysis or by using BioBeads. Also, further freeze–thawing, extrusion, or mild sonication can be performed to obtain more homogeneous and unilamellar proteoliposomes. It must be noted that the described method for IMP reconstitution in liposomes is rather challenging and requires optimization for each particular IMP. Currently, the most widely used method to obtain GUVs is electroformation [260]. This method has been utilized to incorporate IMPs as well—for example, the reconstitution of sarcoplasmic reticulum Ca^{2+}-ATPase and H^+ pump bacteriorhodopsin GUVs preserved these proteins' activity [261]. Recently, a method to reconstitute an IMP into liposomes using native lipid binding without detergent solubilization was illustrated [248]. To do so, cytochrome c oxidase (CytcO) was first solubilized and purified in SMA nanodiscs (Lipodisqs) and then the protein–nanodisc complexes were fused with preformed liposomes, a methodology previously used for IMP delivery and integration into planar lipid membranes [262].

2.4.3. Applications of Liposomes in Functional Studies of Integral Membrane Proteins

As noted above, proteoliposomes (IMP–liposome complexes) are similar to isolated cells to a certain extent: distinct environments of compounds, ions, or pH can be created inside and outside of liposomes, and in addition transmembrane potential can be generated [263–267]. This is a great advantage for the design and implementation of in vitro functional assays of IMPs. Typically, in these assays, the IMP liposomes, also known as unilamellar vesicles, are filled with the desired buffer, with or without IMP ligands, and aliquots of these proteoliposomes are then transferred to a bath buffer with significantly greater volume than that inside of the liposome. Thus, the reconstituted IMPs sense the difference between the buffers inside and outside the liposome. Such experimental setups are used, for example, to quantify the uptake of substrates by membrane transporters or channels, if the bath buffer contains a labeled substrate, e.g., radioactively labeled substrate [28,268,269], or the proteoliposomes are prefilled with a fluorescent dye whose intensity depends on the presence of substrate [270–272] (Figure 5C). In such experiments, the uptake of radioactive ^{86}Rb into liposomes was utilized to measure the activity of channels reconstituted in these liposomes [268]. Radioactively labeled substrates (typically ^{3}H-labeled, but other radioactive atoms can be used as well) have been widely used in liposome-based functional studies of membrane transporters, e.g., Na^+-dependent dicarboxylate transporter [273] and Na^+-dependent aspartate transporter GltPh [274]. A fluorescence-based method using Magnesium Green, a Mg^{2+}-sensitive dye, was used to evaluate ATP/ADP exchange via mitochondrial adenine nucleotide translocase [271]. In a similar assay, either Ca^{2+}- or Na^+-sensitive fluorescent probes entrapped in liposomes containing connexin 26 hemichannels were used to demonstrate for the first time the translocation of Ca^{2+} by the connexin channel [270]. Inhibitors of IMPs have also been tested in liposome-based assays [263]. Using different lipid mixtures to prepare liposomes was also exploited to study specific IMP–lipid interactions. Thus, the activity of mammalian glucose transporter depends upon anionic (phosphatidic acid, phosphatidylserine, phosphatidylglycerol, and phosphatidylinositol) and conical phospholipids (phosphatidylethanolamine and diacylglycerol) [265].

2.4.4. Applications of Liposomes in Studies of Integral Membrane Proteins Using Biophysical and Structural Biology Methods

Due to their complexity, attempting to determine the high-resolution structure of IMPs in proteoliposomes is usually not a researcher's first choice. Still, liposomes have been used to crystallize IMPs incorporated in the bilayer, and the obtained 2D crystals were analyzed by EM [258,275]. Although using EM to characterize the structure of IMPs from 2D crystals formed in flattened liposomes is a difficult task due to varying liposome morphology and other factors, success was achieved. Electron cryotomography, subtomogram averaging, and electron crystallographic image processing were successfully applied to analyze the structure of bovine F1Fo ATP synthase in 2D membrane crystals [276]. Another advancement in determining the structure of IMPs using 2D crystallization of liposomes is to generate buffer gradient from the inside to the outside of the liposome, which activates the IMP. Then, the 2D crystals are quickly frozen under liposome gradient conditions and snapshots of active protein are taken. This technique has contributed to the detailed characterization of IMP functional conformations in lipid bilayers [258].

Conformational dynamics underlying IMPs' function in liposomes have been extensively studied using EPR spectroscopy [27–30,32,119,132]. This technique can be applied to IMPs in both unilamellar and multilamellar vesicles and is not restricted based on the size of proteins in the liposome. In many cases, EPR studies were conducted on the same proteins in detergent and in liposome, revealing distinct membrane-mimetic dependent conformational behavior. Using DEER spectroscopy for the GltPh transporter, Georgieva et al. [28] found that although the subunits in this homotrimeric protein occupy the outward- and inward-facing conformations independently, the population of protomers in an outward-facing state increases for proteins in liposomes. Also, the lipid bilayer affects the assembly of the M2 proton channel from influenza A virus as deduced from DEER modulation depth measurements on spin-labeled M2 transmembrane domain in MLVs compared to detergent (β-DDM)—the dissociation constant (K_d) of M2 tetramer is significantly smaller than that in detergent, therefore the lipid bilayer environment facilitates M2 functional channel formation [29,132]. These studies are extremely important in elucidating the role of lipid bilayers in sculpting and stabilizing the functional states of IMPs.

Single-molecule fluorescence spectroscopy and microscopy have also been used to study conformations of IMPs in liposomes. This technique was used to successfully assess the dimerization of fluorescently labeled IMPs [277,278] and the conformational dynamics of membrane transporters in real time [137,279].

2.5. Other Membrane Mimetics in Studies of Integral Membrane Proteins

2.5.1. Amphipols

The concept of amphipols—amphipathic polymers that can solubilize and stabilize IMPs in their native state without the need for detergent—emerged in 1994. Amphipols' mechanism was validated in a study of four IMPs: bacteriorhodopsin, a bacterial photosynthetic reaction center, cytochrome b6f, and matrix porin [280]. Amphipols were developed to facilitate studies of membrane proteins in an aqueous environment by providing enhanced protein stability compared to that of detergent [281,282]. Functionalized amphipols can be used to trap membrane proteins after purification in detergent, during cell-free synthesis, or during folding [281]. Because of their mild nature, amphipols provide an excellent environment for refolding denatured IMPs, like those produced as inclusion bodies [283]. The stability of IMP–amphipol complexes upon dilution in an aqueous environment is another advantage of these membrane mimetics. Thus, amphipols have been used in numerous IMP studies to monitor the binding of ligands and/or determine structures [280,284]. Still, they have some disadvantages. Their solubility can be affected by changes in pH and the addition of multivalent cations, which neutralize their intrinsic negative charge and lead to low solubility [284,285].

2.5.2. Lipid Cubic Phases

Lipidic cubic phase (LCP) is a liquid crystalline phase that forms spontaneously upon mixing of lipids and water under specific conditions [286,287]. It was introduced as membrane mimetic in 1996 for crystallization of IMPs [18]. Since then, numerous IMP structures that had been difficult or even impossible to crystalize in other mimetic environments were solved in LPC [19,288]. The first structure of GPCR as a fusion construct with T4 lysozyme was solved in LPC by Kobilka et al. [289] LCP can be described as highly curved continuous lipid bilayer made of monoacylglycerol (MAG) lipids, which is surrounded by water-based mesophase. Thus, the whole system forms continuous highly curved channels, in which IMPs are incorporated. Generally, LCPs maintain the IMPs functional conformations and activity. For crystallization in LCPs, the detergent-solubilized IMP is mixed with the LCP-forming lipid, to which specific lipids can be added as well. The addition of precipitant to this system affects the LCP in terms of phases transition and separation, so some of these phases become enriched in IMP leading to nucleation and 3D crystals growth. In addition to crystallography, functional assays have been performed on LPC-reconstituted IMPs as well [290]. Due to space limitations, we do not provide further details of this highly advantageous for X-ray crystallography and protein structure determination. More details can be found in specialized reviews elsewhere [286,291].

3. Conclusions

Due to the important roles of IMPs in cells' and organisms' normal physiology as well as in diseases, there is a need to comprehensively understand the functional mechanisms of these proteins at the molecular level. To this end, in vitro studies on isolated proteins using diverse biochemical and biophysical approaches provide invaluable information. However, studies of IMPs are challenging due to these proteins' hydrophobic nature, low expression levels in heterologous hosts, and low stability when transferred out of the native membrane to a membrane-mimetic platform. To overcome these challenges, progress has been made in multiple directions. We summarized the developments of lipid membrane mimetics in functional and structural studies of IMPs over the past several decades. Indeed, the diversity of these systems grew significantly, and the widely ranging lipid membrane-mimetic platforms now available provide high solubility, stability, more or less lipid-bilayer environments, and other specific properties that are utilized in studies featuring NMR, X-ray crystallography, EM, EPR, fluorescence spectroscopy assays, ligand binding and translocation assays, etc. This has resulted in the continuous expansion of knowledge about IMPs. In Table 1, we provide concise information about the most-widely used membrane mimetics to study IMPs, selected applicable techniques, along with some of their advantages and disadvantages.

The fast development of lipid membrane mimetics and the great expansion of their diversity also provides a great promise for the successful future research to uncover the mechanisms of IMPs, which, to date, have been difficult to stabilize and study. Besides, combining the information from studies of IMPs in different membrane mimetics and by different techniques will help to more completely understand the structure and function of these proteins and avoid possible biases due to the selection of membrane environment.

Table 1. Summary of most widely used lipid membrane mimetics in functional and structural studies of IMPs.

System/Type	Applicable Techniques to Study IMPs	Advantages	Disadvantages
Detergent micelles Ionic detergents Zwitterionic detergents Non-ionic detergents	X-ray crystallography Single-particle cryoEM Solution NMR EPR spectroscopy Fluorescence spectroscopy smFRET Isothermal titration calorimetry (ITC) for ligand binding/protein interactions Functional assays	Easy handling Starting point for downstream applications Availability of large variety of detergents	Propensity of IMP denaturation Chances of non-physiological IMP conformations due to mismatched 'IMP-micelle' hydrophobic thicknesses CMC of the detergent must be considered
Bicelles	Solution NMR Solid-state NMR X-ray crystallography EPR spectroscopy	Easy preparation Homogeneous and translucent suspensions Provide true lipid environment physiological conditions Diverse types of lipids can be incorporated to match Bicelles of different sizes can be prepared	Total lipid concentration can affect size and geometry of bicelle Risk of IMP perturbation in case of insufficient bilayer size
Nanodisc MSP nanodiscs SMALP/Lipodisq® Synthetic peptide-based nanodiscs Saposin nanoparticles	Single particle cryoEM Solution NMR Fluorescence spectroscopy and microscopy smFRET EPR spectroscopy ITC for ligand binding/protein interactions Functional assays	Maintain integrity and shape even upon dilution Easy accessibility of soluble domains in IMPs Possibility of size adjustment to accommodate a monomeric IMP or larger IMP complex	Optimization of assembly conditions can be time consuming Not suitable for large MP oligomers Dynamics of lipids affected by protein 'belt' Limited size range
Liposomes Small unilamellar vesicles (SUVs) Large unilamellar vesicles (LUVs) Giant unilamellar vesicles (GUVs) Multilamellar vesicles (MLVs)	Electron crystallography Solid-state NMR EPR spectroscopy smFRET Functional assays/substrate uptake Electrophysiology	Large size can accommodate large and multicomponent systems Represent continuous membrane providing closer to native environment for IMPs Diffusion behavior similar to native phospholipid membrane Broad range of possible lipid compositions	The orientation of IMP is often non-native Expensive compared to the traditional systems Low solubility
Amphipols	Single-particle cryoEM Solid-state NMR	Assist IMPs study in aqueous environment Stability of IMP-amphipol complex stable on dilution Provides better IMP stability compared to micelle Facilitate refolding of denatured IMPs	Commercially evaluability of only one amphipol type Too difficult to maintain the IMP-amphipol complex sometimes Multivalent cations- and pH-dependent solubility
Lipidic cubic phase	X-ray crystallography Functional studies	More native-like environment for IMPs facilitating their crystallization	Relatively expensive

Author Contributions: S.M., E.R.G., A.B.A. and U.S. data curation; S.M. and E.R.G. manuscript writing and visualization; E.R.G., S.M., A.B.A. and U.S. manuscript finalization; E.R.G. conception, design, supervision and funds acquisition. All authors have read and agreed to the published version of the manuscript.

Funding: This research received no external funding.

Institutional Review Board Statement: Not Applicable.

Informed Consent Statement: Not Applicable.

Acknowledgments: Startup funds from the Department of Chemistry and Biochemistry at TTU to ERG are acknowledged. We thank the Reviewers for their useful suggestions to improve the quality of this manuscript.

Conflicts of Interest: The authors declare no conflict of interest.

References

1. Fagerberg, L.; Jonasson, K.; von Heijne, G.; Uhlén, M.; Berglund, L. Prediction of the human membrane proteome. *Proteomics* **2010**, *10*, 1141–1149. [CrossRef]
2. Wallin, E.; von Heijne, G. Genome-wide analysis of integral membrane proteins from eubacterial, archaean, and eukaryotic organisms. *Protein Sci.* **1998**, *7*, 1029–1038. [CrossRef]
3. Cournia, Z.; Allen, T.W.; Andricioaei, I.; Antonny, B.; Baum, D.; Brannigan, G.; Buchete, N.V.; Deckman, J.T.; Delemotte, L.; Del Val, C.; et al. Membrane Protein Structure, Function, and Dynamics: A Perspective from Experiments and Theory. *J. Membr. Biol.* **2015**, *248*, 611–640. [CrossRef]
4. Whitelegge, J.P. Integral membrane proteins and bilayer proteomics. *Anal. Chem.* **2013**, *85*, 2558–2568. [CrossRef] [PubMed]
5. Robertson, J.L. The lipid bilayer membrane and its protein constituents. *J. Gen. Physiol.* **2018**, *150*, 1472–1483. [CrossRef] [PubMed]
6. Sanders, C.R.; Myers, J.K. Disease-related misassembly of membrane proteins. *Annu. Rev. Biophys. Biomol. Struct.* **2004**, *33*, 25–51. [CrossRef] [PubMed]
7. Marinko, J.T.; Huang, H.; Penn, W.D.; Capra, J.A.; Schlebach, J.P.; Sanders, C.R. Folding and Misfolding of Human Membrane Proteins in Health and Disease: From Single Molecules to Cellular Proteostasis. *Chem. Rev.* **2019**, *119*, 5537–5606. [CrossRef] [PubMed]
8. Drews, J. Drug discovery: A historical perspective. *Science* **2000**, *287*, 1960–1964. [CrossRef]
9. Overington, J.P.; Al-Lazikani, B.; Hopkins, A.L. How many drug targets are there? *Nat. Rev. Drug Discov.* **2006**, *5*, 993–996. [CrossRef] [PubMed]
10. Xie, H. Activity assay of membrane transport proteins. *Acta Biochim. Biophys. Sin.* **2008**, *40*, 269–277. [CrossRef]
11. Obergrussberger, A.; Stolzle-Feix, S.; Becker, N.; Bruggemann, A.; Fertig, N.; Moller, C. Novel screening techniques for ion channel targeting drugs. *Channels* **2015**, *9*, 367–375. [CrossRef]
12. Nozeret, K.; Boucharlat, A.; Agou, F.; Buddelmeijer, N. A sensitive fluorescence-based assay to monitor enzymatic activity of the essential integral membrane protein Apolipoprotein N-acyltransferase (Lnt). *Sci. Rep.* **2019**, *9*, 15978. [CrossRef] [PubMed]
13. Quick, M.; Javitch, J.A. Monitoring the function of membrane transport proteins in detergent-solubilized form. *Proc. Natl. Acad. Sci. USA* **2007**, *104*, 3603–3608. [CrossRef] [PubMed]
14. Loll, P.J. Membrane proteins, detergents and crystals: What is the state of the art? *Acta Crystallogr. F Struct. Biol. Commun.* **2014**, *70*, 1576–1583. [CrossRef]
15. Bonnete, F.; Loll, P.J. Characterization of New Detergents and Detergent Mimetics by Scattering Techniques for Membrane Protein Crystallization. *Methods Mol. Biol.* **2017**, *1635*, 169–193. [CrossRef]
16. Nikolaev, M.; Round, E.; Gushchin, I.; Polovinkin, V.; Balandin, T.; Kuzmichev, P.; Shevchenko, V.; Borshchevskiy, V.; Kuklin, A.; Round, A.; et al. Integral Membrane Proteins Can Be Crystallized Directly from Nanodiscs. *Cryst. Growth Des.* **2017**, *17*, 945–948. [CrossRef]
17. Broecker, J.; Eger, B.T.; Ernst, O.P. Crystallogenesis of Membrane Proteins Mediated by Polymer-Bounded Lipid Nanodiscs. *Structure* **2017**, *25*, 384–392. [CrossRef]
18. Landau, E.M.; Rosenbusch, J.P. Lipidic cubic phases: A novel concept for the crystallization of membrane proteins. *Proc. Natl. Acad. Sci. USA* **1996**, *93*, 14532–14535. [CrossRef]
19. Cherezov, V. Lipidic cubic phase technologies for membrane protein structural studies. *Curr. Opin. Struct. Biol.* **2011**, *21*, 559–566. [CrossRef] [PubMed]
20. Agah, S.; Faham, S. Crystallization of membrane proteins in bicelles. *Methods Mol. Biol.* **2012**, *914*, 3–16. [CrossRef] [PubMed]
21. Zorzi, R.D.; Mi, W.; Liao, M.; Walz, T. Single-particle electron microscopy in the study of membrane protein structure. *Microscopy* **2016**, *65*, 81–96. [CrossRef] [PubMed]
22. Vinothkumar, K.R. Membrane protein structures without crystals, by single particle electron cryomicroscopy. *Curr. Opin. Struct. Biol.* **2015**, *33*, 103–114. [CrossRef]
23. Matthies, D.; Bae, C.; Toombes, G.E.; Fox, T.; Bartesaghi, A.; Subramaniam, S.; Swartz, K.J. Single-particle cryo-EM structure of a voltage-activated potassium channel in lipid nanodiscs. *Elife* **2018**, *7*, e37558. [CrossRef]

24. Wang, S.; Brettmann, J.B.; Nichols, C.G. Studying Structural Dynamics of Potassium Channels by Single-Molecule FRET. *Methods Mol. Biol.* **2018**, *1684*, 163–180. [CrossRef]
25. Gregorio, G.G.; Masureel, M.; Hilger, D.; Terry, D.S.; Juette, M.; Zhao, H.; Zhou, Z.; Perez-Aguilar, J.M.; Hauge, M.; Mathiasen, S.; et al. Single-molecule analysis of ligand efficacy in beta2AR-G-protein activation. *Nature* **2017**, *547*, 68–73. [CrossRef] [PubMed]
26. Asher, W.B.; Geggier, P.; Holsey, M.D.; Gilmore, G.T.; Pati, A.K.; Meszaros, J.; Terry, D.S.; Mathiasen, S.; Kaliszewski, M.J.; McCauley, M.D.; et al. Single-molecule FRET imaging of GPCR dimers in living cells. *Nat. Methods* **2021**, *18*, 397–405. [CrossRef] [PubMed]
27. Claxton, D.P.; Kazmier, K.; Mishra, S.; McHaourab, H.S. Navigating Membrane Protein Structure, Dynamics, and Energy Landscapes Using Spin Labeling and EPR Spectroscopy. *Methods Enzymol.* **2015**, *564*, 349–387. [CrossRef] [PubMed]
28. Georgieva, E.R.; Borbat, P.P.; Ginter, C.; Freed, J.H.; Boudker, O. Conformational ensemble of the sodium-coupled aspartate transporter. *Nat. Struct. Mol. Biol.* **2013**, *20*, 215–221. [CrossRef]
29. Georgieva, E.R.; Borbat, P.P.; Norman, H.D.; Freed, J.H. Mechanism of influenza A M2 transmembrane domain assembly in lipid membranes. *Sci. Rep.* **2015**, *5*, 11757. [CrossRef]
30. Georgieva, E.R. Nanoscale lipid membrane mimetics in spin-labeling and electron paramagnetic resonance spectroscopy studies of protein structure and function. *Nanotechnol. Rev.* **2017**, *6*, 75–92. [CrossRef]
31. Jeschke, G. DEER distance measurements on proteins. *Annu. Rev. Phys. Chem.* **2012**, *63*, 419–446. [CrossRef]
32. Borbat, P.P.; Surendhran, K.; Bortolus, M.; Zou, P.; Freed, J.H.; McHaourab, H.S. Conformational motion of the ABC transporter MsbA induced by ATP hydrolysis. *PLoS Biol.* **2007**, *5*, e271. [CrossRef]
33. Altenbach, C.; Marti, T.; Khorana, H.G.; Hubbell, W.L. Transmembrane protein structure: Spin labeling of bacteriorhodopsin mutants. *Science* **1990**, *248*, 1088–1092. [CrossRef] [PubMed]
34. Gluck, J.M.; Wittlich, M.; Feuerstein, S.; Hoffmann, S.; Willbold, D.; Koenig, B.W. Integral membrane proteins in nanodiscs can be studied by solution NMR spectroscopy. *J. Am. Chem. Soc.* **2009**, *131*, 12060–12061. [CrossRef] [PubMed]
35. Kang, C.; Li, Q. Solution NMR study of integral membrane proteins. *Curr. Opin. Chem. Biol.* **2011**, *15*, 560–569. [CrossRef]
36. Danmaliki, G.I.; Hwang, P.M. Solution NMR spectroscopy of membrane proteins. *Biochim. Biophys. Acta BioMembr.* **2020**, *1862*, 183356. [CrossRef]
37. Patching, S.G. Solid-state NMR structures of integral membrane proteins. *Mol. Membr. Biol.* **2015**, *32*, 156–178. [CrossRef]
38. Liang, B.; Tamm, L.K. NMR as a tool to investigate the structure, dynamics and function of membrane proteins. *Nat. Struct. Mol. Biol.* **2016**, *23*, 468–474. [CrossRef]
39. Brown, L.S.; Ladizhansky, V. Membrane proteins in their native habitat as seen by solid-state NMR spectroscopy. *Protein Sci.* **2015**, *24*, 1333–1346. [CrossRef] [PubMed]
40. Balana, B.; Taylor, N.; Slesinger, P.A. Mutagenesis and functional analysis of ion channels heterologously expressed in mammalian cells. *J. Vis. Exp.* **2010**, e2189. [CrossRef] [PubMed]
41. Lueck, J.D.; Mackey, A.L.; Infield, D.T.; Galpin, J.D.; Li, J.; Roux, B.; Ahern, C.A. Atomic mutagenesis in ion channels with engineered stoichiometry. *Elife* **2016**, *5*, e18976. [CrossRef] [PubMed]
42. Bezanilla, F. The voltage sensor in voltage-dependent ion channels. *Physiol. Rev.* **2000**, *80*, 555–592. [CrossRef] [PubMed]
43. Chavent, M.; Duncan, A.L.; Sansom, M.S. Molecular dynamics simulations of membrane proteins and their interactions: From nanoscale to mesoscale. *Curr. Opin. Struct. Biol.* **2016**, *40*, 8–16. [CrossRef] [PubMed]
44. Jiang, T.; Han, W.; Maduke, M.; Tajkhorshid, E. Molecular Basis for Differential Anion Binding and Proton Coupling in the Cl(-)/H(+) Exchanger ClC-ec1. *J. Am. Chem. Soc.* **2016**, *138*, 3066–3075. [CrossRef] [PubMed]
45. Muller, M.P.; Jiang, T.; Sun, C.; Lihan, M.; Pant, S.; Mahinthichaichan, P.; Trifan, A.; Tajkhorshid, E. Characterization of Lipid-Protein Interactions and Lipid-Mediated Modulation of Membrane Protein Function through Molecular Simulation. *Chem. Rev.* **2019**, *119*, 6086–6161. [CrossRef]
46. Ahmad, I.; Nawaz, N.; Darwesh, N.M.; Ur Rahman, S.; Mustafa, M.Z.; Khan, S.B.; Patching, S.G. Overcoming challenges for amplified expression of recombinant proteins using Escherichia coli. *Protein Expr. Purif.* **2018**, *144*, 12–18. [CrossRef]
47. Jensen, H.M.; Eng, T.; Chubukov, V.; Herbert, R.A.; Mukhopadhyay, A. Improving membrane protein expression and function using genomic edits. *Sci. Rep.* **2017**, *7*, 13030. [CrossRef] [PubMed]
48. Mathieu, K.; Javed, W.; Vallet, S.; Lesterlin, C.; Candusso, M.P.; Ding, F.; Xu, X.N.; Ebel, C.; Jault, J.M.; Orelle, C. Functionality of membrane proteins overexpressed and purified from E. coli is highly dependent upon the strain. *Sci. Rep.* **2019**, *9*, 2654. [CrossRef] [PubMed]
49. Heimpel, S.; Basset, G.; Odoy, S.; Klingenberg, M. Expression of the mitochondrial ADP/ATP carrier in Escherichia coli. Renaturation, reconstitution, and the effect of mutations on 10 positive residues. *J. Biol. Chem.* **2001**, *276*, 11499–11506. [CrossRef]
50. Angiulli, G.; Dhupar, H.S.; Suzuki, H.; Wason, I.S.; Duong Van Hoa, F.; Walz, T. New approach for membrane protein reconstitution into peptidiscs and basis for their adaptability to different proteins. *Elife* **2020**, *9*, e53530. [CrossRef]
51. Chien, C.H.; Helfinger, L.R.; Bostock, M.J.; Solt, A.; Tan, Y.L.; Nietlispach, D. An Adaptable Phospholipid Membrane Mimetic System for Solution NMR Studies of Membrane Proteins. *J. Am. Chem. Soc.* **2017**, *139*, 14829–14832. [CrossRef] [PubMed]
52. Schmidpeter, P.A.M.; Sukomon, N.; Nimigean, C.M. Reconstitution of Membrane Proteins into Platforms Suitable for Biophysical and Structural Analyses. *Methods Mol. Biol.* **2020**, *2127*, 191–205. [CrossRef]
53. Georgieva, E.R.; Fanouraki, C.; Borbat, P. Purification and Biophysical Characterization Of LysE Membrane Exporter from Mycobacterium tuberculosis in Lipodiscs Made of Native E. coli Membranes and Detergent. *Biophys. J.* **2020**, *118*, 503a–504a. [CrossRef]

54. Thoma, J.; Burmann, B.M. Fake It 'Till You Make It—The Pursuit of Suitable Membrane Mimetics for Membrane Protein Biophysics. *Int. J. Mol. Sci.* **2021**, *22*, 50. [CrossRef]
55. Lorsch, J. *Laboratory Methods in Enzymology: Protein Part C*; Elsevier Science: Amsterdam, The Netherlands, 2014.
56. Garavito, R.M.; Ferguson-Miller, S. Detergents as tools in membrane biochemistry. *J. Biol. Chem.* **2001**, *276*, 32403–32406. [CrossRef]
57. Popot, J.-L. Alternatives to Detergents for Handling Membrane Proteins in Aqueous Solutions. In *Membrane Proteins in Aqueous Solutions*; Springer: Berlin/Heidelberg, Germany, 2018; pp. 97–149.
58. Anson, M. The denaturation of proteins by detergents and bile salts. *Science* **1939**, *90*, 256–257. [CrossRef]
59. Brooks, M.M. Comparative studies on respiration: XV. The Effect of Bile Salts and of Saponin upon Respiration. *J. Gen. Physiol.* **1921**, *3*, 527–532. [CrossRef] [PubMed]
60. Jones, A.J.; Gabriel, F.; Tandale, A.; Nietlispach, D. Structure and Dynamics of GPCRs in Lipid Membranes: Physical Principles and Experimental Approaches. *Molecules* **2020**, *25*, 4729. [CrossRef] [PubMed]
61. Seddon, A.M.; Curnow, P.; Booth, P.J. Membrane proteins, lipids and detergents: Not just a soap opera. *Biochim. Biophys. Acta (BBA)-Biomembr.* **2004**, *1666*, 105–117. [CrossRef]
62. Koehler, J.; Sulistijo, E.S.; Sakakura, M.; Kim, H.J.; Ellis, C.D.; Sanders, C.R. Lysophospholipid micelles sustain the stability and catalytic activity of diacylglycerol kinase in the absence of lipids. *Biochemistry* **2010**, *49*, 7089–7099. [CrossRef] [PubMed]
63. Chipot, C.; Dehez, F.; Schnell, J.R.; Zitzmann, N.; Pebay-Peyroula, E.; Catoire, L.J.; Miroux, B.; Kunji, E.R.S.; Veglia, G.; Cross, T.A.; et al. Perturbations of Native Membrane Protein Structure in Alkyl Phosphocholine Detergents: A Critical Assessment of NMR and Biophysical Studies. *Chem. Rev.* **2018**, *118*, 3559–3607. [CrossRef] [PubMed]
64. Almgren, M. Mixed micelles and other structures in the solubilization of bilayer lipid membranes by surfactants. *Biochim. Biophys. Acta (BBA)-Biomembr.* **2000**, *1508*, 146–163. [CrossRef]
65. Ilgü, H.; Jeckelmann, J.-M.; Gachet, M.S.; Boggavarapu, R.; Ucurum, Z.; Gertsch, J.; Fotiadis, D. Variation of the detergent-binding capacity and phospholipid content of membrane proteins when purified in different detergents. *Biophys. J.* **2014**, *106*, 1660–1670. [CrossRef]
66. Rigaud, J.-L.; Levy, D.; Mosser, G.; Lambert, O. Detergent removal by non-polar polystyrene beads. *Eur. Biophys. J.* **1998**, *27*, 305–319. [CrossRef]
67. Tulumello, D.V.; Deber, C.M. Efficiency of detergents at maintaining membrane protein structures in their biologically relevant forms. *Biochim. Biophys. Acta (BBA)-Biomembr.* **2012**, *1818*, 1351–1358. [CrossRef]
68. Wiener, M.C. A pedestrian guide to membrane protein crystallization. *Methods* **2004**, *34*, 364–372. [CrossRef]
69. Klöpfer, K.; Hagn, F. Beyond detergent micelles: The advantages and applications of non-micellar and lipid-based membrane mimetics for solution-state NMR. *Prog. Nucl. Magn. Reson. Spectrosc.* **2019**, *114*, 271–283. [CrossRef]
70. Zhou, H.-X.; Cross, T.A. Influences of membrane mimetic environments on membrane protein structures. *Annu. Rev. Biophys.* **2013**, *42*, 361–392. [CrossRef] [PubMed]
71. Lee, S.; Mao, A.; Bhattacharya, S.; Robertson, N.; Grisshammer, R.; Tate, C.G.; Vaidehi, N. How do short chain nonionic detergents destabilize G-protein-coupled receptors? *J. Am. Chem. Soc.* **2016**, *138*, 15425–15433. [CrossRef]
72. Rajarathnam, K.; Rosgen, J. Isothermal titration calorimetry of membrane proteins—Progress and challenges. *Biochim. Biophys. Acta* **2014**, *1838*, 69–77. [CrossRef] [PubMed]
73. Draczkowski, P.; Matosiuk, D.; Jozwiak, K. Isothermal titration calorimetry in membrane protein research. *J. Pharm Biomed. Anal.* **2014**, *87*, 313–325. [CrossRef]
74. Hariharan, P.; Guan, L. Thermodynamic cooperativity of cosubstrate binding and cation selectivity of Salmonella typhimurium MelB. *J. Gen. Physiol.* **2017**, *149*, 1029–1039. [CrossRef]
75. Amin, A.; Hariharan, P.; Chae, P.S.; Guan, L. Effect of Detergents on Galactoside Binding by Melibiose Permeases. *Biochemistry* **2015**, *54*, 5849–5855. [CrossRef]
76. Reich-Slotky, R.; Panagiotidis, C.; Reyes, M.; Shuman, H.A. The detergent-soluble maltose transporter is activated by maltose binding protein and verapamil. *J. Bacteriol.* **2000**, *182*, 993–1000. [CrossRef]
77. Schneider, E.; Hunke, S. ATP-binding-cassette (ABC) transport systems: Functional and structural aspects of the ATP-hydrolyzing subunits/domains. *FEMS MicroBiol. Rev.* **1998**, *22*, 1–20. [CrossRef]
78. Infed, N.; Hanekop, N.; Driessen, A.J.; Smits, S.H.; Schmitt, L. Influence of detergents on the activity of the ABC transporter LmrA. *Biochim. Biophys. Acta* **2011**, *1808*, 2313–2321. [CrossRef]
79. Slotboom, D.J.; Duurkens, R.H.; Olieman, K.; Erkens, G.B. Static light scattering to characterize membrane proteins in detergent solution. *Methods* **2008**, *46*, 73–82. [CrossRef]
80. Kawate, T.; Gouaux, E. Fluorescence-detection size-exclusion chromatography for precrystallization screening of integral membrane proteins. *Structure* **2006**, *14*, 673–681. [CrossRef]
81. Alexandrov, A.I.; Mileni, M.; Chien, E.Y.; Hanson, M.A.; Stevens, R.C. Microscale fluorescent thermal stability assay for membrane proteins. *Structure* **2008**, *16*, 351–359. [CrossRef]
82. Kotov, V.; Bartels, K.; Veith, K.; Josts, I.; Subhramanyam, U.K.T.; Gunther, C.; Labahn, J.; Marlovits, T.C.; Moraes, I.; Tidow, H.; et al. High-throughput stability screening for detergent-solubilized membrane proteins. *Sci. Rep.* **2019**, *9*, 10379. [CrossRef]
83. Miles, A.J.; Wallace, B.A. Circular dichroism spectroscopy of membrane proteins. *Chem. Soc. Rev.* **2016**, *45*, 4859–4872. [CrossRef] [PubMed]
84. Ireland, S.M.; Sula, A.; Wallace, B.A. Thermal melt circular dichroism spectroscopic studies for identifying stabilising amphipathic molecules for the voltage-gated sodium channel NavMs. *Biopolymers* **2018**, *109*, e23067. [CrossRef] [PubMed]

85. Mancusso, R.; Karpowich, N.K.; Czyzewski, B.K.; Wang, D.N. Simple screening method for improving membrane protein thermostability. *Methods* **2011**, *55*, 324–329. [CrossRef] [PubMed]
86. Kunji, E.R.; Harding, M.; Butler, P.J.; Akamine, P. Determination of the molecular mass and dimensions of membrane proteins by size exclusion chromatography. *Methods* **2008**, *46*, 62–72. [CrossRef] [PubMed]
87. Parker, J.L.; Newstead, S. Membrane Protein Crystallisation: Current Trends and Future Perspectives. *Adv. Exp. Med. Biol.* **2016**, *922*, 61–72. [CrossRef] [PubMed]
88. Newstead, S.; Ferrandon, S.; Iwata, S. Rationalizing alpha-helical membrane protein crystallization. *Protein Sci.* **2008**, *17*, 466–472. [CrossRef]
89. Newstead, S.; Hobbs, J.; Jordan, D.; Carpenter, E.P.; Iwata, S. Insights into outer membrane protein crystallization. *Mol. Membr. Biol.* **2008**, *25*, 631–638. [CrossRef]
90. Sonoda, Y.; Cameron, A.; Newstead, S.; Omote, H.; Moriyama, Y.; Kasahara, M.; Iwata, S.; Drew, D. Tricks of the trade used to accelerate high-resolution structure determination of membrane proteins. *FEBS Lett.* **2010**, *584*, 2539–2547. [CrossRef]
91. Durr, K.L.; Chen, L.; Stein, R.A.; De Zorzi, R.; Folea, I.M.; Walz, T.; McHaourab, H.S.; Gouaux, E. Structure and dynamics of AMPA receptor GluA2 in resting, pre-open, and desensitized states. *Cell* **2014**, *158*, 778–792. [CrossRef]
92. Krishnamurthy, H.; Gouaux, E. X-ray structures of LeuT in substrate-free outward-open and apo inward-open states. *Nature* **2012**, *481*, 469–474. [CrossRef]
93. Quick, M.; Winther, A.M.; Shi, L.; Nissen, P.; Weinstein, H.; Javitch, J.A. Binding of an octylglucoside detergent molecule in the second substrate (S2) site of LeuT establishes an inhibitor-bound conformation. *Proc. Natl. Acad. Sci. USA* **2009**, *106*, 5563–5568. [CrossRef]
94. Perez, C.; Koshy, C.; Yildiz, O.; Ziegler, C. Alternating-access mechanism in conformationally asymmetric trimers of the betaine transporter BetP. *Nature* **2012**, *490*, 126–130. [CrossRef]
95. Birch, J.; Cheruvara, H.; Gamage, N.; Harrison, P.J.; Lithgo, R.; Quigley, A. Changes in Membrane Protein Structural Biology. *Biology* **2020**, *9*, 401. [CrossRef] [PubMed]
96. Gewering, T.; Januliene, D.; Ries, A.B.; Moeller, A. Know your detergents: A case study on detergent background in negative stain electron microscopy. *J. Struct. Biol.* **2018**, *203*, 242–246. [CrossRef]
97. Hauer, F.; Gerle, C.; Fischer, N.; Oshima, A.; Shinzawa-Itoh, K.; Shimada, S.; Yokoyama, K.; Fujiyoshi, Y.; Stark, H. GraDeR: Membrane Protein Complex Preparation for Single-Particle Cryo-EM. *Structure* **2015**, *23*, 1769–1775. [CrossRef] [PubMed]
98. Autzen, H.E.; Julius, D.; Cheng, Y. Membrane mimetic systems in CryoEM: Keeping membrane proteins in their native environment. *Curr. Opin. Struct. Biol.* **2019**, *58*, 259–268. [CrossRef]
99. Zhang, Z.; Chen, J. Atomic Structure of the Cystic Fibrosis Transmembrane Conductance Regulator. *Cell* **2016**, *167*, 1586–1597.e9. [CrossRef] [PubMed]
100. Basak, S.; Gicheru, Y.; Samanta, A.; Molugu, S.K.; Huang, W.; Fuente, M.; Hughes, T.; Taylor, D.J.; Nieman, M.T.; Moiseenkova-Bell, V.; et al. Cryo-EM structure of 5-HT3A receptor in its resting conformation. *Nat. Commun.* **2018**, *9*, 514. [CrossRef]
101. Basak, S.; Gicheru, Y.; Rao, S.; Sansom, M.S.P.; Chakrapani, S. Cryo-EM reveals two distinct serotonin-bound conformations of full-length 5-HT3A receptor. *Nature* **2018**, *563*, 270–274. [CrossRef]
102. Basak, S.; Kumar, A.; Ramsey, S.; Gibbs, E.; Kapoor, A.; Filizola, M.; Chakrapani, S. High-resolution structures of multiple 5-HT3AR-setron complexes reveal a novel mechanism of competitive inhibition. *Elife* **2020**, *9*, e57870. [CrossRef]
103. Mineev, K.S.; Nadezhdin, K.D. Membrane mimetics for solution NMR studies of membrane proteins. *Nanotechnol. Rev.* **2017**, *6*, 15–32. [CrossRef]
104. Meissner, A.; Sorensen, O.W. Optimization of three-dimensional TROSY-type HCCH NMR correlation of aromatic (1)H-(13)C groups in proteins. *J. Magn. Reson.* **1999**, *139*, 447–450. [CrossRef]
105. Tugarinov, V.; Hwang, P.M.; Kay, L.E. Nuclear magnetic resonance spectroscopy of high-molecular-weight proteins. *Annu. Rev. BioChem.* **2004**, *73*, 107–146. [CrossRef]
106. Columbus, L.; Lipfert, J.; Jambunathan, K.; Fox, D.A.; Sim, A.Y.; Doniach, S.; Lesley, S.A. Mixing and matching detergents for membrane protein NMR structure determination. *J. Am. Chem. Soc.* **2009**, *131*, 7320–7326. [CrossRef]
107. Hiller, S.; Garces, R.G.; Malia, T.J.; Orekhov, V.Y.; Colombini, M.; Wagner, G. Solution structure of the integral human membrane protein VDAC-1 in detergent micelles. *Science* **2008**, *321*, 1206–1210. [CrossRef]
108. Bayrhuber, M.; Meins, T.; Habeck, M.; Becker, S.; Giller, K.; Villinger, S.; Vonrhein, C.; Griesinger, C.; Zweckstetter, M.; Zeth, K. Structure of the human voltage-dependent anion channel. *Proc. Natl. Acad. Sci. USA* **2008**, *105*, 15370–15375. [CrossRef]
109. Liang, B.; Tamm, L.K. Structure of outer membrane protein G by solution NMR spectroscopy. *Proc. Natl. Acad. Sci. USA* **2007**, *104*, 16140–16145. [CrossRef]
110. Etzkorn, M.; Raschle, T.; Hagn, F.; Gelev, V.; Rice, A.J.; Walz, T.; Wagner, G. Cell-free expressed bacteriorhodopsin in different soluble membrane mimetics: Biophysical properties and NMR accessibility. *Structure* **2013**, *21*, 394–401. [CrossRef]
111. Isogai, S.; Deupi, X.; Opitz, C.; Heydenreich, F.M.; Tsai, C.J.; Brueckner, F.; Schertler, G.F.; Veprintsev, D.B.; Grzesiek, S. Backbone NMR reveals allosteric signal transduction networks in the beta1-adrenergic receptor. *Nature* **2016**, *530*, 237–241. [CrossRef]
112. Nygaard, R.; Zou, Y.; Dror, R.O.; Mildorf, T.J.; Arlow, D.H.; Manglik, A.; Pan, A.C.; Liu, C.W.; Fung, J.J.; Bokoch, M.P.; et al. The dynamic process of beta(2)-adrenergic receptor activation. *Cell* **2013**, *152*, 532–542. [CrossRef]
113. Ozawa, S.; Kimura, T.; Nozaki, T.; Harada, H.; Shimada, I.; Osawa, M. Structural basis for the inhibition of voltage-dependent K+ channel by gating modifier toxin. *Sci. Rep.* **2015**, *5*, 14226. [CrossRef]
114. Zhang, Q.; Horst, R.; Geralt, M.; Ma, X.; Hong, W.X.; Finn, M.G.; Stevens, R.C.; Wuthrich, K. Microscale NMR screening of new detergents for membrane protein structural biology. *J. Am. Chem. Soc.* **2008**, *130*, 7357–7363. [CrossRef] [PubMed]

115. Horst, R.; Stanczak, P.; Wuthrich, K. NMR polypeptide backbone conformation of the E. coli outer membrane protein W. *Structure* **2014**, *22*, 1204–1209. [CrossRef] [PubMed]
116. Maslennikov, I.; Klammt, C.; Hwang, E.; Kefala, G.; Okamura, M.; Esquivies, L.; Mors, K.; Glaubitz, C.; Kwiatkowski, W.; Jeon, Y.H.; et al. Membrane domain structures of three classes of histidine kinase receptors by cell-free expression and rapid NMR analysis. *Proc. Natl. Acad. Sci. USA* **2010**, *107*, 10902–10907. [CrossRef]
117. Kucharska, I.; Seelheim, P.; Edrington, T.; Liang, B.; Tamm, L.K. OprG Harnesses the Dynamics of its Extracellular Loops to Transport Small Amino Acids across the Outer Membrane of Pseudomonas aeruginosa. *Structure* **2015**, *23*, 2234–2245. [CrossRef] [PubMed]
118. Altenbach, C.; Greenhalgh, D.A.; Khorana, H.G.; Hubbell, W.L. A collision gradient method to determine the immersion depth of nitroxides in lipid bilayers: Application to spin-labeled mutants of bacteriorhodopsin. *Proc. Natl. Acad. Sci. USA* **1994**, *91*, 1667–1671. [CrossRef]
119. Bordignon, E. Site-directed spin labeling of membrane proteins. *Top. Curr. Chem.* **2012**, *321*, 121–157. [CrossRef] [PubMed]
120. Hubbell, W.L.; McHaourab, H.S.; Altenbach, C.; Lietzow, M.A. Watching proteins move using site-directed spin labeling. *Structure* **1996**, *4*, 779–783. [CrossRef]
121. Klug, C.S.; Feix, J.B. Methods and applications of site-directed spin labeling EPR spectroscopy. *Methods Cell Biol.* **2008**, *84*, 617–658. [CrossRef] [PubMed]
122. McHaourab, H.S.; Steed, P.R.; Kazmier, K. Toward the fourth dimension of membrane protein structure: Insight into dynamics from spin-labeling EPR spectroscopy. *Structure* **2011**, *19*, 1549–1561. [CrossRef]
123. Sahu, I.D.; Lorigan, G.A. Site-Directed Spin Labeling EPR for Studying Membrane Proteins. *BioMed Res. Int.* **2018**, *2018*, 3248289. [CrossRef]
124. Georgieva, E.R.; Fanouraki, C.; Borbat, P.P. Expression, purification and initial characterization of LysE membrane exporter from Mycobacterium tuberculosis: Towards comprehensive functional and structural study. *FASEB J.* **2020**, *34*, 1. [CrossRef]
125. Hanelt, I.; Wunnicke, D.; Bordignon, E.; Steinhoff, H.J.; Slotboom, D.J. Conformational heterogeneity of the aspartate transporter Glt(Ph). *Nat. Struct. Mol. Biol.* **2013**, *20*, 210–214. [CrossRef]
126. Hilger, D.; Polyhach, Y.; Padan, E.; Jung, H.; Jeschke, G. High-resolution structure of a Na+/H+ antiporter dimer obtained by pulsed electron paramagnetic resonance distance measurements. *Biophys. J.* **2007**, *93*, 3675–3683. [CrossRef]
127. Joseph, B.; Jeschke, G.; Goetz, B.A.; Locher, K.P.; Bordignon, E. Transmembrane gate movements in the type II ATP-binding cassette (ABC) importer BtuCD-F during nucleotide cycle. *J. Biol. Chem.* **2011**, *286*, 41008–41017. [CrossRef]
128. Nicklisch, S.C.; Wunnicke, D.; Borovykh, I.V.; Morbach, S.; Klare, J.P.; Steinhoff, H.J.; Kramer, R. Conformational changes of the betaine transporter BetP from Corynebacterium glutamicum studied by pulse EPR spectroscopy. *Biochim. Biophys. Acta* **2012**, *1818*, 359–366. [CrossRef] [PubMed]
129. Smirnova, I.; Kasho, V.; Choe, J.Y.; Altenbach, C.; Hubbell, W.L.; Kaback, H.R. Sugar binding induces an outward facing conformation of LacY. *Proc. Natl. Acad. Sci. USA* **2007**, *104*, 16504–16509. [CrossRef]
130. Yardeni, E.H.; Bahrenberg, T.; Stein, R.A.; Mishra, S.; Zomot, E.; Graham, B.; Tuck, K.L.; Huber, T.; Bibi, E.; McHaourab, H.S.; et al. Probing the solution structure of the E. coli multidrug transporter MdfA using DEER distance measurements with nitroxide and Gd(III) spin labels. *Sci. Rep.* **2019**, *9*, 12528. [CrossRef]
131. Dastvan, R.; Mishra, S.; Peskova, Y.B.; Nakamoto, R.K.; McHaourab, H.S. Mechanism of allosteric modulation of P-glycoprotein by transport substrates and inhibitors. *Science* **2019**, *364*, 689–692. [CrossRef]
132. Georgieva, E.R.; Borbat, P.P.; Grushin, K.; Stoilova-McPhie, S.; Kulkarni, N.J.; Liang, Z.; Freed, J.H. Conformational Response of Influenza A M2 Transmembrane Domain to Amantadine Drug Binding at Low pH (pH 5.5). *Front. Physiol.* **2016**, *7*, 317. [CrossRef]
133. Cuello, L.G.; Jogini, V.; Cortes, D.M.; Sompornpisut, A.; Purdy, M.D.; Wiener, M.C.; Perozo, E. Design and characterization of a constitutively open KcsA. *FEBS Lett.* **2010**, *584*, 1133–1138. [CrossRef]
134. Zhu, S.; Stein, R.A.; Yoshioka, C.; Lee, C.H.; Goehring, A.; McHaourab, H.S.; Gouaux, E. Mechanism of NMDA Receptor Inhibition and Activation. *Cell* **2016**, *165*, 704–714. [CrossRef]
135. Wingler, L.M.; Elgeti, M.; Hilger, D.; Latorraca, N.R.; Lerch, M.T.; Staus, D.P.; Dror, R.O.; Kobilka, B.K.; Hubbell, W.L.; Lefkowitz, R.J. Angiotensin Analogs with Divergent Bias Stabilize Distinct Receptor Conformations. *Cell* **2019**, *176*, 468–478.e11. [CrossRef] [PubMed]
136. Castell, O.K.; Dijkman, P.M.; Wiseman, D.N.; Goddard, A.D. Single molecule fluorescence for membrane proteins. *Methods* **2018**, *147*, 221–228. [CrossRef]
137. Bartels, K.; Lasitza-Male, T.; Hofmann, H.; Low, C. Single-Molecule FRET of Membrane Transport Proteins. *ChembioChem* **2021**. [CrossRef]
138. Zhao, Y.; Terry, D.; Shi, L.; Weinstein, H.; Blanchard, S.C.; Javitch, J.A. Single-molecule dynamics of gating in a neurotransmitter transporter homologue. *Nature* **2010**, *465*, 188–193. [CrossRef]
139. Raghuraman, H.; Chatterjee, S.; Das, A. Site-Directed Fluorescence Approaches for Dynamic Structural Biology of Membrane Peptides and Proteins. *Front. Mol. BioSci.* **2019**, *6*, 96. [CrossRef] [PubMed]
140. Ram, P.; Prestegard, J.H. Magnetic field induced ordering of bile salt/phospholipid micelles: New media for NMR structural investigations. *Biochim. Biophys. Acta* **1988**, *940*, 289–294. [CrossRef]
141. Sanders, C.R.; Landis, G.C. Reconstitution of membrane proteins into lipid-rich bilayered mixed micelles for NMR studies. *Biochemistry* **1995**, *34*, 4030–4040. [CrossRef]
142. Sanders, C.R.; Prosser, R.S. Bicelles: A model membrane system for all seasons? *Structure* **1998**, *6*, 1227–1234. [CrossRef]
143. Lucyanna, B.B.; Gelen, R.; Merce, C.; Laia, R.; Carmen, L.I.; de Alfons, L.M.; Olga, L. Structural Versatility of Bicellar Systems and Their Possibilities as Colloidal Carriers. *Pharmaceutics* **2011**, *3*, 636–664. [CrossRef]

144. Nusair, N.A.; Tiburu, E.K.; Dave, P.C.; Lorigan, G.A. Investigating fatty acids inserted into magnetically aligned phospholipid bilayers using EPR and solid-state NMR spectroscopy. *J. Magn. Reson.* **2004**, *168*, 228–237. [CrossRef] [PubMed]
145. Lu, J.X.; Caporini, M.A.; Lorigan, G.A. The effects of cholesterol on magnetically aligned phospholipid bilayers: A solid-state NMR and EPR spectroscopy study. *J. Magn. Reson.* **2004**, *168*, 18–30. [CrossRef] [PubMed]
146. Piai, A.; Fu, Q.; Dev, J.; Chou, J.J. Optimal bicelle q for solution NMR studies of protein transmembrane partition. *Chemistry* **2017**, *23*, 1361. [CrossRef] [PubMed]
147. Poget, S.F.; Girvin, M.E. Solution NMR of membrane proteins in bilayer mimics: Small is beautiful, but sometimes bigger is better. *Biochim. Biophys. Acta (BBA)-Biomembr.* **2007**, *1768*, 3098–3106. [CrossRef]
148. Puthenveetil, R.; Vinogradova, O. Solution NMR: A powerful tool for structural and functional studies of membrane proteins in reconstituted environments. *J. Biol. Chem.* **2019**, *294*, 15914–15931. [CrossRef] [PubMed]
149. Georgieva, E.R.; Ramlall, T.F.; Borbat, P.P.; Freed, J.H.; Eliezer, D. Membrane-bound alpha-synuclein forms an extended helix: Long-distance pulsed ESR measurements using vesicles, bicelles, and rodlike micelles. *J. Am. Chem. Soc.* **2008**, *130*, 12856–12857. [CrossRef]
150. Glover, K.J.; Whiles, J.A.; Wu, G.; Yu, N.; Deems, R.; Struppe, J.O.; Stark, R.E.; Komives, E.A.; Vold, R.R. Structural evaluation of phospholipid bicelles for solution-state studies of membrane-associated biomolecules. *Biophys. J.* **2001**, *81*, 2163–2171. [CrossRef]
151. Yeh, V.; Goode, A.; Bonev, B.B. Membrane protein structure determination and characterisation by solution and solid-state nmr. *Biology* **2020**, *9*, 396. [CrossRef]
152. Prosser, R.S.; Evanics, F.; Kitevski, J.L.; Al-Abdul-Wahid, M.S. Current applications of bicelles in NMR studies of membrane-associated amphiphiles and proteins. *Biochemistry* **2006**, *45*, 8453–8465. [CrossRef]
153. Vold, R.R.; Prosser, R.S. Magnetically oriented phospholipid bilayered micelles for structural studies of polypeptides. Does the ideal bicelle exist? *J. Magn. Reson.* **1996**, *113*, 267–271. [CrossRef]
154. Smrt, S.T.; Draney, A.W.; Singaram, I.; Lorieau, J.L. Structure and Dynamics of Membrane Proteins and Membrane Associated Proteins with Native Bicelles from Eukaryotic Tissues. *Biochemistry* **2017**, *56*, 5318–5327. [CrossRef]
155. Cardon, T.B.; Tiburu, E.K.; Padmanabhan, A.; Howard, K.P.; Lorigan, G.A. Magnetically aligned phospholipid bilayers at the parallel and perpendicular orientations for X-band spin-label EPR studies. *J. Am. Chem. Soc.* **2001**, *123*, 2913–2914. [CrossRef]
156. Tiburu, E.K.; Moton, D.M.; Lorigan, G.A. Development of magnetically aligned phospholipid bilayers in mixtures of palmitoylstearoylphosphatidylcholine and dihexanoylphosphatidylcholine by solid-state NMR spectroscopy. *Biochim. Biophys. Acta* **2001**, *1512*, 206–214. [CrossRef]
157. Fanucci, G.E.; Lee, J.Y.; Cafiso, D.S. Membrane mimetic environments alter the conformation of the outer membrane protein BtuB. *J. Am. Chem. Soc.* **2003**, *125*, 13932–13933. [CrossRef]
158. Uchida, N.; Nishizawa Horimoto, N.; Yamada, K.; Hikima, T.; Ishida, Y. Kinetically Stable Bicelles with Dilution Tolerance, Size Tunability, and Thermoresponsiveness for Drug Delivery Applications. *ChembioChem* **2018**, *19*, 1922–1926. [CrossRef] [PubMed]
159. Dürr, U.H.; Gildenberg, M.; Ramamoorthy, A. The magic of bicelles lights up membrane protein structure. *Chem. Rev.* **2012**, *112*, 6054–6074. [CrossRef]
160. Morrison, E.A.; Henzler-Wildman, K.A. Reconstitution of integral membrane proteins into isotropic bicelles with improved sample stability and expanded lipid composition profile. *Biochim. Biophys. Acta* **2012**, *1818*, 814–820. [CrossRef]
161. Graslund, A.; Maler, L. Testing membrane interactions of CPPs. *Methods Mol. Biol.* **2011**, *683*, 33–40. [CrossRef]
162. Maler, L. Solution NMR studies of peptide-lipid interactions in model membranes. *Mol. Membr. Biol.* **2012**, *29*, 155–176. [CrossRef]
163. Gautier, A.; Mott, H.R.; Bostock, M.J.; Kirkpatrick, J.P.; Nietlispach, D. Structure determination of the seven-helix transmembrane receptor sensory rhodopsin II by solution NMR spectroscopy. *Nat. Struct. Mol. Biol.* **2010**, *17*, 768–774. [CrossRef] [PubMed]
164. Morrison, E.A.; DeKoster, G.T.; Dutta, S.; Vafabakhsh, R.; Clarkson, M.W.; Bahl, A.; Kern, D.; Ha, T.; Henzler-Wildman, K.A. Antiparallel EmrE exports drugs by exchanging between asymmetric structures. *Nature* **2012**, *481*, 45–50. [CrossRef] [PubMed]
165. Bocharov, E.V.; Mayzel, M.L.; Volynsky, P.E.; Goncharuk, M.V.; Ermolyuk, Y.S.; Schulga, A.A.; Artemenko, E.O.; Efremov, R.G.; Arseniev, A.S. Spatial structure and pH-dependent conformational diversity of dimeric transmembrane domain of the receptor tyrosine kinase EphA1. *J. Biol. Chem.* **2008**, *283*, 29385–29395. [CrossRef] [PubMed]
166. Opella, S.J.; Marassi, F.M. Structure determination of membrane proteins by NMR spectroscopy. *Chem. Rev.* **2004**, *104*, 3587–3606. [CrossRef]
167. Carlotti, C.; Aussenac, F.; Dufourc, E.J. Towards high-resolution 1H-NMR in biological membranes: Magic angle spinning of bicelles. *Biochim. Biophys. Acta* **2002**, *1564*, 156–164. [CrossRef]
168. Ujwal, R.; Bowie, J.U. Crystallizing membrane proteins using lipidic bicelles. *Methods* **2011**, *55*, 337–341. [CrossRef]
169. Ujwal, R.; Abramson, J. High-throughput crystallization of membrane proteins using the lipidic bicelle method. *J. Vis. Exp.* **2012**, e3383. [CrossRef]
170. Faham, S.; Bowie, J.U. Bicelle crystallization: A new method for crystallizing membrane proteins yields a monomeric bacteriorhodopsin structure. *J. Mol. Biol.* **2002**, *316*, 1–6. [CrossRef]
171. Rasmussen, S.G.; Choi, H.J.; Rosenbaum, D.M.; Kobilka, T.S.; Thian, F.S.; Edwards, P.C.; Burghammer, M.; Ratnala, V.R.; Sanishvili, R.; Fischetti, R.F.; et al. Crystal structure of the human beta2 adrenergic G-protein-coupled receptor. *Nature* **2007**, *450*, 383–387. [CrossRef]
172. Vinothkumar, K.R. Structure of rhomboid protease in a lipid environment. *J. Mol. Biol.* **2011**, *407*, 232–247. [CrossRef]
173. Ujwal, R.; Cascio, D.; Colletier, J.P.; Faham, S.; Zhang, J.; Toro, L.; Ping, P.; Abramson, J. The crystal structure of mouse VDAC1 at 2.3 A resolution reveals mechanistic insights into metabolite gating. *Proc. Natl. Acad. Sci. USA* **2008**, *105*, 17742–17747. [CrossRef]

174. Sahu, I.D.; Mayo, D.J.; Subbaraman, N.; Inbaraj, J.J.; McCarrick, R.M.; Lorigan, G.A. Probing topology and dynamics of the second transmembrane domain (M2delta) of the acetyl choline receptor using magnetically aligned lipid bilayers (bicelles) and EPR spectroscopy. *Chem. Phys. Lipids* **2017**, *206*, 9–15. [CrossRef]
175. Nusair, N.A.; Mayo, D.J.; Dorozenski, T.D.; Cardon, T.B.; Inbaraj, J.J.; Karp, E.S.; Newstadt, J.P.; Grosser, S.M.; Lorigan, G.A. Time-resolved EPR immersion depth studies of a transmembrane peptide incorporated into bicelles. *Biochim. Biophys. Acta* **2012**, *1818*, 821–828. [CrossRef]
176. Bountra, K.; Hagelueken, G.; Choudhury, H.G.; Corradi, V.; El Omari, K.; Wagner, A.; Mathavan, I.; Zirah, S.; Yuan Wahlgren, W.; Tieleman, D.P.; et al. Structural basis for antibacterial peptide self-immunity by the bacterial ABC transporter McjD. *EMBO J.* **2017**, *36*, 3062–3079. [CrossRef]
177. Bayburt, T.H.; Grinkova, Y.V.; Sligar, S.G. Self-assembly of discoidal phospholipid bilayer nanoparticles with membrane scaffold proteins. *Nano Lett.* **2002**, *2*, 853–856. [CrossRef]
178. Bayburt, T.H.; Carlson, J.W.; Sligar, S.G. Reconstitution and imaging of a membrane protein in a nanometer-size phospholipid bilayer. *J. Struct. Biol.* **1998**, *123*, 37–44. [CrossRef] [PubMed]
179. Denisov, I.G.; Sligar, S.G. Nanodiscs in membrane biochemistry and biophysics. *Chem. Rev.* **2017**, *117*, 4669–4713. [CrossRef] [PubMed]
180. Civjan, N.R.; Bayburt, T.H.; Schuler, M.A.; Sligar, S.G. Direct solubilization of heterologously expressed membrane proteins by incorporation into nanoscale lipid bilayers. *Biotechniques* **2003**, *35*, 556–563. [CrossRef]
181. Frauenfeld, J.; Löving, R.; Armache, J.-P.; Sonnen, A.F.; Guettou, F.; Moberg, P.; Zhu, L.; Jegerschöld, C.; Flayhan, A.; Briggs, J.A. A saposin-lipoprotein nanoparticle system for membrane proteins. *Nat. Methods* **2016**, *13*, 345–351. [CrossRef] [PubMed]
182. Carlson, M.L.; Young, J.W.; Zhao, Z.; Fabre, L.; Jun, D.; Li, J.; Li, J.; Dhupar, H.S.; Wason, I.; Mills, A.T. The Peptidisc, a simple method for stabilizing membrane proteins in detergent-free solution. *Elife* **2018**, *7*, e34085. [CrossRef]
183. Juarez, J.F.B.; Harper, A.J.; Judge, P.J.; Tonge, S.R.; Watts, A. From polymer chemistry to structural biology: The development of SMA and related amphipathic polymers for membrane protein extraction and solubilisation. *Chem. Phys. Lipids* **2019**, *221*, 167–175. [CrossRef]
184. Hagn, F.; Etzkorn, M.; Raschle, T.; Wagner, G. Optimized phospholipid bilayer nanodiscs facilitate high-resolution structure determination of membrane proteins. *J. Am. Chem. Soc.* **2013**, *135*, 1919–1925. [CrossRef]
185. Hiller, S.; Wagner, G. Solution NMR spectroscopy of integral membrane proteins. In *Comprehensive Biophysics*; Elsevier: Amsterdam, The Netherlands, 2012.
186. Boldog, T.; Grimme, S.; Li, M.; Sligar, S.G.; Hazelbauer, G.L. Nanodiscs separate chemoreceptor oligomeric states and reveal their signaling properties. *Proc. Natl. Acad. Sci. USA* **2006**, *103*, 11509–11514. [CrossRef]
187. Bibow, S.; Polyhach, Y.; Eichmann, C.; Chi, C.N.; Kowal, J.; Albiez, S.; McLeod, R.A.; Stahlberg, H.; Jeschke, G.; Güntert, P. Solution structure of discoidal high-density lipoprotein particles with a shortened apolipoprotein AI. *Nat. Struct. Mol. Biol.* **2017**, *24*, 187–193. [CrossRef] [PubMed]
188. Denisov, I.G.; Grinkova, Y.V.; Lazarides, A.A.; Sligar, S.G. Directed self-assembly of monodisperse phospholipid bilayer Nanodiscs with controlled size. *J. Am. Chem. Soc.* **2004**, *126*, 3477–3487. [CrossRef]
189. Grinkova, Y.V.; Denisov, I.G.; Sligar, S.G. Engineering extended membrane scaffold proteins for self-assembly of soluble nanoscale lipid bilayers. *Protein Eng. Des. Sel.* **2010**, *23*, 843–848. [CrossRef]
190. Ritchie, T.K.; Grinkova, Y.V.; Bayburt, T.H.; Denisov, I.G.; Zolnerciks, J.K.; Atkins, W.M.; Sligar, S.G. Chapter 11—Reconstitution of membrane proteins in phospholipid bilayer nanodiscs. *Methods Enzymol.* **2009**, *464*, 211–231. [CrossRef]
191. Nasr, M.L.; Baptista, D.; Strauss, M.; Sun, Z.-Y.J.; Grigoriu, S.; Huser, S.; Plückthun, A.; Hagn, F.; Walz, T.; Hogle, J.M. Covalently circularized nanodiscs for studying membrane proteins and viral entry. *Nat. Methods* **2017**, *14*, 49–52. [CrossRef]
192. Knowles, T.J.; Finka, R.; Smith, C.; Lin, Y.-P.; Dafforn, T.; Overduin, M. Membrane proteins solubilized intact in lipid containing nanoparticles bounded by styrene maleic acid copolymer. *J. Am. Chem. Soc.* **2009**, *131*, 7484–7485. [CrossRef] [PubMed]
193. Hazell, G.; Arnold, T.; Barker, R.D.; Clifton, L.A.; Steinke, N.-J.; Tognoloni, C.; Edler, K.J. Evidence of lipid exchange in styrene maleic acid lipid particle (SMALP) nanodisc systems. *Langmuir* **2016**, *32*, 11845–11853. [CrossRef] [PubMed]
194. Jamshad, M.; Grimard, V.; Idini, I.; Knowles, T.J.; Dowle, M.R.; Schofield, N.; Sridhar, P.; Lin, Y.; Finka, R.; Wheatley, M. Structural analysis of a nanoparticle containing a lipid bilayer used for detergent-free extraction of membrane proteins. *Nano Res.* **2015**, *8*, 774–789. [CrossRef]
195. Vargas, C.; Arenas, R.C.; Frotscher, E.; Keller, S. Nanoparticle self-assembly in mixtures of phospholipids with styrene/maleic acid copolymers or fluorinated surfactants. *Nanoscale* **2015**, *7*, 20685–20696. [CrossRef]
196. Orwick, M.C.; Judge, P.J.; Procek, J.; Lindholm, L.; Graziadei, A.; Engel, A.; Gröbner, G.; Watts, A. Detergent-Free Formation and Physicochemical Characterization of Nanosized Lipid–Polymer Complexes: Lipodisq. *Angew. Chem. Int. Ed.* **2012**, *51*, 4653–4657. [CrossRef] [PubMed]
197. Ravula, T.; Hardin, N.Z.; Ramadugu, S.K.; Cox, S.J.; Ramamoorthy, A. Formation of pH-Resistant Monodispersed Polymer–Lipid Nanodiscs. *Angew. Chem. Int. Ed.* **2018**, *57*, 1342–1345. [CrossRef] [PubMed]
198. Ravula, T.; Ramadugu, S.K.; Di Mauro, G.; Ramamoorthy, A. Bioinspired, Size-Tunable Self-Assembly of Polymer–Lipid Bilayer Nanodiscs. *Angew. Chem.* **2017**, *129*, 11624–11628. [CrossRef]
199. Oluwole, A.O.; Danielczak, B.; Meister, A.; Babalola, J.O.; Vargas, C.; Keller, S. Solubilization of membrane proteins into functional lipid-bilayer nanodiscs using a diisobutylene/maleic acid copolymer. *Angew. Chem. Int. Ed.* **2017**, *56*, 1919–1924. [CrossRef]

200. Anantharamaiah, G.; Jones, J.; Brouillette, C.; Schmidt, C.; Chung, B.H.; Hughes, T.A.; Bhown, A.; Segrest, J. Studies of synthetic peptide analogs of the amphipathic helix. Structure of complexes with dimyristoyl phosphatidylcholine. *J. Biol. Chem.* **1985**, *260*, 10248–10255. [CrossRef]
201. Mishra, V.K.; Anantharamaiah, G.; Segrest, J.P.; Palgunachari, M.N.; Chaddha, M.; Sham, S.S.; Krishna, N.R. Association of a model class A (apolipoprotein) amphipathic α helical peptide with lipid: High resolution NMR studies of peptide· lipid discoidal complexes. *J. Biol. Chem.* **2006**, *281*, 6511–6519. [CrossRef]
202. Epand, R.; Gawish, A.; Iqbal, M.; Gupta, K.; Chen, C.H.; Segrest, J.; Anantharamaiah, G. Studies of synthetic peptide analogs of the amphipathic helix. Effect of charge distribution, hydrophobicity, and secondary structure on lipid association and lecithin: Cholesterol acyltransferase activation. *J. Biol. Chem.* **1987**, *262*, 9389–9396. [CrossRef]
203. Midtgaard, S.R.; Pedersen, M.C.; Kirkensgaard, J.J.K.; Sørensen, K.K.; Mortensen, K.; Jensen, K.J.; Arleth, L. Self-assembling peptides form nanodiscs that stabilize membrane proteins. *Soft Matter* **2014**, *10*, 738–752. [CrossRef] [PubMed]
204. Park, S.H.; Berkamp, S.; Cook, G.A.; Chan, M.K.; Viadiu, H.; Opella, S.J. Nanodiscs versus macrodiscs for NMR of membrane proteins. *Biochemistry* **2011**, *50*, 8983–8985. [CrossRef] [PubMed]
205. Carlson, M. Development of Methods for Study of Membrane Proteins in the Absence of Detergents. Doctoral Dissertation, University of British Columbia, Vancouver, AB, Canada, 2018.
206. Kondo, H.; Ikeda, K.; Nakano, M. Formation of size-controlled, denaturation-resistant lipid nanodiscs by an amphiphilic self-polymerizing peptide. *Colloids Surf. B Biointerfaces* **2016**, *146*, 423–430. [CrossRef] [PubMed]
207. Lyons, J.A.; Boggild, A.; Nissen, P.; Frauenfeld, J. Saposin-Lipoprotein Scaffolds for Structure Determination of Membrane Transporters. *Methods Enzymol.* **2017**, *594*, 85–99. [CrossRef] [PubMed]
208. Flayhan, A.; Mertens, H.D.; Ural-Blimke, Y.; Molledo, M.M.; Svergun, D.I.; Löw, C. Saposin lipid nanoparticles: A highly versatile and modular tool for membrane protein research. *Structure* **2018**, *26*, 345–355.e345. [CrossRef]
209. Kolter, T.; Sandhoff, K. Lysosomal degradation of membrane lipids. *FEBS Lett.* **2010**, *584*, 1700–1712. [CrossRef]
210. Hausler, E.; Fredriksson, K.; Goba, I.; Peters, C.; Raltchev, K.; Sperl, L.; Steiner, A.; Weinkauf, S.; Hagn, F. Quantifying the insertion of membrane proteins into lipid bilayer nanodiscs using a fusion protein strategy. *Biochim. Biophys. Acta BioMembr.* **2020**, *1862*, 183190. [CrossRef] [PubMed]
211. Shirzad-Wasei, N.; van Oostrum, J.; Bovee-Geurts, P.H.; Kusters, L.J.; Bosman, G.J.; DeGrip, W.J. Rapid transfer of overexpressed integral membrane protein from the host membrane into soluble lipid nanodiscs without previous purification. *Biol. Chem.* **2015**, *396*, 903–915. [CrossRef] [PubMed]
212. Overduin, M.; Esmaili, M. Native Nanodiscs and the Convergence of Lipidomics, Metabolomics, Interactomics and Proteomics. *Appl. Sci.* **2019**, *9*, 1230. [CrossRef]
213. Fiori, M.C.; Zheng, W.; Kamilar, E.; Simiyu, G.; Altenberg, G.A.; Liang, H. Extraction and reconstitution of membrane proteins into lipid nanodiscs encased by zwitterionic styrene-maleic amide copolymers. *Sci. Rep.* **2020**, *10*, 9940. [CrossRef]
214. Malhotra, K.; Alder, N.N. Reconstitution of Mitochondrial Membrane Proteins into Nanodiscs by Cell-Free Expression. *Methods Mol. Biol.* **2017**, *1567*, 155–178. [CrossRef] [PubMed]
215. Lyukmanova, E.N.; Shenkarev, Z.O.; Khabibullina, N.F.; Kopeina, G.S.; Shulepko, M.A.; Paramonov, A.S.; Mineev, K.S.; Tikhonov, R.V.; Shingarova, L.N.; Petrovskaya, L.E.; et al. Lipid-protein nanodiscs for cell-free production of integral membrane proteins in a soluble and folded state: Comparison with detergent micelles, bicelles and liposomes. *Biochim. Biophys. Acta* **2012**, *1818*, 349–358. [CrossRef]
216. Roos, C.; Kai, L.; Proverbio, D.; Ghoshdastider, U.; Filipek, S.; Dotsch, V.; Bernhard, F. Co-translational association of cell-free expressed membrane proteins with supplied lipid bilayers. *Mol. Membr. Biol.* **2013**, *30*, 75–89. [CrossRef]
217. Ly, S.; Bourguet, F.; Fischer, N.O.; Lau, E.Y.; Coleman, M.A.; Laurence, T.A. Quantifying interactions of a membrane protein embedded in a lipid nanodisc using fluorescence correlation spectroscopy. *Biophys. J.* **2014**, *106*, L05–L08. [CrossRef]
218. Malhotra, K.; Alder, N.N. Advances in the use of nanoscale bilayers to study membrane protein structure and function. *Biotechnol. Genet. Eng. Rev.* **2014**, *30*, 79–93. [CrossRef] [PubMed]
219. Nasr, M.L.; Singh, S.K. Radioligand binding to nanodisc-reconstituted membrane transporters assessed by the scintillation proximity assay. *Biochemistry* **2014**, *53*, 4–6. [CrossRef] [PubMed]
220. Kawai, T.; Caaveiro, J.M.; Abe, R.; Katagiri, T.; Tsumoto, K. Catalytic activity of MsbA reconstituted in nanodisc particles is modulated by remote interactions with the bilayer. *FEBS Lett.* **2011**, *585*, 3533–3537. [CrossRef]
221. Gardill, B.; Huang, J.; Tu, L.; Van Petegem, F.; Oxenoid, K.; Thomson, C.A. Nanodisc technology facilitates identification of monoclonal antibodies targeting multi-pass membrane proteins. *Sci. Rep.* **2020**, *10*, 1130. [CrossRef]
222. Swainsbury, D.J.; Scheidelaar, S.; van Grondelle, R.; Killian, J.A.; Jones, M.R. Bacterial reaction centers purified with styrene maleic acid copolymer retain native membrane functional properties and display enhanced stability. *Angew. Chem. Int. Ed. Engl.* **2014**, *53*, 11803–11807. [CrossRef] [PubMed]
223. Swainsbury, D.J.K.; Scheidelaar, S.; Foster, N.; van Grondelle, R.; Killian, J.A.; Jones, M.R. The effectiveness of styrene-maleic acid (SMA) copolymers for solubilisation of integral membrane proteins from SMA-accessible and SMA-resistant membranes. *Biochim. Biophys. Acta BioMembr.* **2017**, *1859*, 2133–2143. [CrossRef]
224. Kalienkova, V.; Alvadia, C.; Clerico Mosina, V.; Paulino, C. Single-Particle Cryo-EM of Membrane Proteins in Lipid Nanodiscs. *Methods Mol. Biol.* **2020**, *2127*, 245–273. [CrossRef]

225. Efremov, R.G.; Gatsogiannis, C.; Raunser, S. Lipid Nanodiscs as a Tool for High-Resolution Structure Determination of Membrane Proteins by Single-Particle Cryo-EM. *Methods Enzymol.* **2017**, *594*, 1–30. [CrossRef]
226. Qiu, W.; Fu, Z.; Xu, G.G.; Grassucci, R.A.; Zhang, Y.; Frank, J.; Hendrickson, W.A.; Guo, Y. Structure and activity of lipid bilayer within a membrane-protein transporter. *Proc. Natl. Acad. Sci. USA* **2018**, *115*, 12985–12990. [CrossRef] [PubMed]
227. Kumar, A.; Basak, S.; Rao, S.; Gicheru, Y.; Mayer, M.L.; Sansom, M.S.P.; Chakrapani, S. Mechanisms of activation and desensitization of full-length glycine receptor in lipid nanodiscs. *Nat. Commun.* **2020**, *11*, 3752. [CrossRef]
228. Rheinberger, J.; Gao, X.; Schmidpeter, P.A.; Nimigean, C.M. Ligand discrimination and gating in cyclic nucleotide-gated ion channels from apo and partial agonist-bound cryo-EM structures. *Elife* **2018**, *7*, e39775. [CrossRef]
229. Nygaard, R.; Kim, J.; Mancia, F. Cryo-electron microscopy analysis of small membrane proteins. *Curr. Opin. Struct. Biol.* **2020**, *64*, 26–33. [CrossRef] [PubMed]
230. Kim, J.; Tan, Y.Z.; Wicht, K.J.; Erramilli, S.K.; Dhingra, S.K.; Okombo, J.; Vendome, J.; Hagenah, L.M.; Giacometti, S.I.; Warren, A.L.; et al. Structure and drug resistance of the Plasmodium falciparum transporter PfCRT. *Nature* **2019**, *576*, 315–320. [CrossRef] [PubMed]
231. Puthenveetil, R.; Vinogradova, O. Optimization of the design and preparation of nanoscale phospholipid bilayers for its application to solution NMR. *Proteins Struct. Funct. Bioinform.* **2013**, *81*, 1222–1231. [CrossRef] [PubMed]
232. Kijac, A.Z.; Li, Y.; Sligar, S.G.; Rienstra, C.M. Magic-angle spinning solid-state NMR spectroscopy of nanodisc-embedded human CYP3A4. *Biochemistry* **2007**, *46*, 13696–13703. [CrossRef]
233. Lavington, S.; Watts, A. Lipid nanoparticle technologies for the study of G protein-coupled receptors in lipid environments. *Biophys. Rev.* **2020**, *12*, 1287–1302. [CrossRef]
234. Viennet, T.; Bungert-Plumke, S.; Elter, S.; Viegas, A.; Fahlke, C.; Etzkorn, M. Reconstitution and NMR Characterization of the Ion-Channel Accessory Subunit Barttin in Detergents and Lipid-Bilayer Nanodiscs. *Front. Mol. BioSci.* **2019**, *6*, 13. [CrossRef]
235. Martens, C.; Stein, R.A.; Masureel, M.; Roth, A.; Mishra, S.; Dawaliby, R.; Konijnenberg, A.; Sobott, F.; Govaerts, C.; McHaourab, H.S. Lipids modulate the conformational dynamics of a secondary multidrug transporter. *Nat. Struct. Mol. Biol.* **2016**, *23*, 744–751. [CrossRef]
236. Sahu, I.D.; Dixit, G.; Reynolds, W.D.; Kaplevatsky, R.; Harding, B.D.; Jaycox, C.K.; McCarrick, R.M.; Lorigan, G.A. Characterization of the Human KCNQ1 Voltage Sensing Domain (VSD) in Lipodisq Nanoparticles for Electron Paramagnetic Resonance (EPR) Spectroscopic Studies of Membrane Proteins. *J. Phys. Chem. B* **2020**, *124*, 2331–2342. [CrossRef] [PubMed]
237. Samanta, D.; Widom, J.; Borbat, P.P.; Freed, J.H.; Crane, B.R. Bacterial Energy Sensor Aer Modulates the Activity of the Chemotaxis Kinase CheA Based on the Redox State of the Flavin Cofactor. *J. Biol. Chem.* **2016**, *291*, 25809–25814. [CrossRef]
238. Nath, A.; Trexler, A.J.; Koo, P.; Miranker, A.D.; Atkins, W.M.; Rhoades, E. Single-molecule fluorescence spectroscopy using phospholipid bilayer nanodiscs. *Methods Enzymol.* **2010**, *472*, 89–117. [CrossRef] [PubMed]
239. Nath, A.; Koo, P.K.; Rhoades, E.; Atkins, W.M. Allosteric effects on substrate dissociation from cytochrome P450 3A4 in nanodiscs observed by ensemble and single-molecule fluorescence spectroscopy. *J. Am. Chem. Soc.* **2008**, *130*, 15746–15747. [CrossRef]
240. Sadler, E.E.; Kapanidis, A.N.; Tucker, S.J. Solution-Based Single-Molecule FRET Studies of K(+) Channel Gating in a Lipid Bilayer. *Biophys. J.* **2016**, *110*, 2663–2670. [CrossRef]
241. Overduin, M.; Esmaili, M. Structures and Interactions of Transmembrane Targets in Native Nanodiscs. *SLAS Discov.* **2019**, *24*, 943–952. [CrossRef]
242. Bangham, A.D.; Horne, R.W. Negative Staining of Phospholipids and Their Structural Modification by Surface-Active Agents as Observed in the Electron Microscope. *J. Mol. Biol.* **1964**, *8*, 660–668. [CrossRef]
243. Akbarzadeh, A.; Rezaei-Sadabady, R.; Davaran, S.; Joo, S.W.; Zarghami, N.; Hanifehpour, Y.; Samiei, M.; Kouhi, M.; Nejati-Koshki, K. Liposome: Classification, preparation, and applications. *Nanoscale Res. Lett.* **2013**, *8*, 102. [CrossRef] [PubMed]
244. Sessa, G.; Weissmann, G. Phospholipid spherules (liposomes) as a model for biological membranes. *J. Lipid Res.* **1968**, *9*, 310–318. [CrossRef]
245. Hickey, K.D.; Buhr, M.M. Lipid bilayer composition affects transmembrane protein orientation and function. *J. Lipids* **2011**, *2011*, 208457. [CrossRef]
246. Daraee, H.; Etemadi, A.; Kouhi, M.; Alimirzalu, S.; Akbarzadeh, A. Application of liposomes in medicine and drug delivery. *Artif. Cells Nanomed. Biotechnol.* **2016**, *44*, 381–391. [CrossRef] [PubMed]
247. Österlund, N.; Luo, J.; Wärmländer, S.K.; Gräslund, A. Membrane-mimetic systems for biophysical studies of the amyloid-β peptide. *Biochim. Biophys. Acta (BBA)-Proteins Proteom.* **2019**, *1867*, 492–501. [CrossRef]
248. Smirnova, I.A.; Adelroth, P.; Brzezinski, P. Extraction and liposome reconstitution of membrane proteins with their native lipids without the use of detergents. *Sci. Rep.* **2018**, *8*, 14950. [CrossRef] [PubMed]
249. Skrzypek, R.; Iqbal, S.; Callaghan, R. Methods of reconstitution to investigate membrane protein function. *Methods* **2018**, *147*, 126–141. [CrossRef]
250. Johnston, M.J.; Semple, S.C.; Klimuk, S.K.; Ansell, S.; Maurer, N.; Cullis, P.R. Characterization of the drug retention and pharmacokinetic properties of liposomal nanoparticles containing dihydrosphingomyelin. *Biochim. Biophys. Acta* **2007**, *1768*, 1121–1127. [CrossRef] [PubMed]
251. Allen, T.M. Liposomes. Opportunities in drug delivery. *Drugs* **1997**, *54* (Suppl. 4), 8–14. [CrossRef]
252. Parmar, M.M.; Edwards, K.; Madden, T.D. Incorporation of bacterial membrane proteins into liposomes: Factors influencing protein reconstitution. *Biochim. Biophys. Acta* **1999**, *1421*, 77–90. [CrossRef]

253. Winterstein, L.M.; Kukovetz, K.; Rauh, O.; Turman, D.L.; Braun, C.; Moroni, A.; Schroeder, I.; Thiel, G. Reconstitution and functional characterization of ion channels from nanodiscs in lipid bilayers. *J. Gen. Physiol.* **2018**, *150*, 637–646. [CrossRef]
254. Jones, A.J.; Blaza, J.N.; Bridges, H.R.; May, B.; Moore, A.L.; Hirst, J. A Self-Assembled Respiratory Chain that Catalyzes NADH Oxidation by Ubiquinone-10 Cycling between Complex I and the Alternative Oxidase. *Angew. Chem. Int. Ed.* **2016**, *55*, 728–731. [CrossRef] [PubMed]
255. Mouritsen, O.G. Model answers to lipid membrane questions. *Cold Spring Harb. Perspect. Biol.* **2011**, *3*, a004622. [CrossRef]
256. Plummer, A.M.; Fleming, K.G. BamA alone accelerates outer membrane protein folding in vitro through a catalytic mechanism. *Biochemistry* **2015**, *54*, 6009–6011. [CrossRef]
257. Teucher, M.; Zhang, H.; Bader, V.; Winklhofer, K.F.; Garcia-Saez, A.J.; Rajca, A.; Bleicken, S.; Bordignon, E. A new perspective on membrane-embedded Bax oligomers using DEER and bioresistant orthogonal spin labels. *Sci. Rep.* **2019**, *9*, 13013. [CrossRef]
258. Sejwal, K.; Chami, M.; Baumgartner, P.; Kowal, J.; Müller, S.A.; Stahlberg, H. Proteoliposomes—A system to study membrane proteins under buffer gradients by cryo-EM. *Nanotechnol. Rev.* **2017**, *6*, 57–74. [CrossRef]
259. Rideau, E.; Dimova, R.; Schwille, P.; Wurm, F.R.; Landfester, K. Liposomes and polymersomes: A comparative review towards cell mimicking. *Chem. Soc. Rev.* **2018**, *47*, 8572–8610. [CrossRef]
260. Angelova, M.I.; Dimitrov, D.S. Liposome electroformation. *Faraday Discuss. Chem. Soc.* **1986**, *81*, 303–311. [CrossRef]
261. Girard, P.; Pecreaux, J.; Lenoir, G.; Falson, P.; Rigaud, J.L.; Bassereau, P. A new method for the reconstitution of membrane proteins into giant unilamellar vesicles. *Biophys. J.* **2004**, *87*, 419–429. [CrossRef]
262. Dorr, J.M.; Koorengevel, M.C.; Schafer, M.; Prokofyev, A.V.; Scheidelaar, S.; van der Cruijsen, E.A.; Dafforn, T.R.; Baldus, M.; Killian, J.A. Detergent-free isolation, characterization, and functional reconstitution of a tetrameric K+ channel: The power of native nanodiscs. *Proc. Natl. Acad. Sci. USA* **2014**, *111*, 18607–18612. [CrossRef] [PubMed]
263. Scalise, M.; Pochini, L.; Giangregorio, N.; Tonazzi, A.; Indiveri, C. Proteoliposomes as tool for assaying membrane transporter functions and interactions with xenobiotics. *Pharmaceutics* **2013**, *5*, 472–497. [CrossRef]
264. Zollmann, T.; Moiset, G.; Tumulka, F.; Tampe, R.; Poolman, B.; Abele, R. Single liposome analysis of peptide translocation by the ABC transporter TAPL. *Proc. Natl. Acad. Sci. USA* **2015**, *112*, 2046–2051. [CrossRef]
265. Hresko, R.C.; Kraft, T.E.; Quigley, A.; Carpenter, E.P.; Hruz, P.W. Mammalian Glucose Transporter Activity Is Dependent upon Anionic and Conical Phospholipids. *J. Biol. Chem.* **2016**, *291*, 17271–17282. [CrossRef]
266. Peterson, E.; Ryser, T.; Funk, S.; Inouye, D.; Sharma, M.; Qin, H.; Cross, T.A.; Busath, D.D. Functional reconstitution of influenza A M2(22-62). *Biochim. Biophys. Acta* **2011**, *1808*, 516–521. [CrossRef]
267. Apell, H.J.; Marcus, M.M.; Anner, B.M.; Oetliker, H.; Lauger, P. Optical study of active ion transport in lipid vesicles containing reconstituted Na,K-ATPase. *J. Membr. Biol.* **1985**, *85*, 49–63. [CrossRef]
268. Nimigean, C.M. A radioactive uptake assay to measure ion transport across ion channel-containing liposomes. *Nat. Protoc.* **2006**, *1*, 1207–1212. [CrossRef]
269. Johnson, Z.L.; Lee, S.Y. Liposome reconstitution and transport assay for recombinant transporters. *Methods Enzymol.* **2015**, *556*, 373–383. [CrossRef]
270. Fiori, M.C.; Figueroa, V.; Zoghbi, M.E.; Saez, J.C.; Reuss, L.; Altenberg, G.A. Permeation of calcium through purified connexin 26 hemichannels. *J. Biol. Chem.* **2012**, *287*, 40826–40834. [CrossRef] [PubMed]
271. Kreiter, J.; Beitz, E.; Pohl, E.E. A Fluorescence-Based Method to Measure ADP/ATP Exchange of Recombinant Adenine Nucleotide Translocase in Liposomes. *Biomolecules* **2020**, *10*, 685. [CrossRef] [PubMed]
272. Fitzgerald, G.A.; Terry, D.S.; Warren, A.L.; Quick, M.; Javitch, J.A.; Blanchard, S.C. Quantifying secondary transport at single-molecule resolution. *Nature* **2019**, *575*, 528–534. [CrossRef]
273. Mulligan, C.; Fitzgerald, G.A.; Wang, D.N.; Mindell, J.A. Functional characterization of a Na+-dependent dicarboxylate transporter from Vibrio cholerae. *J. Gen. Physiol.* **2014**, *143*, 745–759. [CrossRef] [PubMed]
274. Ryan, R.M.; Compton, E.L.; Mindell, J.A. Functional characterization of a Na+-dependent aspartate transporter from Pyrococcus horikoshii. *J. Biol. Chem.* **2009**, *284*, 17540–17548. [CrossRef]
275. Stokes, D.L.; Rice, W.J.; Hu, M.; Kim, C.; Ubarretxena-Belandia, I. Two-dimensional crystallization of integral membrane proteins for electron crystallography. *Methods Mol. Biol.* **2010**, *654*, 187–205. [CrossRef]
276. Jiko, C.; Davies, K.M.; Shinzawa-Itoh, K.; Tani, K.; Maeda, S.; Mills, D.J.; Tsukihara, T.; Fujiyoshi, Y.; Kuhlbrandt, W.; Gerle, C. Bovine F1Fo ATP synthase monomers bend the lipid bilayer in 2D membrane crystals. *Elife* **2015**, *4*, e06119. [CrossRef]
277. Chadda, R.; Robertson, J.L. Measuring Membrane Protein Dimerization Equilibrium in Lipid Bilayers by Single-Molecule Fluorescence Microscopy. *Methods Enzymol.* **2016**, *581*, 53–82. [CrossRef] [PubMed]
278. You, M.; Li, E.; Wimley, W.C.; Hristova, K. Forster resonance energy transfer in liposomes: Measurements of transmembrane helix dimerization in the native bilayer environment. *Anal. BioChem.* **2005**, *340*, 154–164. [CrossRef]
279. Wang, Y.; Liu, Y.; Deberg, H.A.; Nomura, T.; Hoffman, M.T.; Rohde, P.R.; Schulten, K.; Martinac, B.; Selvin, P.R. Single molecule FRET reveals pore size and opening mechanism of a mechano-sensitive ion channel. *Elife* **2014**, *3*, e01834. [CrossRef]
280. Tribet, C.; Audebert, R.; Popot, J.-L. Amphipols: Polymers that keep membrane proteins soluble in aqueous solutions. *Proc. Natl. Acad. Sci. USA* **1996**, *93*, 15047–15050. [CrossRef] [PubMed]
281. Popot, J.-L.; Althoff, T.; Bagnard, D.; Banères, J.-L.; Bazzacco, P.; Billon-Denis, E.; Catoire, L.; Champeil, P.; Charvolin, D.; Cocco, M. Amphipols from A to Z. *Annu. Rev. Biophys.* **2011**, *40*, 379–408. [CrossRef] [PubMed]

282. Zoonens, M.; Catoire, L.J.; Giusti, F.; Popot, J.-L. NMR study of a membrane protein in detergent-free aqueous solution. *Proc. Natl. Acad. Sci. USA* **2005**, *102*, 8893–8898. [CrossRef] [PubMed]
283. Bazzacco, P.; Billon-Denis, E.; Sharma, K.S.; Catoire, L.J.; Mary, S.; Le Bon, C.; Point, E.; Banères, J.-L.; Durand, G.g.; Zito, F. Nonionic homopolymeric amphipols: Application to membrane protein folding, cell-free synthesis, and solution nuclear magnetic resonance. *Biochemistry* **2012**, *51*, 1416–1430. [CrossRef] [PubMed]
284. Zoonens, M.; Popot, J.-L. Amphipols for each season. *J. Membr. Biol.* **2014**, *247*, 759–796. [CrossRef] [PubMed]
285. Gohon, Y.; Giusti, F.; Prata, C.; Charvolin, D.; Timmins, P.; Ebel, C.; Tribet, C.; Popot, J.L. Well-defined nanoparticles formed by hydrophobic assembly of a short and polydisperse random terpolymer, amphipol A8-35. *Langmuir* **2006**, *22*, 1281–1290. [CrossRef] [PubMed]
286. Gaffrey, M. A comprehensive review of the lipid cubic phase or in meso method for crystallizing membrane and soluble proteins and complexes. *Acta Crystallogr. Sect. F* **2015**, *F71*, 3–18. [CrossRef]
287. Available online: https://cherezov.usc.edu/resources.htm (accessed on 30 August 2021).
288. Zhang, H.; Luginina, A.; Mishin, A.; Baidya, M.; Shukla, A.K.; Cherezov, V. Structural insights into ligand recognition and activation of angiotensin receptors. *Trends Pharmacol. Sci.* **2021**, *42*, 577–587. [CrossRef] [PubMed]
289. Rosenbaum, D.M.; Cherezov, V.; Hanson, M.A.; Rasmussen, S.G.; Thian, F.S.; Kobilka, T.S.; Choi, H.J.; Yao, X.J.; Weis, W.I.; Stevens, R.C.; et al. GPCR engineering yields high-resolution structural insights into beta2-adrenergic receptor function. *Science* **2007**, *318*, 1266–1273. [CrossRef]
290. Li, D.; Caffrey, M. Lipid cubic phase as a membrane mimetic for integral membrane protein enzymes. *Proc. Natl. Acad. Sci. USA* **2011**, *108*, 8639–8644. [CrossRef] [PubMed]
291. Li, D.; Caffrey, M. Structure and Functional Characterization of Membrane Integral Proteins in the Lipid Cubic Phase. *J. Mol. Biol.* **2020**, *432*, 5104–5123. [CrossRef] [PubMed]

membranes

MDPI

Article

Structural Arrangement Produced by Concanavalin A Binding to Homomeric GluK2 Receptors

Cuauhtemoc U. Gonzalez [1,2], Elisa Carrillo [1], Vladimir Berka [1] and Vasanthi Jayaraman [1,2,*]

[1] Center for Membrane Biology, Department of Biochemistry and Molecular Biology, University of Texas Health Science Center at Houston, 6431 Fannin St., Houston, TX 77030, USA; Cuauhtemoc.Gonzalez@uth.tmc.edu (C.U.G.); Elisa.CarrilloFlores@uth.tmc.edu (E.C.); Vladimir.Berka@uth.tmc.edu (V.B.)

[2] MD Anderson Cancer Center UTHealth Graduate School of Biomedical Sciences, University of Texas Health Science Center at Houston, 6431 Fannin St., Houston, TX 77030, USA

* Correspondence: vasanthi.jayaraman@uth.tmc.edu; Tel.: +1-(713)-500-6236

Abstract: Kainate receptors are members of the ionotropic glutamate receptor family. They form cation-specific transmembrane channels upon binding glutamate that desensitize in the continued presence of agonists. Concanavalin A (Con-A), a lectin, stabilizes the active open-channel state of the kainate receptor and reduces the extent of desensitization. In this study, we used single-molecule fluorescence resonance energy transfer (smFRET) to investigate the conformational changes underlying kainate receptor modulation by Con-A. These studies showed that Con-A binding to GluK2 homomeric kainate receptors resulted in closer proximity of the subunits at the dimer–dimer interface at the amino-terminal domain as well as between the subunits at the dimer interface at the agonist-binding domain. Additionally, the modulation of receptor functions by monovalent ions, which bind to the dimer interface at the agonist-binding domain, was not observed in the presence of Con-A. Based on these results, we conclude that Con-A modulation of kainate receptor function is mediated by a shift in the conformation of the kainate receptor toward a tightly packed extracellular domain.

Keywords: kainate receptor; Concanavalin A; smFRET; electrophysiology

1. Introduction

Kainate receptors belong to the family of cation-permeable channels known as ionotropic glutamate receptors (iGluRs). These receptors are activated by glutamate and are important in excitatory neuronal transmission throughout the central nervous system [1–4]. Kainate receptors, like other iGluRs, are tetrameric proteins arranged as a dimer of dimers in their extracellular segments. Furthermore, each subunit of the receptor can be separated into four major domains: the two extracellular domains, which comprise the amino-terminal domain (ATD) and the agonist-binding domain (ABD); the transmembrane segments; and the cytosolic carboxy-terminal domain [3,5,6]. It is through induced conformational changes occurring at these domains that agonists and other small molecules regulate channel activity [7–9].

Structural and spectroscopic measurements have provided significant insight into the conformational changes underlying kainate receptor activation and desensitization [6,7,9–14]. In the resting, apo state of the kainate receptor, when there is no glutamate bound, the cleft within the bi-lobed, agonist-binding domain is open, and there are extensive interactions (coupling) between the subunits forming the dimer of the ABD. The binding of glutamate leads to closure of the ABD cleft [13,15]. In the event of coupling across the ABD dimers, the cleft closure conformational change pulls on the transmembrane segments, moving them apart, resulting in opening of the channel pore and thus, allowing the influx of cations. Decoupling of the ABD dimer interface, on the other hand, allows for the release of stress at the transmembrane segments. This release in stress allows for the transmembrane segments to close, preventing further influx of cations and leading to receptor desensitization [6,7,9].

Exogenous jack-bean extract, Concanavalin A (Con-A), is a lectin which is known to bind to carbohydrates attached to proteins through *N*-glycosylation [16–18]. Con-A has been found to modulate the function of kainate receptors by changing the extent of desensitization [16,19]. Using site-directed mutagenesis, it has been shown that the glycosylation sites on kainate receptors play a role in Con-A modulation [17,18]. These glycosylation sites are spread throughout the extracellular ATD and ABD domains, and it has not been possible to draw direct correlations between binding site, conformation, and functional modulation [17,18,20,21]. Functional studies have shown that Con-A binding and modulation is state-dependent. Specifically, preincubation with Con-A prior to glutamate application produces larger effects on loss of desensitization than upon co-application with glutamate simultaneously. Furthermore, the preincubation effects are less pronounced with partial agonists [22,23]. The differential effects due to pre-incubation provide indirect evidence for the state-dependent effect of Con-A. However, apart from these indirect measurements, the specific conformational mechanism underlying Con-A modulation is still unknown.

Here we used single-molecule fluorescence resonance energy transfer (smFRET) to study the conformational changes that underlie the reduction of desensitization in kainate receptors due to Con-A. FRET is ideally suited for such studies, as it can measure distances like a molecular ruler by detecting the extent of energy transfer between a donor and acceptor fluorophore, which are attached to specific sites on the protein of interest. Using the FRET methodology at the single-molecule level allows us to understand the conformational heterogeneity in the protein associated with each of the functional states [24–34]. We have previously used smFRET to investigate the conformations associated with the apo (unliganded), channel open/active, and desensitized states of the GluK2 homomeric kainate receptors [34]. These studies identified the ABD dimer interface as a site showing large-scale changes between active and desensitized states with no significant changes at the ATD interface across the subunits. The ABD dimer interface showed multiples states with varying degrees of decoupling, with the dimer interface being primarily tightly coupled in the resting and active states and shifting toward decoupled states upon desensitization [34]. This was consistent with the mechanism that was proposed based on cryo-EM structures of the resting and desensitized states [6,7,9]. Given that the subunit interfaces played a role in conformational control of function, here we studied the conformational changes underlying the reduction in desensitization of kainate receptors by Con-A by investigating the changes at the ATD and ABD subunit interfaces. Our studies show that Con-A alters the subunit interface at both the ATD and ABD favoring states that stabilize the active state of the receptor.

Prior structural and functional studies have shown that Na^+ ions stabilize the dimer interface at the agonist-binding domain [10,35–39]. Removal of these ions, or their replacement with Cs^+, leads to an increase in population of receptor conformations that are decoupled at the dimer interface, and this, in turn, favors inactivation and stabilization of the desensitized state of the receptor [34,37]. Given that Con-A binding favors the coupled dimer interface at the agonist-binding domain, we investigated the ion modulation function of kainate receptors in the presence of Con-A to determine if the two modulators had overlapping conformational control of function.

2. Methods

2.1. Generation of FRET Constructs

GluK2 constructs containing native glutamine at site 590 from *Rattus norvegicus* previously used in cryo-EM and smFRET studies were used. GluK2 insertion into pcDNA3.1; replacement of cysteine to serine at sites C91, C199, and C432; and mutation of site

S266 (ATD) and A479 (LBD) to cysteine has been previously described in prior studies [6,7,33,34].

2.2. Electrophysiology

HEK 293T cells were transfected using Lipofectamine 2000 (GluK2-wt, GluK2-ABD, and GluK2-ATD) and co-transfected with GFP at a microgram ratio of 3:1. Whole-cell patch clamp recordings were performed 24–48 h after transfection using fire-polished borosilicate glass (Sutter instruments, Novato, CA, USA) pipettes with 3–5 megaohms resistance that were filled with internal solution: 110 mM CsF, 30 mM CsCl, 4 mM NaCl, 0.5 mM $CaCl_2$, 10 mM HEPES, and 5 mM EGTA (adjusted to pH 7.4 with CsOH). The extracellular solution consisted of 150 mM NaCl or CsCl, 2.8 mM KCl, 1.8 mM $CaCl_2$, 1.0 mM $MgCl_2$, and 10 mM HEPES and was adjusted to pH 7.4 with NaOH or CsOH. External solutions were locally applied to lifted cells using a SF-77B perfusion fast-step (Warner Instruments, Holliston, MA, USA) in the presence or absence of 1 mM glutamate. Recordings were performed using an Axopatch 200B amplifier (Molecular Devices, San Jose, CA, USA) at −60 mV hold potential, acquired at 10 kHz using pCLAMP10 software (Molecular Devices), and filtered online at 5 kHz. Single-channel recordings were performed in the outside-out patch-clamp configuration 24 h after transfection (GluK2-wt). Patch pipettes had a resistance of 8 to 15 megaohms when filled with an internal solution. Buffers and solution concentrations were like those used for whole cell recordings. Data were acquired at 50 kHz and low pass filtered at 10 kHz (Axon 200B and Digidata 1550A; Molecular Devices). The holding potential was −100 mV. Data were further filtered at 1 Hz. All recordings were idealized using the segmental *k*-means algorithm of QuB [40,41].

2.3. smFRET Sample Preparation

HEK 293T cells were transiently transfected using the JetPrime protocol at 10 μg per 10-cm plate. One day after transfection, the cells from two 10-cm dishes were harvested; washed with extracellular buffer (ECB) containing 135 mM NaCl or CsCl, 3 mM KCl, 2 mM $CaCl_2$, 20 mM glucose, and 20 mM HEPES; and adjusted to pH 7.4 with NaOH or CsOH. Post-wash, the cells were labeled for 1h at room temperature with 600 nM of donor fluorophore Alexa 555 maleimide (ThermoFisher, Waltham, MA, USA) and 2.4 μM of acceptor fluorophore Alexa 647 maleimide (ThermoFisher) in 3 mL of ECB. Post-labeling, the cells were washed and solubilized for 1 h at 4 °C. Solubilization buffer contained 1% lauryl maltose neopentyl glycol (Anatrace, Maumee, OH, USA), 2 mM cholesteryl hydrogen succinate (MP Biomedicals, Irvine, CA, USA), and $\frac{1}{4}$ protease inhibitor tablet (Pierce™) in phosphate-buffer saline containing either NaCl and Na_2HPO_4 for experiments in the presence of Na^+ or CsCl and K_2HPO_4 for experiments in the absence of Cs^+. Solubilized cells were filtered from unsolubilized debris by ultracentrifugation at 100,000× *g* for 1 h at 4 °C using a TLA 100.3 rotor. The supernatant was collected and kept on ice until used for smFRET samples.

2.4. smFRET Slide Preparation

Glass slides (20 × 20 mm) were washed using bath sonication in Liquinox phosphate-free detergent (Alconox, Inc., New York, NY, USA) followed by washing with 4.3% NH_4OH and 4.3% H_2O_2 solution. Slides were then washed with purified water and dried with nitrogen gas. Post-drying, slides were plasma cleaned using a Harrick Plasma PDC-32G Plasma Cleaner followed by Vectabond (Vector Laboratories, Burlingame, CA, USA) treatment for aminosilanization. Slides were then stored under vacuum. Silicon templates (Grace Bio-Labs, Bend, OR, USA) were cleaned by sonication in methanol for 20 min followed by vortexing. Silicon templates were stored in clean methanol and dried with nitrogen gas when applied over Vectabond-treated slides. Slides were then treated with 50 μL of PEG solution containing 0.25% *w/w* biotinylated PEG, 25% *w/w* mPEG-succinimidyl carbonate, and either 0.1 M $NaHCO_3$ or 0.1 m Tris for Na^+ free conditions and stored overnight in a dark, moist environment. The following day, the slides were cleaned with purified water and dried with nitrogen gas prior to treatment with short-chain PEG solution containing 25 mM short-chain 333 Da MS(PEG)$_4$ Methyl-PEG-NHS-Ester Reagent and either 0.1 M $NaHCO_3$ or 0.1 M Tris, followed by room temperature incubation for 2–3 h. After incubation, slides were washed with purified water and dried with nitrogen gas. Silicon templates were removed, and Hybridwell chambers and press-fit tubing connectors (Grace Bio-Labs) were applied. Streptavidin solution containing 0.2 mg/mL streptavidin, 1X smFRET imaging buffer (1 mM DDM (n-dodecyl-β-D-maltoside), 0.2 mM CHS (cholesteryl hydrogen succinate), and 1X phosphate-buffer saline containing either NaCl and Na_2HPO_4 or CsCl and K_2HPO_4 was applied inside the chamber and incubated for 10 min, followed by washing with the appropriate 1X phosphate-buffer saline with or without Na^+. 10 nM of biotinylated Goat Anti-Mouse IgG (H + L) secondary antibody (Jackson Immunoresearch Laboratories, Inc., West Grove, PA, USA, catalog number 115-065-003) was then flowed through the chamber and incubated for 20 min, followed by washing with the appropriate 1x phosphate-buffer saline. Next, 10 nM of 6X-His Tag mouse anti-Tag, Clone: HIS.H8 primary antibody (Invitrogen, Waltham, MA, USA, MA1213151MG) was introduced through the chamber and incubated for 20 min, followed by washing with the appropriate 1x phosphate-buffer saline. Bovine serum albumin (1 mg/mL) was introduced into the chamber and incubated for 15 min. Detergent-solubilized purified proteins containing 8X-His tagged homomeric GluK2 receptors were attached to the glass slide using an in situ immuno-precipitation (SiMPull) method by applying 50 μL of sample three times through the chamber and incubating for 20 min. 90 μL of oxygen-scavenging solution buffer system (ROXS) was applied inside the chamber containing 1 mM methyl viologen, 1 mM ascorbic acid, 0.01% *w/w* pyranose oxidase, 0.001% *w/v* catalase, 3.3% *w/w* glucose (all from Sigma-Aldrich, Inc., St. Louis, MO, USA), 1 mM DDM (Chem-Impex, Wood Dale, IL, USA), and 0.2 mM CHS (MP Biomedicals, LLC, Santa Ana, CA, USA) in phosphate-buffer saline containing either NaCl and Na_2HPO_4 or CsCl and K_2HPO_4, pH 7.4. In the glutamate-treated condition, 1 mM glutamate, 1 mM of $MnCl_2$, and 1 mM of $CaCl_2$ were introduced into the ROXS. In the Con-A treated condition, 40 μM Con-A (MilliporeSigma, Burlington, MA, USA), 1 mM of $MnCl_2$, and 1 mM of $CaCl_2$ were introduced into the ROXS and incubated for 10 min before application of glutamate-containing ROXS.

2.5. smFRET Data Acquisition

smFRET measurements were collected using a custom-built PicoQuant Microtime 200 Fluorescence Lifetime Microscope. To characterize the fluorescent behavior of both donor and acceptor fluorophores as well as the efficiency of the energy transfer between donors and acceptors, acquisition was conducted using both 532 nm (LDH-D-TA-530; Picoquant, Berlin, Germany) and 637 nm (LDH-D-C-640; Picoquant) lasers using pulsed interleaved excitation at 80 MHz. During scanning, the slide was immobilized on a scanning x-y-z piezo stage (P-733.2CD; Physik Instrumente) and observed through a 100× oil-immersed lens (100 × 1.4 NA; Olympus, Tokyo, Japan). Photons from samples post-excitation were then collected back through the objective and separated through a dual band dichroic beam splitter (Zt532/640rpc-UF3; AHF/Chroma, Bellows Falls, VT, USA). Prior to detection, photons were then filtered through emission filters (550 nm (FF01-582/64; AHF/Semrock, Rochester, NY, USA) for the donor or 650 nm (2XH690/70; AHF, Tübingen-Pfrondorf, Germany) for the acceptor) and into two SPAD photodiodes (SPCM CD3516H; Excelitas Technologies, Waltham, MA, USA).

2.6. smFRET Data Analysis

Molecules exhibiting a donor–acceptor anticorrelation and with a single photobleaching step were used in the analysis. FRET efficiencies were calculated based on the donor and acceptor intensities. MATLAB (Mathworks, Natick, MA, USA) was used to denoise donor and acceptor traces using wavelet-based denoising [42]. Origin (OriginLab Corp, Northampton, MA, USA) was used to create smFRET histograms and traces. The total numbers of molecules used for each condition were as follows: Con-A treated under Na^+ condition, 40 molecules for GluK2-ATD and 38 molecules for GluK2-LBD; Con-A treated under Cs^+ condition, 36 molecules for GluK2-LBD.

3. Results

We used maleimide-thiol reaction chemistry to attach donor and acceptor fluorophores at specific sites on the GluK2 receptor via mutated cysteines. To introduce cysteines at specific sites, we used a Cys-light GluK2 receptor, where the extracellular accessible cysteines at sites C91, C199, and C432 were mutated to serines [32–34]. For investigating changes at the dimer–dimer interface at the ATD we introduced a cysteine at site 266, henceforth referred to as GluK2-ATD. To study the interface within the dimer at the ABD, we introduced a cysteine at site 479 in the Cys-light background, henceforth referred to as GluK2-ABD. These sites were optimal for the donor (Alexa 555) and acceptor (Alexa 647) fluorophore pair, as FRET was only observed for the interface of interest, and all other distances had much lower FRET efficiencies (the R_0 for Alexa 555 and Alexa 647 was 51 Å) (Figure 1). In order to attach the receptor to the imaging cover slip, we introduced an octa-histidine tag on the C-terminus of the receptor, allowing us to perform in situ pulldown of solubilized receptors that were expressed in HEK-293T cells [43]. Electrophysiological measurements were performed with the smFRET constructs, GluK2-ATD and GluK2-ABD, expressed in HEK-293T cells. The glutamate-mediated currents, as well as the reduction in the extent of desensitization by Con-A in the glutamate-mediated currents for these constructs, are shown in Figure 1 and were similar to that of the wild-type GluK2 receptor. This established that these smFRET constructs served as good models for wild-type GluK2 receptors.

To perform smFRET measurements, we transfected HEK 293T cells with GluK2-ATD or GluK2-ABD and labeled them with donor and acceptor fluorophores. Detergent-solubilized membrane preparations of these cells were used for the in situ pull down of the receptor onto the coverslip. Fluorescence intensities of donor and acceptor fluorophores measured from single molecules using donor excitation (for determining FRET) were used to determine FRET efficiency between the donor and acceptor fluorophores. To ensure that only a single FRET distance was being measured, only traces exhibiting a single donor and a single acceptor photobleaching step and showing anticorrelation after the photobleaching of the acceptor were used to obtain FRET efficiencies. Representative single molecule traces and the ensemble FRET efficiency histograms from at least 30 molecules are shown in Figure 2 for GluK2-ATD and in Figures 3 and 4 for GluK2-ABD. Wavelet-based denoising was performed on the donor and acceptor traces, and these denoised traces were used to determine FRET efficiencies of the different conformational states. Histograms showing the denoised efficiencies are overlaid on the histograms from the observed traces [42].

In the presence of Con-A and glutamate, the denoised smFRET efficiency histogram for GluK2-ATD showed that the receptor exhibited a primary population that had a FRET efficiency of 0.95, corresponding to the distance of 31 Å (Figure 2), with smaller fractions at FRET efficiencies of 0.88, 0.80, and 0.72, corresponding to distances of 37 Å, 40 Å, and 44 Å, respectively (Figure 2). The primary FRET efficiency of 0.95 for the ATD dimer distance, at site 266, was significantly higher than the primary FRET efficiency of 0.83 that we previously observed for GluK2-ATD receptors bound to glutamate in the absence of Con-A. This shift in the FRET efficiency population histogram is shown in the overlay of the observed FRET efficiency histogram for GluK2-ATD bound to glutamate in the presence and absence of Con-A in Figure 2. Based on these data showing conformations with higher FRET efficiency corresponding to shorter distances between the dimer–dimer subunits, we could conclude that Con-A binding to GluK2 receptors led to tighter packing at the ATD.

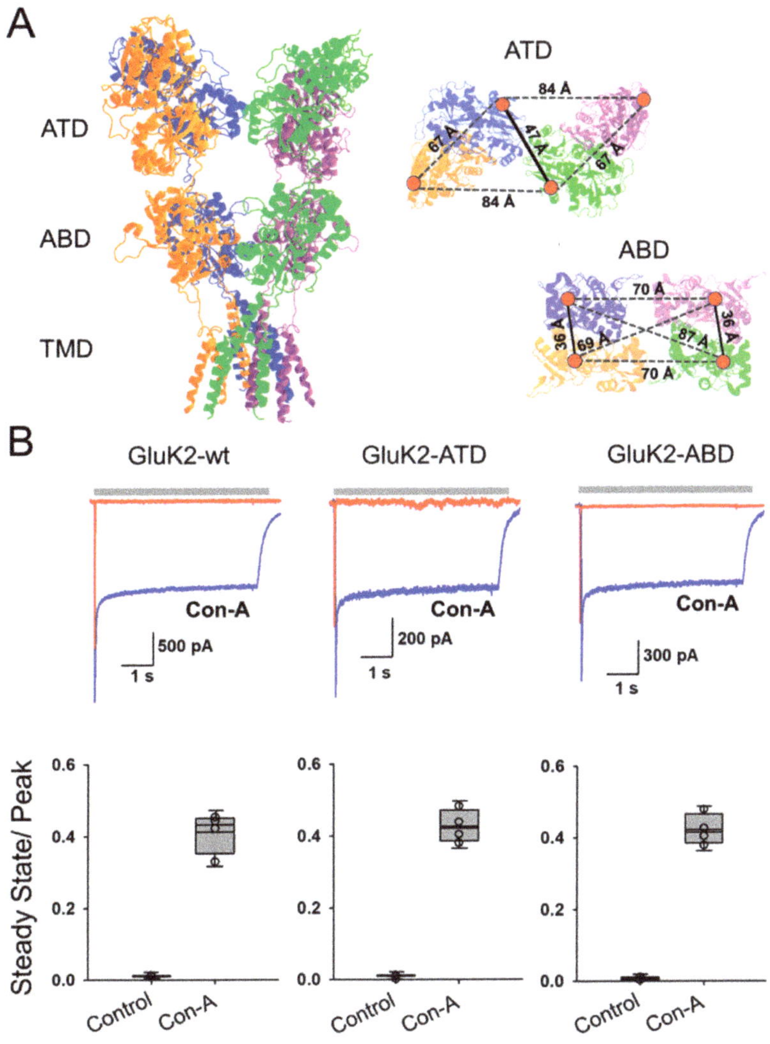

Figure 1. Depiction of the structural arrangement of homomeric GluK2 receptor and fluorophore conjugation sites. (**A**) Cryo-EM structure of a homomeric GluK2 receptor (PDB:5KUF) displaying the amino-terminal domain (ATD), agonist-binding domain (ABD), and transmembrane domain (TMD) and sites of fluorophore conjugation at the ATD (266) and ABD (479); top-down view shown for ATD (PDB:5KUF) and ABD (PDB:5KUH) with distances between sites 266 and 479, respectively. (**B**) Representative whole-cell recordings with 1 mM of glutamate for GluK2-wt, GluK2-ATD, and GluK2-ABD in the presence (blue) or absence (red) of 40 μM of Con-A (top) along with the ratio of steady-state to peak currents obtained from at least three different cells (bottom). These data show higher steady-state currents in the presence of Con-A.

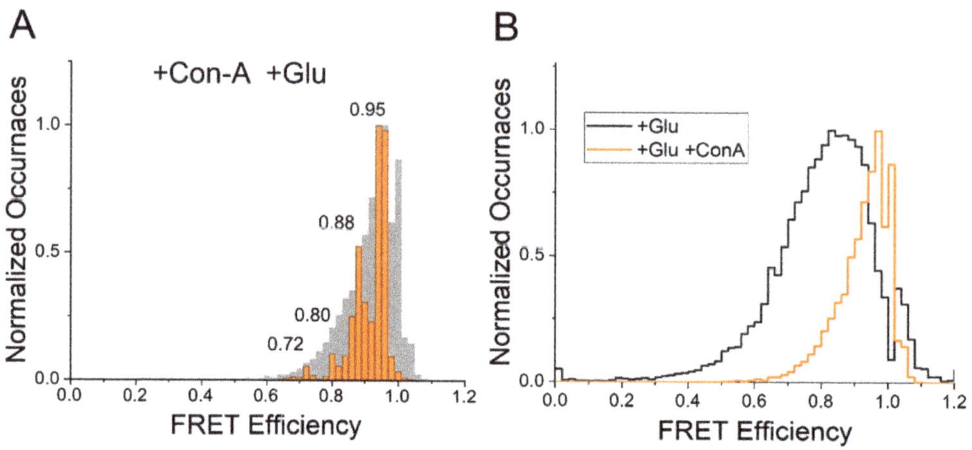

Figure 2. Conformational landscape of the distance across the dimer–dimer interface at the amino-terminal domain (ATD) of the homomeric GluK2 receptor (measured at site 266). (**A**) Denoised FRET efficiency histogram, in orange, overlapped onto an observed FRET efficiency histogram, in gray, depicting FRET efficiency occurrences in the presence of Con-A and glutamate. (**B**) Comparison between observed FRET efficiency histograms for glutamate-bound homomeric GluK2 receptors in the absence (in black) [34] and presence (in orange) of Con-A, illustrating a shift toward higher FRET efficiencies in the presence of Con-A.

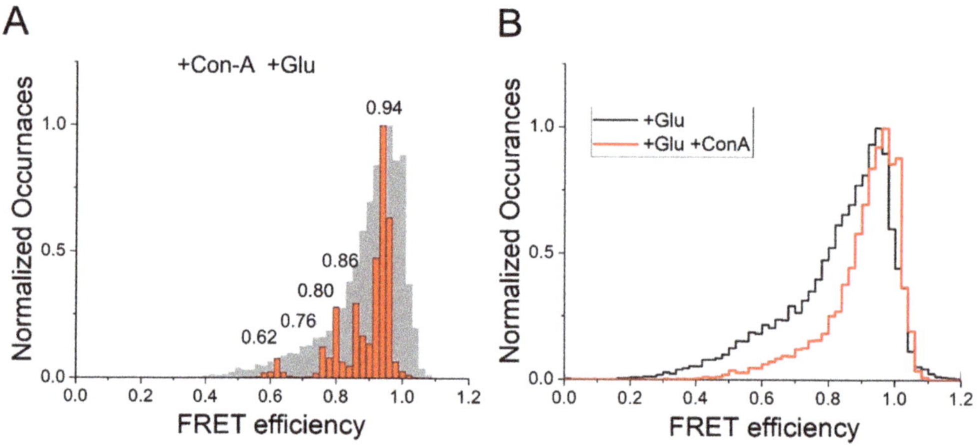

Figure 3. Conformational landscape of the distance between subunits at the dimer interface of the agonist-binding domain (ABD) of a homomeric GluK2 receptor (measured at site 479). (**A**) Denoised FRET efficiency histogram, in red, overlapped over observed FRET efficiency histogram, in gray, depicting FRET efficiency occurrences in the presence of Con-A and glutamate. (**B**) Comparison between observed FRET efficiency histograms of glutamate-bound homomeric GluK2 receptors in the absence (in black) [34] and presence (in red) of Con-A, illustrating a decrease in the low-FRET efficiency population in the presence of Con-A.

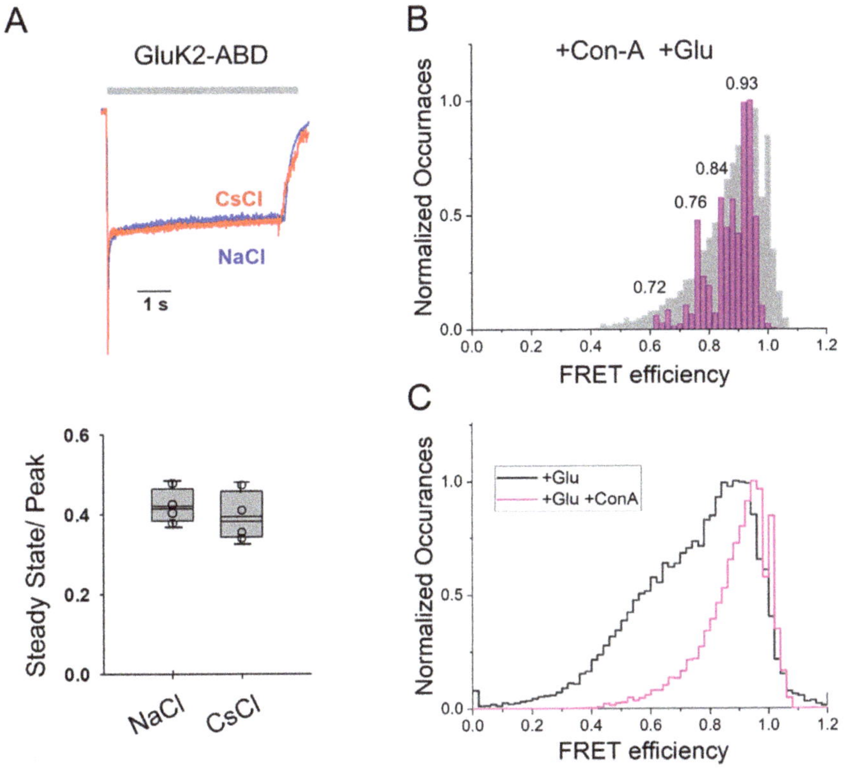

Figure 4. Effect of Con-A on ion modulation of GluK2 receptors. (**A**) Representative whole-cell recording with 1mM of glutamate and 40μM of Con-A for GluK2-ABD in the presence of either CsCl (red) or NaCl (blue) (top) along with the ratio of steady-state to peak currents obtained from at least three cells (bottom), showing no difference in steady-state response between NaCl- and CsCl-treated. (**B**) Denoised FRET efficiency histogram, in magenta, overlapped over the observed FRET efficiency histogram, in gray, depicting FRET efficiency occurrences in the presence of Con-A and glutamate obtained with CsCl replacing NaCl. (**C**) Comparison between observed FRET efficiency histograms of the homomeric, glutamate-bound, GluK2 receptor obtained with CsCl replacing NaCl, in the absence (in black) [34] and presence of Con-A (in magenta), illustrating a decrease in the low-FRET efficiency population in the presence of Con-A.

At the ABD dimer interface, smFRET histograms with GluK2-ABD receptors showed a major peak region at 0.94 FRET efficiency, corresponding to a distance of 32 Å (Figure 3) in the presence of Con-A and glutamate. Additional, less populous peaks were observed at 0.86, 0.80, 0.76, and 0.62, corresponding to 38 Å, 40 Å, 42 Å, and 47 Å, respectively (Figure 3). While conformations with similar FRET efficiencies were observed previously for the GluK2 receptors bound to glutamate in the absence of Con-A, the fraction of receptors in the high FRET states was significantly lower than that observed in the presence of Con-A. This shift in population to higher FRET efficiency states in the presence of Con-A can be seen in the overlay of the FRET efficiency histogram of GluK2-ABD bound to glutamate in the presence and absence of Con-A (Figure 3B). Prior structural and spectroscopic investigations of the α-amino-3-hydroxy-5-methyl-4-isoxazolepropionic acid receptor (AMPAR) and kainate receptors have shown that decoupling of the agonist-binding dimer drives desensitization in these receptors [6,9,44]. Thus, our data, which show a shift toward a more coupled dimer interface at the agonist-binding domain in the presence of Con-A, suggest that Con-A favors activation and reduces desensitization by allosterically stabilizing the dimer interface at the agonist-binding domain.

To further investigate the changes at the dimer interface of the ABD caused by Con-A binding, we studied the effect of Con-A on the Na^+ modulation of kainate receptor function. For investigating Na^+ ion modulation, we compared the whole-cell currents from GluK2-ABD homomeric receptors mediated by 10 mM glutamate in the presence of Con-A with Na^+ ions in the extracellular buffer to those measured in the extracellular buffer when Na^+ was replaced with Cs^+ (Figure 4A). No significant differences were observed in the currents when Na^+ was replaced with Cs^+ in the presence of Con-A. These results in the presence of Con-A are in contrast with prior investigations showing a large decrease in currents when Na^+ was replaced with Cs^+ in the absence of Con-A [37]. Additionally, the extent of the residual current (non-desensitizing) due to stabilization with Con-A was similar in Cs^+ (Figure 4A) and in Na^+ (Figure 1). These investigations showed that in the presence of Con-A, the Na^+ ion modulation of GluK2 receptor function was not observed and was consistent with the smFRET observations that showed that Con-A stabilized the ABD dimer interface and thus, removed the need for stabilization by Na^+ ions. To further confirm that the stabilization of the dimer interface in the presence of Con-A was maintained in the presence of Cs^+ ions, we investigated the distances at the dimer interface using smFRET. The smFRET histogram obtained with GluK2-ABD receptors when Na^+ ions were replaced with Cs^+ ions (Figure 4B) was similar to that obtained with Na^+ ions (Figure 3). The shift to higher efficiency FRET conformations in the presence of Con-A relative to those previously reported in its absence in measurements performed when Na^+ ions were replaced with Cs^+ ions is shown by overlapping the FRET efficiency histograms (Figure 4C). Overall, the smFRET data show that Con-A favors a more coupled dimer interface even when Na^+ ions are replaced with Cs^+ ions, and this, in turn, can be related to the similar whole cell currents observed in both Na^+ ions and Cs^+ ions.

4. Discussion

Lectins such as Con-A are proteins that bind specifically to carbohydrates, glycoproteins, and glycolipids. The pharmacological effects of Con-A on kainate receptors have been studied extensively in both native and heterologous expression systems [16–19,36,45,46]. These studies have shown that Con-A increases the steady-state response of kainate receptors while not altering the kinetics for desensitization. To account for these electrophysiological observations, a kinetic mechanism involving multiple open states has been proposed, with Con-A shifting the population between these states. Additionally, mutational studies have shown that Con-A binds irreversibly to nine *N*-glycosylated amino acid residues located near the agonist-binding domain [18]. Given the multiple Con-A binding sites and the fact that tetrameric Con-A is more effective than the dimeric succinyl Con-A [38], it has been proposed that Con-A may act by restricting protein movements in the ABD. Consistent with the hypothesis that Con-A alters the agonist-binding domain, Con-A has been shown to have differential effects on currents mediated by agonists of differing efficacy [23]. These biochemical and functional studies provide indirect evidence of a shift in the conformational landscape of the agonist-binding domain as the mechanism underlying Con-A modulation of kainate receptors, and our smFRET investigations presented here provide insight into the specific conformations and shifts induced by Con-A binding.

Previous smFRET investigations allowed us to characterize the conformational landscape of the extracellular domain of GluK2 receptors in the apo (unliganded), glutamate-bound form (where the receptor was primarily desensitized) as well as the glutamate-bound D776K mutant receptor (where the receptor was primarily in the open active state) [34]. These studies showed that the dimer interface at the agonist-binding domain exhibits varying degrees of decoupling when the GluK2 receptor is bound to glutamate, while in the glutamate-bound form of the D776K mutant, the dimer interface of the agonist-binding domain exists primarily in a tightly coupled conformation. These studies suggested that the lower FRET (longer distance) decoupled states at the dimer interface of the ABD correlated to the desensitized state conformations, while the high FRET (shorter distance) coupled state at the dimer interface of the ABD represented the active, open-channel state of the receptor. The smFRET histograms for GluK2 receptors with Con-A at the dimer interface of the agonist-binding domain studied here showed a shift toward a higher FRET population, and the shift was consistent with the higher residual currents observed in the whole-cell recordings. These results suggest that modulation via Con-A is through a conformational selection process wherein conformation with the coupled dimer interface of the agonist-binding domain is favored.

This mechanism of Con-A modulation is further confirmed by the loss in Na^+ ion modulation in the presence of Con-A. Na^+ ions bind to the agonist-binding domain dimer interface and stabilize the interface. Replacing Na^+ with Cs^+ ions, which bind with a significantly lower affinity at the dimer interface, leads to lower currents and decreased stability of desensitized states in GluK2 receptors. This functional change is reflected in the higher population of lower FRET states in the presence of Cs^+, further supporting the correlation between the decoupling at the dimer interface and the desensitization of the receptor. In this study, we showed that in the presence of Con-A, both function and conformational spread was unaltered when Na^+ was replaced with Cs^+ ions, as Na^+ was no longer required for stabilization of the agonist-binding domain interface. Overall, our smFRET data showed a tighter packing at both the ATD and ABD subunit interfaces in the GluK2 receptor in the presence of Con-A that underlay the observed functional changes.

All ionotropic glutamate receptors are N-glycosylated, and deficiencies associated with glycosylation have been implicated in a number of diseases [47]. Interestingly, it has been shown that receptors without glycosylation still maintain their primary glutamate-gated ion channel function [17]. However, these studies were done using isolated receptors. The present study, along with prior functional studies, shows that lectins such as Con-A can modulate function by stabilizing extracellular domains upon binding to glycosylation sites. This raises the question as to whether endogenous proteins present in synapses can modulate ionotropic glutamate receptors by binding to glycosylation sites in a manner similar to what is seen for Con-A. This would imply that deficiencies in glycosylation might not affect isolated receptor function, but they would alter the modulatory effect of these proteins on receptor function in the context of the interacting proteins and thus, alter excitatory transmission. Recent studies have shown that several endogenous proteins, such as those that belong to the C1q/TNF superfamily, have a similar beta sheet fold as Con-A [48], and these have been shown to bind to ionotropic glutamate receptors. Among the most studied and better understood is cerebellin, an endogenous soluble protein, which forms a tripartite complex with ionotropic glutamate delta receptors and the presynaptic transmembrane protein, neurexin1β [49,50]. The modulatory mechanism through which these proteins work is still largely unknown. Future studies on the modulatory mechanism mediated by such C1q/TNF superfamily proteins, on the ionotropic glutamate receptors, and on the possible role of glycosylation on such modulation will shed light on whether these share common pathways to those observed here for Con-A binding to kainate receptors.

Author Contributions: V.J. and C.U.G. contributed toward the conceptualization; E.C. acquired the electrophysiology data; C.U.G. and V.B. acquired and analyzed smFRET data; V.J., C.U.G., E.C. and V.B. wrote and edited the manuscript. All authors have read and agreed to the published version of the manuscript.

Funding: This work was funded by Training Interdisciplinary Pharmacology Scientists (TIPS) Program Grant Fellowship T32 GM120011 to C.U.G., National Institutes of Health Grant R35 GM122528 to V.J., and American Heart Association Fellowship 18POST34030189 to E.C.

Institutional Review Board Statement: No humans or animals were used.

Informed Consent Statement: Not applicable.

Data Availability Statement: The data generated, analyzed, and presented in this study are available from the corresponding author on reasonable request.

Acknowledgments: We would like to thank Drew Dolino and Ryan Durham for providing comments and editing the manuscript.

Conflicts of Interest: Authors declare that they have no conflict of interest.

References

1. Dingledine, R.; Borges, K.; Bowie, D.; Traynelis, S.F. The Glutamate Receptor Ion Channels. *Pharmacol. Rev.* **1999**, *51*, 7–61.
2. Traynelis, S.F.; Wollmuth, L.P.; Mcbain, C.J.; Menniti, F.S.; Vance, K.M., Ogden, K.K.; Hansen, K.B.; Yuan, H.; Myers, S.J.; Dingledine, R. Glutamate Receptor Ion Channels: Structure, Regulation, and Function. *Pharmacol. Rev.* **2010**, *62*, 405–496. [CrossRef]
3. Contractor, A.; Mulle, C.; Swanson, G.T. Kainate Receptors Coming of Age: Milestones of Two Decades of Research. *Trends Neurosci.* **2011**, *34*, 154–163. [CrossRef]
4. Popescu, G.K. Modes of Glutamate Receptor Gating. *J. Physiol.* **2012**, *590*, 73–91. [CrossRef]
5. Hollmann, M.; Maron, C.; Heinemann, S. N-Glycosylation Site Tagging Suggests a Three Transmembrane Domain Topology for the Glutamate Receptor GluR1. *Neuron* **1994**, *13*, 1331–1343. [CrossRef]
6. Meyerson, J.; Chittori, S.; Merk, A.; Rao, P.; Hee Han, T.; Serpe, M.; Mayer, M.L.; Subramaniam, S. Structural Basis of Kainate Subtype Glutamate Receptor Desensitization. *Nature* **2016**, *537*, 567–571. [CrossRef]
7. Meyerson, J.R.; Kumar, J.; Chittori, S.; Rao, P.; Pierson, J.; Bartesaghi, A.; Mayer, M.L.; Subramaniam, S. Structural Mechanism of Glutamate Receptor Activation and Desensitization. *Nature* **2014**, *514*, 328–334. [CrossRef] [PubMed]
8. Nayeem, N.; Mayans, O.; Green, T. Conformational Flexibility of the Ligand-Binding Domain Dimer in Kainate Receptor Gating and Desensitization. *J. Neurosci.* **2011**, *31*, 2916–2964. [CrossRef] [PubMed]
9. Schauder, D.M.; Kuybeda, O.; Zhang, J.; Klymko, K.; Bartesaghi, A.; Borgnia, M.J.; Mayer, M.L.; Subramaniam, S. Glutamate Receptor Desensitization Is Mediated by Changes in Quaternary Structure of the Ligand Binding Domain. *Proc. Natl. Acad. Sci. USA* **2013**, *110*, 5921–5926. [CrossRef] [PubMed]
10. Dawe, G.B.; Musgaard, M.; Andrews, E.D.; Daniels, B.A.; Aurousseau, M.R.P.; Biggin, P.C.; Bowie, D. Defining the Structural Relationship between Kainate-Receptor Deactivation and Desensitization. *Nat. Struct. Mol. Biol.* **2013**, *20*, 1054–1061. [CrossRef]
11. Twomey, E.C.; Yelshanskaya, M.V.; Grassucci, R.A.; Frank, J.; Sobolevsky, A.I. Structural Bases of Desensitization in AMPA Receptor-Auxiliary Subunit Complexes. *Neuron* **2017**, *94*, 569–580. [CrossRef]
12. Twomey, E.C.; Yelshanskaya, M.V.; Grassucci, R.A.; Frank, J.; Sobolevsky, A.I. Channel Opening and Gating Mechanism in AMPA-Subtype Glutamate Receptors. *Nature* **2017**, *549*, 60–65. [CrossRef]
13. Naur, P.; Vestergaard, B.; Skov, L.K.; Egebjerg, J.; Gajhede, M.; Kastrup, J.S. Crystal structure of the kainate receptor GluR5 ligand-binding core in complex with (S)-glutamate. *FEBS Lett.* **2005**, *579*, 1154–1160. [CrossRef] [PubMed]
14. Yelshanskaya, M.V.; Li, M.; Sobolevsky, A.I. Structure of an Agonist-Bound Ionotropic Glutamate Receptor. *Science* **2014**, *345*, 1070–1074. [CrossRef] [PubMed]
15. Mayer, M.L. Crystal Structures of the GluR5 and GluR6 Ligand Binding Cores: Molecular Mechanisms Underlying Kainate Receptor Selectivity. *Neuron* **2005**, *45*, 539–552. [CrossRef] [PubMed]
16. Yue, K.-T.; Macdonald, J.F.; Pekhletski, R.; Hampson, D.R. Differential Effects of Lectins on Recombinant Glutamate Receptors. *Eur. J. Pharmacol. Mol. Pharmacol.* **1995**, *291*, 229–235. [CrossRef]
17. Everts, I.; Villmann, C.; Hollmann, M. N-Glycosylation Is Not a Prerequisite for Glutamate Receptor Function but Is Essential for Lectin Modulation. *Mol. Pharmacol.* **1997**, *52*, 861–873. [CrossRef]
18. Everts, I.; Petroski, R.; Kizelsztein, P.; Teichberg, V.I.; Heinemann, S.F.; Hollmann, M. Lectin-Induced Inhibition of Desensitization of the Kainate Receptor GluR6 Depends on the Activation State and Can Be Mediated by a Single Native or Ectopic N-Linked Carbohydrate Side Chain. *J. Neurosci.* **1999**, *19*, 916–927. [CrossRef] [PubMed]
19. Mayer, M.L.; Vyklicky, L. Concanavalin A Selectively Reduces Desensitization of Mammalian Neuronal Quisqualate Receptors. *Proc. Natl. Acad. Sci. USA* **1989**, *86*, 1411–1415. [CrossRef]
20. Galen Wo, Z.; Oswald, R.E. A Topological Analysis of Goldfish Kainate Receptors Predicts Three Transmembrane Segments. *J. Biol. Chem.* **2000**, *270*, 2000–2009. [CrossRef]
21. Taverna, F.A.; Wang, L.Y.; MacDonald, J.F.; Hampson, D.R. A Transmembrane Model for an Ionotropic Glutamate Receptor Predicted on the Basis of the Location of Asparagine-Linked Oligosaccharides. *J. Biol. Chem.* **1994**, *269*, 14159–14164. [CrossRef]
22. Sahara, Y.; Noro, N.; Iida, Y.; Soma, K.; Nakamura, Y. Glutamate Receptor Subunits GluR5 and KA-2 Are Coexpressed in Rat Trigeminal Ganglion Neurons. *J. Neurosci.* **1997**, *17*, 6611–6620. [CrossRef]
23. Fay, A.-M.L.; Bowie, D. Concanavalin—A Reports Agonist-Induced Conformational Changes in the Intact GluR6 Kainate Receptor. *J. Physiol.* **2006**, *572*, 201–213. [CrossRef]
24. Landes, C.F.; Rambbhadran, A.; Taylor, J.N.; Salatan, F.; Jayaraman, V. Structural Landscape of Isolated Agonist-Binding Domains from Single AMPA Receptors. *Nat. Chem. Biol.* **2011**, *7*, 168–173. [CrossRef]

25. Ramaswamy, S.; Cooper, D.; Poddar, N.; MacLean, D.M.; Rambhadran, A.; Taylor, J.N.; Uhm, H.; Landes, C.F.; Jayaraman, V. Role of Conformational Dynamics in α-Amino-3-Hydroxy-5-Methylisoxazole-4-Propionic Acid (AMPA) Receptor Partial Agonism. *J. Biol. Chem.* **2012**, *287*, 43557–43564. [CrossRef] [PubMed]
26. MacLean, D.M.; Ramaswamy, S.S.; Du, M.; Howe, J.R.; Jayaraman, V. Stargazin Promotes Closure of the AMPA Receptor Ligand-Binding Domain. *J. Gen. Physiol.* **2014**, *144*, 503–512. [CrossRef]
27. Dolino, D.M.; Cooper, D.; Ramaswamy, S.; Jaurich, H.; Landes, C.F.; Jayaraman, V. Structural Dynamics of the Glycine-Binding Domain of the N-Methyl-D-Aspartate Receptor. *J. Biol. Chem.* **2014**, *290*, 797–804. [CrossRef] [PubMed]
28. Sirrieh, R.E.; MacLean, D.M.; Jayaraman, V. A Conserved Structural Mechanism of NMDA Receptor Inhibition: A Comparison of Ifenprodil and Zinc. *J. Gen. Physiol.* **2015**, *146*, 173–181. [CrossRef]
29. Dolino, D.M.; Adariani, S.R.; Shaikh, S.A.; Jayaraman, V.; Sanabria, H. Conformational Selection and Submillisecond Dynamics of the Ligand-Binding Domain of the N-Methyl-d-Aspartate Receptor. *J. Biol. Chem.* **2016**, *291*, 16175–16185. [CrossRef]
30. Shaikh, S.A.; Dolino, D.M.; Lee, G.; Chatterjee, S.; MacLean, D.M.; Flatebo, C.; Landes, C.F.; Jayaraman, V. Stargazin Modulation of AMPA Receptors. *Cell Rep.* **2016**, *17*, 328–335. [CrossRef] [PubMed]
31. Dolino, D.M.; Chatterjee, S.; MacLean, D.M.; Flatebo, C.; Bishop, L.D.C.; Shaikh, S.A.; Landes, C.F.; Jayaraman, V. The Structure–Energy Landscape of NMDA Receptor Gating. *Nat. Chem. Biol.* **2017**, *13*, 1232–1238. [CrossRef]
32. Litwin, D.B.; Durham, R.J.; Jayaraman, V. Single-molecule FRET methods to study glutamate receptors. In *Methods in Molecular Biology*; Humana Press Inc.: Totowa, NJ, USA, 2019; Volume 1941, pp. 3–16.
33. Litwin, D.B.; Paudyal, N.; Carrillo, E.; Berka, V.; Jayaraman, V. The Structural Arrangement and Dynamics of the Heteromeric GluK2/GluK5 Kainate Receptor as Determined by SmFRET. *Biomembranes* **2020**, *1862*, 183001. [CrossRef] [PubMed]
34. Litwin, D.B.; Carrillo, E.; Shaikh, S.A.; Berka, V.; Jayaraman, V. The Structural Arrangement at Intersubunit Interfaces in Homomeric Kainate Receptors. *Sci. Rep.* **2019**, *9*, 6969. [CrossRef]
35. Bowie, D. External Anions and Cations Distinguish between AMPA and Kainate Receptor Gating Mechanisms. *J. Physiol.* **2002**, *539*, 725–733. [CrossRef] [PubMed]
36. Wong, A.Y.C.; Fay, A.-M.L.; Bowie, D. External Ions Are Coactivators of Kainate Receptors. *J. Neurosci.* **2006**, *26*, 5750–5755. [CrossRef] [PubMed]
37. Plested, A.J.R.; Vijayan, R.; Biggin, P.C.; Mayer, M.L. Molecular Basis of Kainate Receptor Modulation by Sodium. *Neuron* **2008**, *58*, 720–735. [CrossRef] [PubMed]
38. Bowie, D.; Garcia, E.P.; Marshall, J.; Traynelis, S.F.; Lange, D.G. Allosteric Regulation and Spatial Distribution of Kainate Receptors Bound to Ancillary Proteins. *J. Physiol.* **2003**, *547*, 373–385. [CrossRef]
39. Paramo, T.; Brown, P.M.G.E.; Musgaard, M.; Bowie, D.; Biggin, P.C. Functional Validation of Heteromeric Kainate Receptor Models. *Biophys. J.* **2017**, *113*, 2173–2177. [CrossRef] [PubMed]
40. Qin, F. Restoration of Single-Channel Currents Using the Segmental k-Means Method Based on Hidden Markov Modeling. *Biophys. J.* **2004**, *86*, 1488–1501. [CrossRef]
41. Nicolai, C.; Sachs, F. Solving Ion Channel Kinetics with the QuB Software. *Biophys. Rev. Lett.* **2014**, *8*, 191–211. [CrossRef]
42. Shuang, B.; Cooper, D.; Taylor, J.N.; Kisley, L.; Chen, J.; Wang, W.; Li, C.B.; Komatsuzaki, T.; Landes, C.F. Fast StepTransition and State Identification (STaSI) for Discrete Single-Molecule Data Analysis. *J. Phys. Chem. Lett.* **2014**, *5*, 3157–3161. [CrossRef] [PubMed]
43. Jain, A.; Liu, R.; Ramani, B.; Arauz, E.; Ishitsuka, Y.; Ragunathan, K.; Park, J.; Chen, J.; Xiang, Y.K.; Ha, T. Probing Cellular Protein Complexes Using Single-Molecule Pull-Down. *Nature* **2011**, *473*, 484–488. [CrossRef] [PubMed]
44. Dürr, K.L.; Chen, L.; Stein, R.A.; de Zorzi, R.; Folea, I.M.; Walz, T.; McHaourab, H.S.; Gouaux, E. Structure and Dynamics of AMPA Receptor GluA2 in Resting, Pre-Open, and Desensitized States. *Cell* **2014**, *158*, 778–792. [CrossRef]
45. Huettner, J.E. Glutamate Receptor Channels in Rat DRG Neurons: Activation by Kainate and Quisqualate and Blockade of Desensitization by Con A. *Neuron* **1990**, *5*, 255–266. [CrossRef]
46. Paternain, A.V.; Rodríguez-Moreno, A.; Villarroel, A.; Lerma, J. Activation and Desensitization Properties of Native and Recombinant Kainate Receptors. *Neuropharmacoloy* **1998**, *37*, 1249–1259. [CrossRef]
47. Hennet, T.; Cabalzar, J. Congenital disorders of glycosylation: A concise chart of glycocalyx dysfunction. *Trends Biochem Sci.* **2015**, *40*, 377–384. [CrossRef]
48. Ressl, S.; Vu, B.K.; Vivona, S.; Martinelli, D.C.; Südhof, T.C.; Brunger, A.T. Structures of C1q-like proteins reveal unique features among the C1q/TNF superfamily. *Structure* **2015**, *23*, 688–699. [CrossRef]
49. Matsuda, K.; Yuzaki, M. Cbln1 and the δ2 glutamate receptor–an orphan ligand and an orphan receptor find their partners. *Cerebellum* **2012**, *11*, 78–84. [CrossRef]
50. Matsuda, K. Synapse organization and modulation via C1q family proteins and their receptors in the central nervous system. *Neurosci. Res.* **2017**, *116*, 46–53. [CrossRef]

membranes

Review

Function-Related Dynamics in Multi-Spanning Helical Membrane Proteins Revealed by Solution NMR

Koh Takeuchi [1,*], Yutaka Kofuku [2], Shunsuke Imai [3], Takumi Ueda [2], Yuji Tokunaga [1], Yuki Toyama [2], Yutaro Shiraishi [3] and Ichio Shimada [2,3,*]

1 Cellular and Molecular Biotechnology Research Institute, National Institute of Advanced Industrial Science and Technology, Aomi, Koto, Tokyo 135-0064, Japan; tokunaga.y@aist.go.jp
2 Graduate School of Pharmaceutical Sciences, The University of Tokyo, Hongo, Bunkyo, Tokyo 113-0033, Japan; happy@nmrlab.f.u-tokyo.ac.jp (Y.K.); ueda@nmrlab.f.u-tokyo.ac.jp (T.U.); yuki.toyama@utoronto.ca (Y.T.)
3 Center for Biosystems Dynamics Research, RIKEN, Suehiro, Tsurumi, Yokohama 230-0045, Japan; shunsuke.imai.ku@riken.jp (S.I.); shiraishi@nmrlab.f.u-tokyo.ac.jp (Y.S.)
* Correspondence: koh-takeuchi@aist.go.jp (K.T.); ichio.shimada@riken.jp (I.S.)

Abstract: A primary biological function of multi-spanning membrane proteins is to transfer information and/or materials through a membrane by changing their conformations. Therefore, particular dynamics of the membrane proteins are tightly associated with their function. The semi-atomic resolution dynamics information revealed by NMR is able to discriminate function-related dynamics from random fluctuations. This review will discuss several studies in which quantitative dynamics information by solution NMR has contributed to revealing the structural basis of the function of multi-spanning membrane proteins, such as ion channels, GPCRs, and transporters.

Keywords: membrane protein; NMR; ion channel; GPCR; transporter; dynamics

1. Introduction

Multi-spanning helical membrane proteins, such as GPCRs, ion channels, and transporters, play a critical role in transferring information and materials through a membrane. Since the function and malfunction of multi-spanning helical membrane proteins are critical in many biological processes and their associated diseases, understanding the functional mechanism of these proteins is of great interest from biological, medical, and pharmaceutical aspects. The recent advances in X-ray crystallography and cryo-electron microscopy (Cryo-EM) unveiled a significant number of structures of membrane proteins with biological, medical, and pharmacological importance. The number of published membrane protein structures is currently over 4000 and rapidly increases after the establishment of single-particle analysis of Cryo-EM.

The atomic-resolution structures of membrane proteins have contributed to various research fields as a structural basis for understanding their functions. One of the early landmark examples is the determination of potassium ion channel structure by R. Mackinnon [1]. The structure of the KcsA potassium ion channel revealed the selective ion conductance mechanism of the channel by the formation of square antiprism coordination of carbonyl oxygens around each K^+ binding site as if to mimic the waters of hydration [1,2] (Figure 1A). The following crystal structure of a calcium-dependent potassium channel MthK in the opened state unveiled the structural difference between the closed and opened potassium channels [3,4]. Under the closed conditions, the inner helix adopts straight conformation to form a bundle, closing the ion-conducting path in the intracellular side (Figure 1A). In contrast, the inner helices are in the bent configuration in the open conditions, which sprays out the intracellular gate, allowing the ion conductance (Figure 1B). In addition, the structures of ion channels with different regulation mechanisms have been solved to unveil how ion conductance is regulated by various inputs such as pH, ligands, temperature, membrane voltage, ion concentrations, mechanosensation, and protein–protein interactions, etc. [5–12]. The recent image processing technique in Cryo-EM even allows the classification of open and closed conformations of ion channels in the mixture to quantify the increased fraction of open conformation with the exposure to the inducers [13].

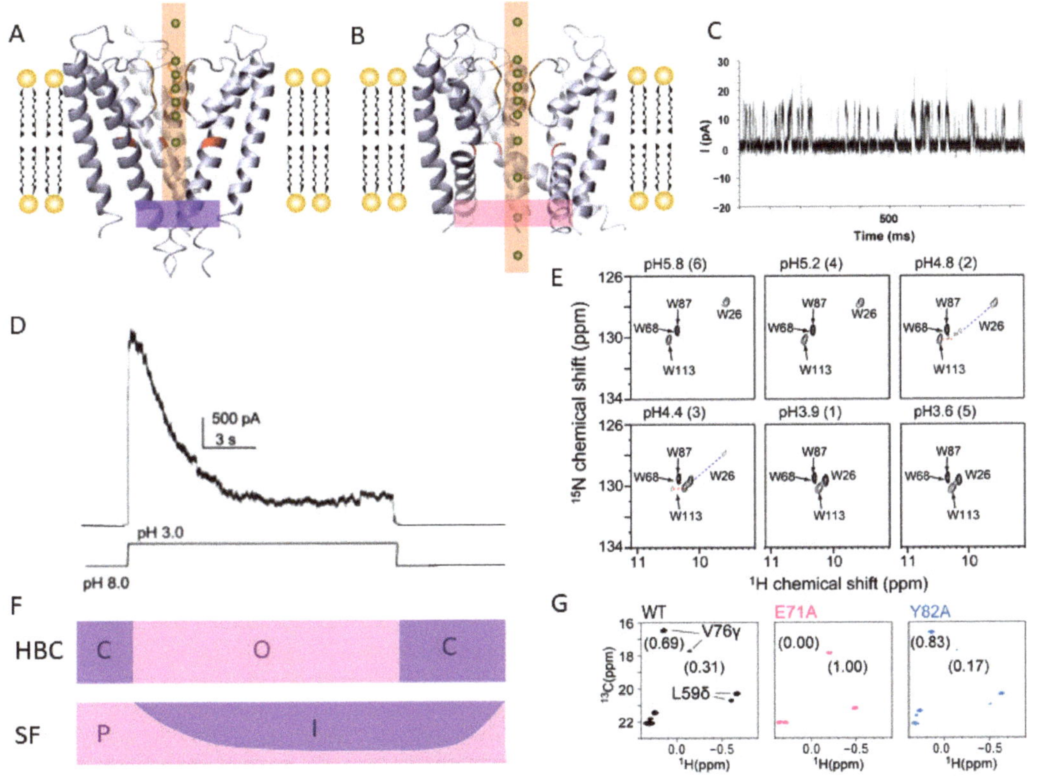

Figure 1. Functional dynamics of an ion channel. (**A**) Crystal structure of KcsA in the closed conformation at neutral pH (PDB ID: 1BL8). (**B**) Crystal structure of pore domain of MthK in the open conformation (PDB ID: 1LNQ). (**C**) Single-channel recording of KcsA in acidic condition [14]. (**D**) A macroscopic current of KcsA after pH jump. (**E**) Close-up view of the Trp indole region of the ^{1}H–^{15}N TROSY HSQC spectra of Hα-partially protonated/^{2}H/^{15}N-labeled Δ125–160 KcsA. The resonances from Trp-68 and Trp-87 overlap with each other. Assignments were established by site-specific mutagenesis. The dashed lines connect corresponding signals. (**F**) The conformational state of HBC and SF in the corresponding condition is shown in panel (**D**). (**G**) Methyl-TROSY spectra of WT (Left), the E71A (Middle), and Y82A (Right) mutants at pH 3.2 and 25 °C. The numbers in parentheses are the relative intensities of the two signals of V76 γ. The figure was reproduced from [14–16] with permission.

The dynamic behavior of membrane proteins is of importance as they are required to change their structure in response to various stimuli and transfer the information in and out of the membrane to exert their functions. For example, the single-channel recording shows that ion channels in the open conditions are not continuously permeating ions. They are instead under the equilibrium between open, closed, and inactive conformations, reflected in the stochastic profile of the ion permeabilization through a channel (Figure 1C). The open dwelling times are mostly in the order of milliseconds, which correspond to the exchange rate 1000 s^{-1}. Since the open probability (i.e., the population of open state per unit time) determined from the open/closed dwell time and the ion conductivity in the open state defines the total ion flux, the activity of the ion channel can be described from the stochastic dynamic behavior of the proteins and is beyond the description by static snapshot structures.

NMR is an optimal spectroscopic technique to reveal such dynamics of proteins. NMR is able to define the function-associated dynamics in a site-specific manner and quantify the population of each conformation under an equilibrium, along with the rate of interconversion. For this purpose, various NMR measurements and stable-isotope labeling strategies are available. The quantitative analysis of the NMR relaxation rate can detect a high-energy functional state that populates less than one percent in an equilibrium. Due to its versatility to the condition, NMR can monitor the changes in structure and dynamics of membrane proteins associated with the functionally important parameters such as temperature, pH, ions, and inducers. The features would allow dissecting function-associated dynamics from random fluctuations. Thus, NMR would be of choice in understanding the functional mechanism of membrane proteins. In this review, we will overview some recent achievements that highlight the strength of solution-state NMR in the structure and dynamics analyses of the multi-spanning helical membrane proteins, such as ion channels, GPCRs, and transporters.

2. Ion Channels

As mentioned above, KcsA potassium channels are a hallmark example where structural studies, especially those from NMR, have contributed to understanding their functional mechanism. KcsA is a pH-dependent potassium channel derived from *Streptomyces lividans*. Structurally, KcsA is one of the simplest potassium channels consisting only of a pore domain harboring an ion-selective filter and an intracellular gate. Two-membrane spanning helices from each monomeric unit form a tetrameric assembly to form the ion conductance path at the rotation symmetry axis (Figure 1A).

At neutral pH, KcsA adopts a closed state and is impermeable to K^+ ions. As depicted by X-ray crystallography, inner transmembrane helices of KcsA form a bundle in the intracellular side to pinch the ion-conducting path. This intracellular helix bundle crossing (HBC) is a common structural feature among ion channels in closed conditions. While in the acidic condition, KcsA transiently and stochastically opens to conduct the K^+ channel (Figure 1B). The microscopic current from patch-clamp multichannel recording showed that KcsA conducts transient maximum current right after reducing pH; however, the current decays exponentially to a certain inactivation level in 1–3 s [14] (Figure 1C). While the activation-coupled inactivation proceeds, the K^+ conductance at the single-channel level remains invariant [17] (Figure 1D). Thus, the activation-coupled inactivation reflects the reduction of the open probability of the ion channel after activation. Indeed, the channels are reportedly open in a burst of activity before entering a long-lived nonconductive state [17]. The activation-coupled inactivation is present in almost all K^+ channels and many other ion channels; thus, it reflects the common structural behavior of ion channels.

While the open probability of KcsA in the inactivated condition is known to be as much as 10% [14], our NMR study revealed that intracellular HBC is fully open even in the inactivated condition [15]. In a series of $^1H^{15}N$-TROSY HSQC spectra measured at various pH values, significant chemical shift changes were detected between pH 3.9 and 5.2, reflecting a conformational rearrangement associated with the gating of the ion channel. Since the 2D NMR measurement takes hours, the acidic condition spectra are those that experience the activation-coupled inactivation. The pH-dependent chemical shift changes were primarily observed in the residues near the intracellular HBC. Especially, the chemical shifts Trp-26 and Trp-113 are distinct at acidic and neutral pH, indicating that the intracellular gate changes to a completely different structure, presumably in open conformation at acidic conditions (Figure 1E). This observation supports the "dual-gate" properties of the K^+ channel, in which the channel becomes permeable to K^+ ions only when both intracellular HBC and the selective filter (SF) are in conductive conformation [15] (Figure 1F). Since the intracellular gate of KcsA is fully open even after a few hours of exposure to the acidic pH, the selectivity filter determines the channel's open probability in the inactivated state. It should also be noted that from a pH-dependent measurement of KcsA, we revealed that the H25A mutation abolished this pH-dependent conformational rearrangement. These results indicate that the residue near the intracellular gate serves as a "pH-sensor" of the channel [15].

Further NMR investigation showed that the signals in the vicinity of the selectivity filter (SF) split into two peaks in acidic conditions [16] (Figure 1G). The relative intensity of the signal changes substantially with the temperature, which strongly suggests that SF in the acidic condition is under the equilibrium between two distinct conformations. Since the open probability of KcsA in the acidic inactivated condition is as much as 10% [17], we consider that the major peaks observed at 25 °C correspond to the impermeable conformation and the minor peak reflects the permeable conformation of the selective filter. From the analysis of the EXSY spectrum, the exchange rate between the two conformational states in SF was determined to be ~1 s^{-1}. The value is closely in agreement with the inactivation kinetics of the channel [18].

To further verify this hypothesis, we compared the NMR spectra of the WT KcsA with those of the E71A and Y82A mutants (Figure 1G). The E71A mutant does not show inactivation. In contrast, the Y82A mutant experiences extensive activation-coupled inactivation, in which the open probability is less than 5%. In the NMR spectrum of the E71A mutant, the only signals that correspond to the permeable conformation (minor peak in WT KcsA) were mostly observed [16]. While in the Y82A mutant, the signal that corresponds to the permeable condition was much less intense than WT. These results support that major and minor signals observed in the WT KcsA are from impermeable and permeable conformational states of SF, respectively.

Furthermore, the K^+ titration experiments revealed that KcsA weakens the affinity to K^+ ion in an acidic condition, and the loss of K^+ ion at SF results in the deformation to an impermeable conformation [16]. Since NOE signals between the methyl group of Val-76 in SF and water were detected only in the impermeable state, we think that a water-binding instead of K^+ ion leads to the blockage of the ion permeabilization at the SF. It should be noted that the reduction of K^+ ion binding to the open state was also confirmed by recent solid-state NMR analysis of the constitutively open H25R/E118A mutant of KcsA embedded in a lipid bilayer at neutral pH [19].

As shown here, NMR successfully identifies the dual gate properties in KcsA, which thoroughly explains the electrophysiological profile of the channel. Independent NMR studies of KcsA by other groups also confirmed the structural and dynamic properties associated with the gating function of the ion channel [20]. It should also be noted that while the initial study was done in n-dodecyl β-d-maltoside (DDM) micelle, we also investigated the conformational equilibrium of KcsA in lipid bilayers [18]. Toward this end, reconstituted high-density lipoprotein (rHDL), also known as nanodisc, was used [21]. The rHDL is a disc-shaped particle of about 10 nm in diameter, in which a membrane scaffold protein with an amphiphilic helix structure holds the lipid bilayer structure inside. Since rHDL is monodisperse and soluble, it is suitable for NMR analysis. We reconstituted KcsA into rHDL and performed NMR observations. The NMR spectra in rHDL were similar to those in DDM, and there was no global structural difference between detergent-solubilized and rHDL-reconstituted KcsA. However, in the conformational equilibrium of the residues constituting SF, the ratio of impermeable state was higher in rHDL than DDM. This result strongly suggests that the lipid bilayer environment affects the structural equilibrium of KcsA and NMR is a suitable technique to dissect the mechanistic feature of ion channels embedded in lipid bilayer [18]. In line with this notion, it is shown that specific lipid composition strongly increases the KcsA open probability, indicating the critical role of the surrounding lipid bilayer in the gating of the KcsA channel [22].

Bax et al. reported the measurements of backbone ^{15}N R_1, $R_{1\rho}$, ^{15}N-(^1H) NOE, and ^{15}N CSA/dipolar cross-correlation for the KcsA channel in SDS micelles under conditions where this channel is in the closed state [23] as well as in the open state [24], and the dynamics at different time scales were successfully described. It should be noted that the study was conducted at a relatively high temperature (323 K) in SDS micelle. As a result, while the channel remains in the tetrameric form in the measurements under the closed condition, the channel dissociated into a monomeric form under the open condition at acidic pH, which might not be fully relevant to its functional state.

In addition, we recently investigated the effect of disease-related mutations on K^+ channel conformational equilibrium [25]. Val-408 in the human Kv1.1 channel is located near intracellular HBC, and an Ala mutation at the position is known to associate with episodic ataxia. Mutational studies using the Shaker voltage-gated K^+ (Kv) channel have revealed that increased sidechain volumes at the corresponding position substantially decrease the K^+-permeability. Our NMR study revealed that the Ala to Val mutation at the corresponding residue (position 111) in KcsA stabilizes the closed conformation of HBC. The A111V mutant forms the structure closely resembling the closed state in the WT KcsA, even under the pH 3.0 condition. The results are consistent with the case of the Shaker Kv channel in that the bulky sidechain at the corresponding position results in a nonconductive phenotype. The Ala-111 contributes to extensive inter-helix van-der-Waals contacts in the open conformation, while the closed conformation is stabilized by introducing mutations with increased sidechain volumes. Thus, we considered that the stability of closed HBC conformation and disfavored sidechain van-der-Waals contacts in the open conformation would be the underlying mechanism of the increased population of the closed conformation in the KcsA mutants. Due to the structural similarity to KcsA, the disease-related mutations should also affect the gating of the human Kv1.1 channel.

We have also clarified the function-associated conformational equilibrium of the ATP-dependent ion channel, P2X [26]. In the α,β-methylene ATP-bound state, NMR signals from the residues in the transmembrane region and the residues that connect the ATP-binding site and the transmembrane region are in a conformational equilibrium between closed and open conformations [27,28]. NMR study also contributed to understanding the regulation mechanism of the NaK channel [29] and G protein-activated inwardly rectifying potassium channel [28,30], along with the structural characterization of ion channels [31–35] and their regulatory domains [36–38]. In addition, the inhibition mechanisms of ion channels by peptide toxins [39–42] were characterized by extensive structural and interaction analysis at residue-specific levels. These indicate the versatility of the NMR strategy to unveil the functional and inhibition mechanism of the class of multi-spanning helical membrane proteins.

3. G-Protein Coupled Receptors (GPCRs)

GPCRs are one of the largest families of membrane proteins in eukaryotes. In humans, there are more than 800 GPCRs, which function as receptors for neurotransmitters, hormones, and cytokines, etc. To date, more than 550 structures of >100 GPCRs have been deposited to PDB. Especially, the recent development of cryo-EM technology allowed us to determine the structure of GPCR in complex with heterotrimeric G proteins. While these structures substantially advanced the structural study of GPCRs, these structures are static and do not necessarily capture the major structures in physiological conditions. Instead, by using NMR and other biophysical methods, it becomes clear that GPCRs are under dynamic structural equilibrium with multiple conformations under physiological conditions to evoke multiple intracellular signaling pathways in various amplitudes. Especially, the development of in situ NMR strategy allows quantitative analyses of the dynamic structural equilibrium in GPCRs, such as the rate and amount ratio of exchange between different structures [43,44].

One of the unique characteristics of GPCR is the presence of basal activity (Figure 2A). With the basal activity, GPCRs are always partially activated intracellular G proteins to a certain level. Upon the ligand binding, GPCRs both upregulate or downregulate the activation level compared to the basal level, depending on the ligand types. The degree to which the ligand activates the GPCR at the saturated concentration is called efficacy

(Figure 2A). The ligands that inhibit the basal activity are called inverse agonists, and the ligands that do not change the basal activity are called neutral antagonists. A ligand that completely activates the target GPCR is called a full agonist. It is also known that there is a ligand called partial agonist that weakly activates the target GPCR. These differences in efficacy are known to affect the biological activity and usage of the drug. For the β2-adrenergic receptor (β2AR), it is reported that a full agonist offers a clinical advantage over a partial agonist in severe acute asthma. In contrast, partial agonists with fewer side effects are preferable in chronic treatment.

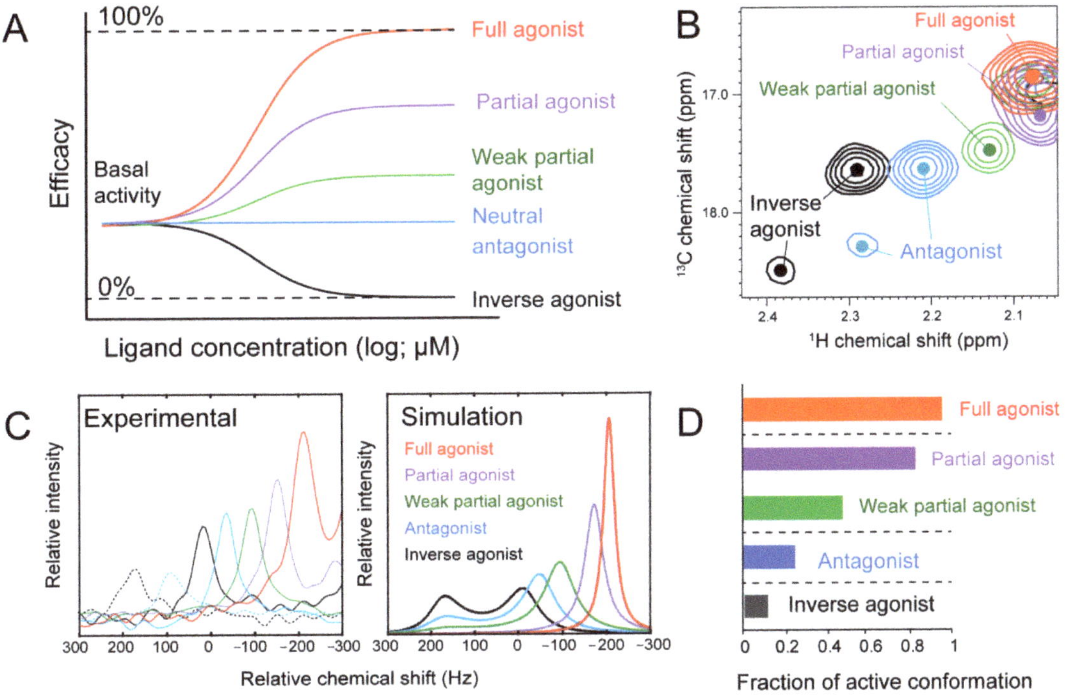

Figure 2. Efficacy of GPCR elucidated by NMR. (**A**) Pharmacological concepts of efficacy. (**B**) Overlay of the ^1H–^{13}C SOFAST-HMQC spectra of [α,β,β-$^2H_3$-, methyl-^{13}C-Met]β2AR at 298 K in complex with an inverse agonist, carazolol (black), a neutral antagonist, alprenolol (cyan), a weak partial agonist, tulobuterol (green), a partial agonist, clenbuterol (violet), and a full agonist, formoterol (red). Only the regions with M82 resonances are shown. The signals are color-coded in the same way as in panel (**A**). (**C**) Experimental (left) and simulation (right) spectra of the ligand-dependent shift of the M82 resonances. Overlay of the ^1H slice of M82 resonances in complexes with indicated ligands with different efficacy at 298 K is shown. (**D**) Conformational equilibrium quantified by NMR explains the efficacy of β2AR ligands. The fraction of active conformation that was obtained by fitting, shown in panel (**C**), was used. The figure was reproduced from [45] with permission.

In the structure of the inverse agonist-bound β2AR, the fifth and sixth transmembrane helices (TM5 and TM6) interact with each other in the closed conformation. While, in the ternary complex of β2AR, full agonist, and G protein, the TM5 and TM6 of β2AR open outward, and the C-terminal helix of the α subunit of G protein enters the resulting cavity. The difference in the three-dimensional structures explains the mechanism by which a ligand induces the activation of GPCRs. However, the difference in the efficacy of each ligand and the basal activity of β2AR cannot be explained by a simple two-state conformational change from inactive to active conformations.

In order to unveil the structural mechanism underlying the difference in the efficacy of each ligand, we subjected β2AR for NMR analyses. For this purpose, NMR signals of

Met-82, which undergoes a conformational change in conjunction with the P/I/F motif residues was used. The structure change in the P/I/F motif (Pro-211, Ile-121, and Phe-282 in β2AR) is known to trigger conformational changes in the intracellular side. In the inverse agonist-bound condition, two signals derived from Met-82 were observed [45] (Figure 2B). On the other hand, in the full agonist-bound condition, only one signal of Met-82 with a chemical shift distinct from that in the inverse agonist-bound form was observed [45] (Figure 2B). These observations indicate that the differences in the activation states of β2AR are reflected in the NMR spectra.

The ^{1}H chemical shifts of the Met methyl group are known to reflect the surrounding environment, especially the positions of the aromatic rings. In contrast, those of the ^{13}C sidechain mostly reflect the χ^3 dihedral angle of its own [46]. The ^{1}H and ^{13}C chemical shifts of Met-82 in the inverse agonist-bound state corresponded well to those in the crystal structure of β2AR. Similarly, the ^{1}H and ^{13}C chemical shifts of Met-82 in the full agonist-bound state correspond well to those of the full agonist-bound form. Therefore, the observed signals report the structural difference between inactive and active conformations of the GPCR [45]. It should be noted that two signals were observed for Met-82 in the spectrum, indicating that there are at least two distinct inactivation states in β2AR.

In the antagonist-bound state, two Met-82 derived signals are observed as it is similar to those in the inverse agonist-bound state; however, their chemical shifts are slightly different. A single Met-82 signal with a chemical shift between the inverse agonist- and full agonist-bound states was observed in the partial agonist-bound states. Additionally, depending on the strength in the activation level, the chemical shifts of the resultant spectra were changed; the weak antagonist is closer to that of the inverse agonist-bound state. Based on the correlation between the chemical shifts and the efficacy of the binding ligands, we concluded that β2AR is in equilibrium between the two inactive and one active conformation in each ligand-bound state and that the ratio of each state differs depending on the binding ligand (Figure 2C). In this case, the population of the active conformation upon the binding to the saturated concentration of ligands should reflect the efficacy. The ratio of the active conformation calculated from the chemical shifts of Met-82 signal in each ligand-bound state corresponds well to their G-protein activation capacity (Figure 2D) [45]. Similar function-related conformational equilibria of β2AR/β1AR in the transmembrane regions have later been investigated by other groups [47–49]. In addition, several groups investigated the conformational equilibria in the cytoplasmic regions by ^{19}F-NMR, with chemical modifications by CF_3 groups on the cysteine sidechains [50,51]. Interestingly, these reports suggest that the conformational equilibria in the transmembrane and cytoplasmic regions are only partially coupled [43].

The structure model of agonist-bound β2AR in solution was recently proposed by quantifying the paramagnetic relaxation enhancement (PRE) effect [52]. The structure of the agonist-bound β2AR in solution was different from the structure of any GPCR reported thus far. Compared to the structure of the inverse agonist-bound β2AR, TM6 is rotated about 90° clockwise when viewed from the intracellular side. In addition, the intracellular side of TM6 was closed as compared to the structure of the G protein-bound state. Due to the rotation of TM6, the G-protein-interacting residues on TM5 and TM6 cluster on the same surface to facilitate the interaction. Furthermore, the Leu mainchain amide signal broadening indicates the presence of a structural equilibrium in the transmembrane region of β2AR in the agonist-bound state, including the site where the effector binds.

No such structural equilibrium was suggested in previous reports that utilize the extensive introduction of thermostable mutations. Thus, excessive mutation and modification of GPCR introduced to determine the structures by X-ray crystallography and Cryo-EM can be detrimental to the function-associated dynamics of proteins. It should also be noted that inserting fusion proteins in the effector-binding site also causes the loss of the function and associated dynamics of GPCRs. Thus, the observation of functional GPCR dynamics without modification would be critical in understanding the function of proteins with dynamic structural nature.

The measurements of backbone ^{15}N relaxation rates are challenging for GPCR due to its molecular weight and limitation in deuteriation. Grzesiek et al. recently reported the ^{15}N relaxation rates in the β1-adrenergic receptor in its *apo* form and seven ligand complexes [53], while the construct used retains extensive thermostabilizing mutations [48]. The measurements of ^{19}F NMR relaxation times were also conducted for the ^{19}F probes that are introduced to the functionally important sites [54,55]. The results clearly describe the changes in the amplitudes of local motions and the local reorientation frequency upon ligand bindings, which might reflect the exchange between conformational states associated with the GPCR [54,55]. A ^{19}F CPMG relaxation dispersion experiment was conducted to quantitatively describe the rate of interconversion between the conformational states in β_2-AR [51]. The strategy was also applied to obtain mechanistic insights into allosteric regulation of the $A_{2A}R$ by physiological cations [56] and ligand-dependent conformational equilibria in the transmembrane regions in β_1AR [57]. It should also be noted that fast methyl dynamics were quantified in the presence of different ligands in $A_{2A}R$ by using triple-quantum relaxation data [58], as well as in sensory rhodopsin II [59] and bacteriorhodopsin [60].

The NMR analyses of GPCRs were also conducted in a lipid bilayer environment using rHDL. Since GPCR embedded in rHDL is equivalent to a protein with a molecular weight of 200 K, the NMR spectrum of fully protonated β2AR-rHDL is rather poor. Thus, the amino acid selective deuteration was employed for the residues located in the proximity of the observing methionine residues to enhance the methyl TROSY effect. The procedure enhances the sensitivity of the NMR signal by ~5 fold [61]. Similar to that observed in DDM micelle, the NMR spectra of β2AR-rHDL bound to various ligands reflected the distinct efficacy of each ligand. However, the chemical shift of the β2AR in the partial agonist-bound state was closer to that of the full agonist-bound state in rHDL than those in micelles, indicating that the proportion of the active form was slightly higher in the lipid bilayer environment [61]. In addition, the signal in the partial agonist-bound state was significantly broader than that in micelles, and two separated signals were observed in the weak partial agonist-bound state. These results indicate that the exchange rate between the active and inactive forms is slower in the lipid bilayer environment than in micelles.

Although the GPCR dynamics study conducted in a lipid bilayer environment is limited, one more example was recently presented for the leukotriene B4 receptor BLT2 [62]. It is shown that BLT2 explores at least four different conformational states for the unliganded condition, and the relative population of the conformational states is modulated by ligands and the sterol content of the membrane.

We also calculated the G-protein activation capacity using the percentage of active β2AR in micelles and rHDL derived from NMR signals and compared it with the experimental G-protein activation capacity in β2AR-expressing cells. The results showed that the G-protein activation in rHDL corresponded better to the experimental values in cells than in micelles. Therefore, the lipid bilayer environment has an essential effect on the activity of β2AR. From the exchange rate of the structural equilibrium, we can also estimate the time required to activate the downstream signal. The results showed that the activation of β2AR occurred in ~1 ms, followed by the activation of G protein in ~100 ms, and the change in intracellular cAMP concentration in ~1 s. This estimation closely matches the experimentally measured time required for the G protein activation and the change of the intracellular cAMP concentration. The rapid activation of GPCRs is thought to be advantageous for neurotransmission and sensory reception.

The NMR observation of critical methionine residues also explained the structural basis for the selective activation of intracellular signaling. GPCRs not only activate G-protein signaling but also evoke G-protein-independent arrestin signaling after phosphorylation of the intracellular region by GPCR kinases (GRKs). GPCR ligands promote differing degrees of signaling in the G-protein and arrestin pathways called biased signaling [63,64]. Ligands that promote both signaling pathways are called balanced ligands, whereas those that activate one of the signaling pathways are called biased ligands. Although the crystal structure of the rhodopsin–arrestin complex is available, the structure of arrestin-bound rhodopsin was almost identical to that of G protein–mimetic peptide bound form [65]. Thus, the structural basis of biased signaling is largely unknown.

In μ opioid receptor (μOR), Met-245, adjacent to the aforementioned P/I/F motif, showed distinct chemical shifts in the activated form, depending on the signaling bias conditions [66]. The signal corresponding to the activated conformation of the G-protein-biased-partial-agonist (TRV130) bound form, and that of the β-arrestin-biased mutant bound to full-agonist shifted in the opposite direction relative to that of the balanced-full-agonist bound state. In addition, the chemical shifts of Met-245 correlated well with the bias factors, which is the ratio of the β-arrestin signaling efficacy to the G-protein signaling efficacy in each state. These results indicate that μOR is under an equilibrium among distinct activation states that favor the G protein activation and arrestin activation. The G protein-biased ligands induce the activation state that is suitable for G protein signaling, while the arrestin-biased mutant populates more to the activation state that favors the initialization of arrestin signaling. The chemical shifts of Met-163, Met-257, and Met-283, located in the intracellular side of the TM3, TM5, and TM6, respectively, also showed the chemical shift change that reflects the signaling bias. These residues are located in positions that reflect TM3, TM5, TM6, and TM7 conformations. Thus, the conformational equilibrium described above accompanies the coupled conformational changes in these TM helices [66]. The intracellular cavity, which is composed of TM3, TM5, TM6, and TM7, seems to exist in equilibrium between multiple open conformations, including the conformations that preferentially activate either G-protein-mediated signaling or β-arrestin-mediated signaling. Each biased ligand seems to shift the equilibrium between the multiple open conformations toward the conformation that preferentially activates G-protein-mediated signaling or arrestin signaling. The chemical shift derived from the NMR experiment thus provides information to dissect or verify the enhanced selectivity of a ligand to a certain signaling pathway [66].

As described above, phosphorylation of the C-terminal region of GPCRs by GRK initializes the arrestin signaling. However, the structure of GRK-phosphorylated GPCRs has not been described in detail. It is known that the lipid bilayer environment is important for the phosphorylation of GPCRs by GRK; thus, we reconstituted the β2AR reconstituted in the rHDL to obtain phosphorylated β2AR. The NMR spectra of the phosphorylated β2AR-rHDL showed that the signal from Met-215, located in TM5, underwent a significant chemical shift change upon phosphorylation. In addition, the Met-279 signal originating from the TM6 became broader. Therefore, phosphorylation of the C-terminal region affects the structure of the transmembrane region of β2AR. In addition, the Met-215 signal in the unligated phosphorylated β2AR was observed between those of unphosphorylated and arrestin-bound phosphorylated states. Thus, the phosphorylated β2AR partly forms a structure that is similar to the arrestin-bound structure [67].

Furthermore, by utilizing the segmental labeling strategy using the trans-splicing technique, the NMR signals originating from the C-terminal region were selectively detected [67]. The Hα and ^{13}Cα chemical shifts of the C-terminal region of the full agonist-bound state of β2AR reconstituted in the rHDL indicated that the region does not form a specific secondary structure. Upon the phosphorylation by GRK, part of the signals from the C-terminal region were significantly broadened. In particular, the signal originating from the γ2 methyl group of Thr-360 was broadened toward the ^{13}C axis, suggesting the existence of the equilibrium between multiple rotameric states with different χ1 angles on a timescale of μs–ms. In addition, the saturation transfer from the transmembrane region of β2AR to the C-terminal residue, Thr-360, was observed. Therefore, the corresponding region is close to the transmembrane region upon phosphorylation. Since arrestin recognizes both the transmembrane region and the phosphorylated C-terminal region of the GPCR, the proximity of the phosphate group to the transmembrane region would be required to form the surface suitable for arrestin interaction. Since the intracellular side of the transmembrane region of GPCR, including β2AR, is rich in basic residues, these basic residues would contribute to the interaction to the phosphorylated and negatively charged C-terminal region. The interaction analyses of phosphorylated GPCR and arrestin by NMR have also revealed that the distinct receptor phospho-barcodes are translated to specific arrestin conformations and define selective signaling [68].

As described above, NMR revealed the dynamic structural equilibrium of GPCRs in the lipid bilayer environment close to physiological conditions, which is directly related to their activity. The applications of NMR to understand the dynamics of GPCRs are extensively pursued as illustrated in recent reviews [43,69–71]. It should also be noted that the dynamics of the human neuropeptide Y receptor type 2 reconstituted into dimyristoylphosphatidylcholine (DMPC) membrane were recently investigated by solid-state NMR spectroscopy along with molecular dynamics simulations [71,72]. From ^{15}N and ^{2}H NMR spectra and the ^{13}C MAS NMR spectra, a large amplitude of structural fluctuations in a liquid-crystalline membrane was described at physiological temperature, which further indicates the high mobility of the protein in the lipid bilayer membrane. The information would also guide the ligand design to control the efficacy and signal selectivity, distinct from the conventional ligand design that uses the only affinity to the target protein as an indicator.

4. Transporters

Exchanging substances across the membrane to supply nutrients and remove waste products is fundamental for cells to keep biological activities. The exchanges are carried out by membrane proteins such as channels and transporters in a lipid bilayer. While channels passively transport ions through the pore structure, most transporters actively transport their ligands using energy molecules, such as ATP and/or concentration difference of protons, ions, and other substances across the membrane. Since transporters are also important to uptake or excrete drug molecules, they are critically involved in the pharmacokinetics and therapeutic efficacy of drugs. One of the hallmark examples is multidrug efflux transporters, which are also recognized as the most fundamental biological defense mechanism widely distributed in all three kingdoms of life. For the sake of clarity, here we focus on the structural study of multidrug efflux transporters.

There are two distinct types in multidrug efflux transporters; one is the ABC (ATP-binding cassette)-type, which utilizes the energy derived from the hydrolysis of ATP molecules for transport activity. The others are ATP-independent transporters, mostly antiporters, which carry in H+/Na^+ ion and drugs in opposite directions. Recent advances in the structural analysis have revealed how transporters achieve active transport through various mechanisms and work as sophisticated molecular machines. As for P-glycoprotein (P-gp), an ABC-type multidrug efflux transporter, X-ray, and Cryo-EM structures unveiled overall structural architectures, as well as the ligand and inhibitor recognition mechanism of the transporter [73–76]. Most of the P-gp structures are inwardly facing, presumably because the outward-facing state is not stable enough to be crystallized. Recently, the crystal structure of *Cyanidioschyzon merolae*, red algae, P-gp was determined in an outward-facing conformation with bound nucleotide, representing the first case that both inward and outward-facing conformations are determined for the eukaryotic homolog [77]. As for antiporter, the AcrAB has been extensively analyzed to unveil how trimeric AcrB transporter recognizes and eliminates various drugs in a functionally rotating mechanism based on the detailed three-dimensional structures [78,79]. Even a whole architecture of the AcrAB–TolC multidrug efflux pump that is composed of the outer-membrane channel TolC, the secondary transporter AcrB located in the inner membrane, and the periplasmic AcrA, which bridges these two integral membrane proteins, were experimentally visualized by Cryo-EM [79,80].

EmrE is another multidrug efflux transporter that is extensively analyzed by multiple structural methods. EmrE is a 110-amino-acid residue antiporter that belongs to the small multidrug resistance (SMR) transporter family. The protein exports a broad range of cationic aromatic substrates in exchange for two protons in the opposite direction [81]. It is known that the initial structure of EmrE was erroneously determined due to an improper process in the in-house program [82,83]. The current consensus topology of the EmrE structure is an asymmetric dimer with antiparallel orientation for the monomers, which is consistent with initially determined by two-dimensional crystal cryoelectron microscopy (cryo-EM) [84], the revised X-ray structure [85], and NMR as discussed below [86] (Figure 3A). In the ^{1}H–^{15}N transverse relaxation-optimized spectroscopy (TROSY) spectrum of TPP$^+$-bound ^{2}H/^{15}N-EmrE embedded in isotropic bicelles, twice as many peaks as expected for a monomer of EmrE with almost equal intensity was observed [86] (Figure 3B). Two possibilities are consistent with the NMR data: (1) parallel, symmetric EmrE dimers interconverting between inward- and outward-facing states and (2) the unique model of asymmetric antiparallel EmrE dimers interconverting between two states that are identical but open to opposite sides of the membrane. With equal populations, asymmetric antiparallel EmrE dimers can be expected, as symmetric parallel EmrE dimer model requires two distinct conformations with exactly equal free energies. In contrast, asymmetric antiparallel EmrE inherently shows equal populations because each dimer consists of one monomer in each conformation. This expectation was further confirmed by bulk and single-molecule FRET and cross-linking experiments [86].

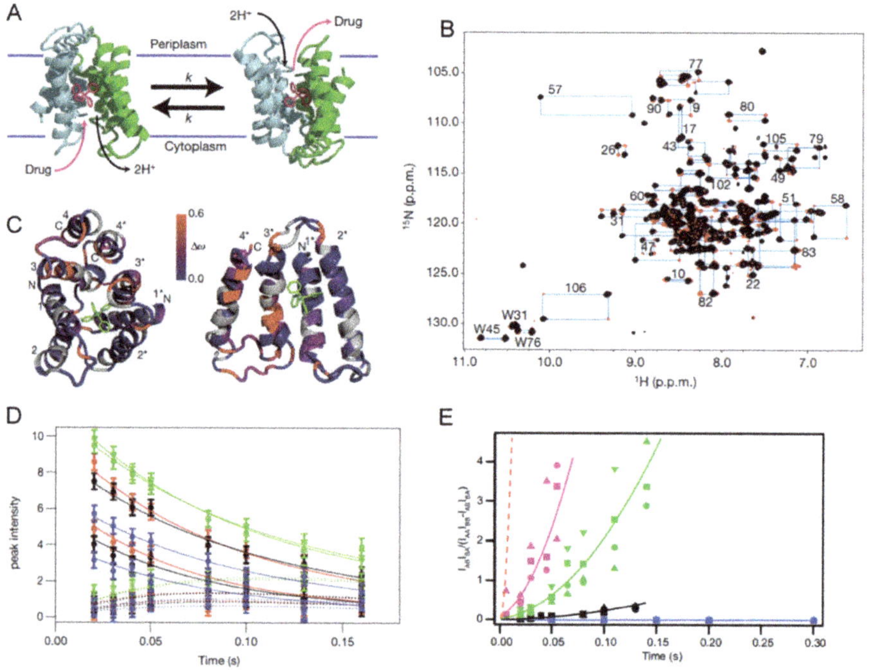

Figure 3. Transportation coupled conformational exchange of EmrE transporter revealed by NMR. (**A**) Proposed transport mechanism of EmrE by antiparallel asymmetrical dimer. The two conformational states adopt an identical structure in opposite orientations with respect to the membrane. Interconversion between the inward-open and outward-open conformations enables EmrE to transport its substrate. (**B**) Overlay of conventional ^{1}H–^{15}N TROSY HSQC spectrum (black) with the TROSY ZZ-exchange spectrum (red, 100 ms mixing time). Blue lines connect exchange peaks and auto-peaks, and residue numbers are indicated. (**C**) Mapping of normalized chemical shift changes ($\Delta\omega$) between auto-peaks in the structure of EmrE. (**D**) The global fit of auto- (solid circles) and cross-peak (open circles) intensities in the ZZ-exchange experiment. (**E**) Composite peak ratio analysis of ZZ-exchange data. Significantly different exchange rates for EmrE bound to TPP$^+$ (black) and the other slow-exchanging derivatives, EtTPP$^+$ (magenta), MBTPP$^+$ (green), and DPhTPP$^+$ (blue). Global fits are depicted with solid lines, along with the experimental data points. The dashed red line corresponds to a simulation of the composite peak ratio of fast-exchanging MeTPP$^+$ using the rate of conformational interconversion determined via line shape analysis. This figure was adapted from [86] with permission.

TROSY-selected ZZ-exchange experiments were performed to quantify the interconversion between the two states. As a result, nearly every resolved TROSY peak was assigned to exchange pairs, suggesting a widespread conformational exchange in the EmrE dimer (Figure 3C). The kinetics of conformational exchange was determined by varying the mixing time and fitting the cross-peak build-up and auto-peak decay as a function of time (Figure 3D). For TPP^+-bound EmrE in isotropic bicelles at 45 °C, a global conformational exchange with a rate constant of $4.8 \pm 0.5\,s^{-1}$ was defined, consistent with the previously observed rate determined by fluorescent measurements. A similar experiment conducted for various ligands showed that the rate of interconversion between the inward- and outward-facing states of EmrE varies significantly depending on the substrate [87] (Figure 3E). Thus, the bound substrate seems to control the rate of this critical step in the transport process. In addition, a cross-link that dramatically reduces the conformational exchange of EmrE impairs ethidium efflux activity. The observation directly confirms that the conformational exchange is vital for substrate transport [88]. Very recently, the structure of drug-bound EmrE in phospholipid bilayers was determined by 214 protein–substrate distances originating from ^{19}F and ^{1}H solid-state NMR experiments using a fluorinated substrate, tetra(4-fluorophenyl) phosphonium (F4-TPP^+). The structure showed that F4-TPP^+ lies closer to the proton-binding residue E14 in one subunit than the other subunit in the dimer, explaining the asymmetric protonation [89]. This structure and dynamics information of EmrE obtained by NMR would contribute to the future design of inhibitors for the multidrug resistance systems.

5. Conclusions and Future Perspectives

As described here, the conformational dynamics in multi-spanning helical membrane proteins quantitatively define their activity. Membrane proteins are under conformational equilibrium between different conformational states that are tightly associated with distinctive functional states. The relative population of the different conformational states can quantitatively predict the functionality of the membrane protein under certain conditions. The exchange rates between the distinct conformational states described by NMR can ultimately define the time scale of the cell response as exemplified for ion channels and GPCRs. Since NMR can provide various methods that allow detecting and quantifying protein dynamics on various timescales at a semi-atomic resolution [90,91], the method is suitable to understand the functional mechanism of the multi-spanning helical membrane proteins [44,92]. The multi-spanning helical membrane proteins are still a challenging target for solution NMR studies, as the apparent molecular weight of the detergent-solubilized and rHDL-embedded states often exceed 100 K. In addition, some of the proteins allow their overexpression only in insect or mammalian cells. These expression systems are not suitable for perdeuteration; thus, conventional proton detection experiments would not be practical. For those systems, low γ detection experiments, such as ^{15}N-detected CRINEPT or FC-TROSY, that are less susceptible to large molecular weight even in non-deuterated conditions would be of interest [93–95]. Further developments of these strategies will fill the molecular weight gap between NMR and other structure methods such as X-ray and Cryo-EM and allow the establishment of the integrated structural biology for membrane proteins, in which NMR plays a pivotal role in directly connecting the dynamics of protein to the function.

Funding: This work was supported by grants from the Japan Agency for Medical Research and Development Grant Number JP18ae010104 (to I.S.); Japan Society for the Promotion of Science, KAKENHI Grant Numbers JP17H06097 (to I.S.) and JP20H04722, JP20H03378, and JP20K21494 (to K.T.).

Institutional Review Board Statement: Not applicable.

Informed Consent Statement: Not applicable.

Conflicts of Interest: The authors declare no conflict of interest.

References

1. Doyle, D.A.; Cabral, J.M.; Pfuetzner, R.A.; Kuo, A.; Gulbis, J.M.; Cohen, S.L.; Chait, B.T.; MacKinnon, R. The Structure of the Potassium Channel: Molecular Basis of K^+ Conduction and Selectivity. *Science* **1998**, *280*, 69. [CrossRef]
2. Zhou, Y.; Morais-Cabral, J.H.; Kaufman, A.; MacKinnon, R. Chemistry of ion coordination and hydration revealed by a K^+ channel-Fab complex at 2.0 A resolution. *Nature* **2001**, *414*, 43–48. [CrossRef] [PubMed]
3. Jiang, Y.; Lee, A.; Chen, J.; Cadene, M.; Chait, B.T.; MacKinnon, R. Crystal structure and mechanism of a calcium-gated potassium channel. *Nature* **2002**, *417*, 515–522. [CrossRef] [PubMed]
4. Jiang, Y.; Lee, A.; Chen, J.; Cadene, M.; Chait, B.T.; MacKinnon, R. The open pore conformation of potassium channels. *Nature* **2002**, *417*, 523–526. [CrossRef] [PubMed]
5. Cao, E.; Liao, M.; Cheng, Y.; Julius, D. TRPV1 structures in distinct conformations reveal activation mechanisms. *Nature* **2013**, *504*, 113–118. [CrossRef] [PubMed]
6. Du, J.; Lü, W.; Wu, S.; Cheng, Y.; Gouaux, E. Glycine receptor mechanism elucidated by electron cryo-microscopy. *Nature* **2015**, *526*, 224–229. [CrossRef]
7. Wu, J.; Yan, Z.; Li, Z.; Yan, C.; Lu, S.; Dong, M.; Yan, N. Structure of the voltage-gated calcium channel Cav1.1 complex. *Science* **2015**, *350*, aad2395. [CrossRef] [PubMed]
8. Saotome, K.; Murthy, S.E.; Kefauver, J.M.; Whitwam, T.; Patapoutian, A.; Ward, A.B. Structure of the mechanically activated ion channel Piezo1. *Nature* **2018**, *554*, 481–486. [CrossRef]
9. Zhao, Q.; Zhou, H.; Chi, S.; Wang, Y.; Wang, J.; Geng, J.; Wu, K.; Liu, W.; Zhang, T.; Dong, M.-Q.; et al. Structure and mechanogating mechanism of the Piezo1 channel. *Nature* **2018**, *554*, 487–492. [CrossRef]
10. Long, S.B.; Tao, X.; Campbell, E.B.; MacKinnon, R. Atomic structure of a voltage-dependent K^+ channel in a lipid membrane-like environment. *Nature* **2007**, *450*, 376–382. [CrossRef]
11. Payandeh, J.; Scheuer, T.; Zheng, N.; Catterall, W.A. The crystal structure of a voltage-gated sodium channel. *Nature* **2011**, *475*, 353–358. [CrossRef] [PubMed]
12. McCusker, E.C.; Bagnéris, C.; Naylor, C.E.; Cole, A.R.; D'Avanzo, N.; Nichols, C.G.; Wallace, B.A. Structure of a bacterial voltage-gated sodium channel pore reveals mechanisms of opening and closing. *Nat. Commun.* **2012**, *3*, 1102. [CrossRef]
13. Hite, R.K.; MacKinnon, R. Structural Titration of Slo2.2, a Na(+)-Dependent K(+) Channel. *Cell* **2017**, *168*, 390–399.e311. [CrossRef] [PubMed]
14. Chakrapani, S.; Cordero-Morales, J.F.; Perozo, E. A Quantitative Description of KcsA Gating I: Macroscopic Currents. *J. Gen. Physiol.* **2007**, *130*, 465–478. [CrossRef]
15. Takeuchi, K.; Takahashi, H.; Kawano, S.; Shimada, I. Identification and Characterization of the Slowly Exchanging pH-dependent Conformational Rearrangement in KcsA. *J. Biol. Chem.* **2007**, *282*, 15179–15186. [CrossRef]
16. Imai, S.; Osawa, M.; Takeuchi, K.; Shimada, I. Structural basis underlying the dual gate properties of KcsA. *Proc. Natl. Acad. Sci. USA* **2010**, *107*, 6216–6221. [CrossRef] [PubMed]
17. Chakrapani, S.; Cordero-Morales, J.F.; Perozo, E. A Quantitative Description of KcsA Gating II: Single-Channel Currents. *J. Gen. Physiol.* **2007**, *130*, 479–496. [CrossRef] [PubMed]
18. Imai, S.; Osawa, M.; Mita, K.; Toyonaga, S.; Machiyama, A.; Ueda, T.; Takeuchi, K.; Oiki, S.; Shimada, I. Functional Equilibrium of the KcsA Structure Revealed by NMR. *J. Biol. Chem.* **2012**, *287*. [CrossRef]
19. Sun, Z.; Xu, Y.; Zhang, D.; McDermott, A.E. Probing allosteric coupling in a constitutively open mutant of the ion channel KcsA using solid-state NMR. *Proc. Natl. Acad. Sci. USA* **2020**, *117*, 7171. [CrossRef] [PubMed]
20. Baker, K.A.; Tzitzilonis, C.; Kwiatkowski, W.; Choe, S.; Riek, R. Conformational dynamics of the KcsA potassium channel governs gating properties. *Nat. Struct. Mol. Biol.* **2007**, *14*, 1089–1095. [CrossRef]
21. Bayburt, T.H.; Grinkova, Y.V.; Sligar, S.G. Self-Assembly of Discoidal Phospholipid Bilayer Nanoparticles with Membrane Scaffold Proteins. *Nano Lett.* **2002**, *2*, 853–856. [CrossRef]
22. van der Cruijsen, E.A.W.; Prokofyev, A.V.; Pongs, O.; Baldus, M. Probing Conformational Changes during the Gating Cycle of a Potassium Channel in Lipid Bilayers. *Biophys. J.* **2017**, *112*, 99–108. [CrossRef]
23. Chill, J.H.; Louis, J.M.; Baber, J.L.; Bax, A. Measurement of 15N relaxation in the detergent-solubilized tetrameric KcsA potassium channel. *J. Biomol. NMR* **2006**, *36*, 123–136. [CrossRef]
24. Chill, J.H.; Louis, J.M.; Delaglio, F.; Bax, A. Local and global structure of the monomeric subunit of the potassium channel KcsA probed by NMR. *Biochim. Biophys. Acta* **2007**, *1768*, 3260–3270. [CrossRef]
25. Iwahashi, Y.; Toyama, Y.; Imai, S.; Itoh, H.; Osawa, M.; Inoue, M.; Shimada, I. Conformational equilibrium shift underlies altered K(+) channel gating as revealed by NMR. *Nat. Commun.* **2020**, *11*, 5168. [CrossRef]
26. Minato, Y.; Suzuki, S.; Hara, T.; Kofuku, Y.; Kasuya, G.; Fujiwara, Y.; Igarashi, S.; Suzuki, E.; Nureki, O.; Hattori, M.; et al. Conductance of P2X4 purinergic receptor is determined by conformational equilibrium in the transmembrane region. *Proc. Natl. Acad. Sci. USA* **2016**, *113*, 4741–4746. [CrossRef]
27. Mase, Y.; Yokogawa, M.; Osawa, M.; Shimada, I. Structural basis for modulation of gating property of G protein-gated inwardly rectifying potassium ion channel (GIRK) by i/o-family G protein α subunit (Gαi/o). *J. Biol. Chem.* **2012**, *287*, 19537–19549. [CrossRef] [PubMed]

28. Toyama, Y.; Kano, H.; Mase, Y.; Yokogawa, M.; Osawa, M.; Shimada, I. Structural basis for the ethanol action on G-protein-activated inwardly rectifying potassium channel 1 revealed by NMR spectroscopy. *Proc. Natl. Acad. Sci. USA* **2018**, *115*, 3858–3863. [CrossRef]
29. Brettmann, J.B.; Urusova, D.; Tonelli, M.; Silva, J.R.; Henzler-Wildman, K.A. Role of protein dynamics in ion selectivity and allosteric coupling in the NaK channel. *Proc. Natl. Acad. Sci. USA* **2015**, *112*, 15366. [CrossRef] [PubMed]
30. Yokogawa, M.; Osawa, M.; Takeuchi, K.; Mase, Y.; Shimada, I. NMR analyses of the Gbetagamma binding and conformational rearrangements of the cytoplasmic pore of G protein-activated inwardly rectifying potassium channel 1 (GIRK1). *J. Biol. Chem.* **2011**, *286*, 2215–2223. [CrossRef] [PubMed]
31. Mandala, V.S.; Loftis, A.R.; Shcherbakov, A.A.; Pentelute, B.L.; Hong, M. Atomic structures of closed and open influenza B M2 proton channel reveal the conduction mechanism. *Nat. Struct. Mol. Biol.* **2020**, *27*, 160–167. [CrossRef]
32. Hiller, S.; Garces, R.G.; Malia, T.J.; Orekhov, V.Y.; Colombini, M.; Wagner, G. Solution Structure of the Integral Human Membrane Protein VDAC-1 in Detergent Micelles. *Science* **2008**, *321*, 1206. [CrossRef]
33. Retel, J.S.; Nieuwkoop, A.J.; Hiller, M.; Higman, V.A.; Barbet-Massin, E.; Stanek, J.; Andreas, L.B.; Franks, W.T.; van Rossum, B.J.; Vinothkumar, K.R.; et al. Structure of outer membrane protein G in lipid bilayers. *Nat. Commun.* **2017**, *8*, 2073. [CrossRef] [PubMed]
34. Oxenoid, K.; Dong, Y.; Cao, C.; Cui, T.; Sancak, Y.; Markhard, A.L.; Grabarek, Z.; Kong, L.; Liu, Z.; Ouyang, B.; et al. Architecture of the mitochondrial calcium uniporter. *Nature* **2016**, *533*, 269. [CrossRef]
35. Ge, L.; Villinger, S.; Mari, S.A.; Giller, K.; Griesinger, C.; Becker, S.; Müller, D.J.; Zweckstetter, M. Molecular Plasticity of the Human Voltage-Dependent Anion Channel Embedded into a Membrane. *Structure* **2016**, *24*, 585–594. [CrossRef] [PubMed]
36. Taylor, K.C.; Kang, P.W.; Hou, P.; Yang, N.D.; Kuenze, G.; Smith, J.A.; Shi, J.; Huang, H.; White, K.M.; Peng, D.; et al. Structure and physiological function of the human KCNQ1 channel voltage sensor intermediate state. *Elife* **2020**, *9*, e53901. [CrossRef]
37. Qasim, A.; Sher, I.; Hirschhorn, O.; Shaked, H.; Qasem, Z.; Ruthstein, S.; Chill, J.H. Investigation of a KcsA Cytoplasmic pH Gate in Lipoprotein Nanodiscs. *ChemBioChem* **2019**, *20*, 813–821. [CrossRef] [PubMed]
38. Boulton, S.; Akimoto, M.; Akbarizadeh, S.; Melacini, G. Free energy landscape remodeling of the cardiac pacemaker channel explains the molecular basis of familial sinus bradycardia. *J. Biol. Chem.* **2017**, *292*, 6414–6428. [CrossRef] [PubMed]
39. Ozawa, S.; Kimura, T.; Nozaki, T.; Harada, H.; Shimada, I.; Osawa, M. Structural basis for the inhibition of voltage-dependent K+ channel by gating modifier toxin. *Sci. Rep.* **2015**, *5*, 14226. [CrossRef]
40. Takeuchi, K.; Yokogawa, M.; Matsuda, T.; Sugai, M.; Kawano, S.; Kohno, T.; Nakamura, H.; Takahashi, H.; Shimada, I. Structural basis of the KcsA K(+) channel and agitoxin2 pore-blocking toxin interaction by using the transferred cross-saturation method. *Structure* **2003**, *11*, 1381–1392. [CrossRef]
41. Matsumura, K.; Shimomura, T.; Kubo, Y.; Oka, T.; Kobayashi, N.; Imai, S.; Yanase, N.; Akimoto, M.; Fukuda, M.; Yokogawa, M.; et al. Mechanism of hERG inhibition by gating-modifier toxin, APETx1, deduced by functional characterization. *BMC Mol. Cell Biol.* **2021**, *22*, 3. [CrossRef] [PubMed]
42. Cao, C.; Wang, S.; Cui, T.; Su, X.-C.; Chou, J.J. Ion and inhibitor binding of the double-ring ion selectivity filter of the mitochondrial calcium uniporter. *Proc. Natl. Acad. Sci. USA* **2017**, *114*, E2846. [CrossRef]
43. Shimada, I.; Ueda, T.; Kofuku, Y.; Eddy, M.T.; Wüthrich, K. GPCR drug discovery: Integrating solution NMR data with crystal and cryo-EM structures. *Nat. Rev. Drug Discov.* **2019**, *18*, 59–82. [CrossRef] [PubMed]
44. Nishida, N.; Osawa, M.; Takeuchi, K.; Imai, S.; Stampoulis, P.; Kofuku, Y.; Ueda, T.; Shimada, I. Functional dynamics of cell surface membrane proteins. *J. Magn. Reson.* **2014**, *241*, 86–96. [CrossRef] [PubMed]
45. Kofuku, Y.; Ueda, T.; Okude, J.; Shiraishi, Y.; Kondo, K.; Maeda, M.; Tsujishita, H.; Shimada, I. Efficacy of the β_2-adrenergic receptor is determined by conformational equilibrium in the transmembrane region. *Nat. Commun.* **2012**, *3*, 1045. [CrossRef] [PubMed]
46. Butterfoss, G.; DeRose, E.; Gabel, S.; Perera, L.; Krahn, J.; Mueller, G.; Zheng, X.; London, R. Conformational dependence of 13C shielding and coupling constants for methionine methyl groups. *J. Biomol. NMR* **2010**, *48*, 31–47. [CrossRef]
47. Nygaard, R.; Zou, Y.; Dror, R.O.; Mildorf, T.J.; Arlow, D.H.; Manglik, A.; Pan, A.C.; Liu, C.W.; Fung, J.J.; Bokoch, M.P.; et al. The Dynamic Process of β2-Adrenergic Receptor Activation. *Cell* **2013**, *152*, 532–542. [CrossRef]
48. Isogai, S.; Deupi, X.; Opitz, C.; Heydenreich, F.M.; Tsai, C.-J.; Brueckner, F.; Schertler, G.F.X.; Veprintsev, D.B.; Grzesiek, S. Backbone NMR reveals allosteric signal transduction networks in the β1-adrenergic receptor. *Nature* **2016**, *530*, 237. [CrossRef]
49. Solt, A.S.; Bostock, M.J.; Shrestha, B.; Kumar, P.; Warne, T.; Tate, C.G.; Nietlispach, D. Insight into partial agonism by observing multiple equilibria for ligand-bound and Gs-mimetic nanobody-bound beta1-adrenergic receptor. *Nat. Commun.* **2017**, *8*, 1795. [CrossRef]
50. Liu, J.J.; Horst, R.; Katritch, V.; Stevens, R.C.; Wüthrich, K. Biased Signaling Pathways in β_2-Adrenergic Receptor Characterized by ^{19}F-NMR. *Science* **2012**, *335*, 1106. [CrossRef]
51. Manglik, A.; Kimm, T.H.; Masureel, M.; Altenbach, C.; Yang, Z.; Hilger, D.; Lerch, M.T.; Kobilka, T.S.; Thian, F.S.; Hubbell, W.L.; et al. Structural Insights into the Dynamic Process of β2-Adrenergic Receptor Signaling. *Cell* **2015**, *161*, 1101–1111. [CrossRef] [PubMed]
52. Imai, S.; Yokomizo, T.; Kofuku, Y.; Shiraishi, Y.; Ueda, T.; Shimada, I. Structural equilibrium underlying ligand-dependent activation of β(2)-adrenoreceptor. *Nat. Chem. Biol.* **2020**, *16*, 430–439. [CrossRef] [PubMed]

53. Grahl, A.; Abiko, L.A.; Isogai, S.; Sharpe, T.; Grzesiek, S. A high-resolution description of β1-adrenergic receptor functional dynamics and allosteric coupling from backbone NMR. *Nat. Commun.* **2020**, *11*, 2216. [CrossRef] [PubMed]
54. Chung, K.Y.; Kim, T.H.; Manglik, A.; Alvares, R.; Kobilka, B.K.; Prosser, R.S. Role of detergents in conformational exchange of a G protein-coupled receptor. *J. Biol. Chem.* **2012**, *287*, 36305–36311. [CrossRef] [PubMed]
55. Kim, T.H.; Chung, K.Y.; Manglik, A.; Hansen, A.L.; Dror, R.O.; Mildorf, T.J.; Shaw, D.E.; Kobilka, B.K.; Prosser, R.S. The Role of Ligands on the Equilibria Between Functional States of a G Protein-Coupled Receptor. *J. Am. Chem. Soc.* **2013**, *135*, 9465–9474. [CrossRef] [PubMed]
56. Ye, L.; Neale, C.; Sljoka, A.; Lyda, B.; Pichugin, D.; Tsuchimura, N.; Larda, S.T.; Pomes, R.; Garcia, A.E.; Ernst, O.P.; et al. Mechanistic insights into allosteric regulation of the A2A adenosine G protein-coupled receptor by physiological cations. *Nat. Commun.* **2018**, *9*, 1372. [CrossRef] [PubMed]
57. Frei, J.N.; Broadhurst, R.W.; Bostock, M.J.; Solt, A.; Jones, A.J.Y.; Gabriel, F.; Tandale, A.; Shrestha, B.; Nietlispach, D. Conformational plasticity of ligand-bound and ternary GPCR complexes studied by 19F NMR of the β1-adrenergic receptor. *Nat. Commun.* **2020**, *11*, 669. [CrossRef]
58. Clark, L.D.; Dikiy, I.; Chapman, K.; Rödström, K.E.J.; Aramini, J.; LeVine, M.V.; Khelashvili, G.; Rasmussen, S.G.F.; Gardner, K.H.; Rosenbaum, D.M. Ligand modulation of sidechain dynamics in a wild-type human GPCR. *eLife* **2017**, *6*, e28505. [CrossRef]
59. O'Brien, E.S.; Fuglestad, B.; Lessen, H.J.; Stetz, M.A.; Lin, D.W.; Marques, B.S.; Gupta, K.; Fleming, K.G.; Wand, A.J. Membrane Proteins Have Distinct Fast Internal Motion and Residual Conformational Entropy. *Angew. Chem. Int. Ed.* **2020**, *59*, 11108–11114. [CrossRef]
60. Kooijman, L.; Schuster, M.; Baumann, C.; Jurt, S.; Löhr, F.; Fürtig, B.; Güntert, P.; Zerbe, O. Dynamics of Bacteriorhodopsin in the Dark-Adapted State from Solution Nuclear Magnetic Resonance Spectroscopy. *Angew. Chem. Int. Ed.* **2020**, *59*, 20965–20972. [CrossRef]
61. Kofuku, Y.; Ueda, T.; Okude, J.; Shiraishi, Y.; Kondo, K.; Mizumura, T.; Suzuki, S.; Shimada, I. Functional dynamics of deuterated β2 -adrenergic receptor in lipid bilayers revealed by NMR spectroscopy. *Angew. Chem. Int. Ed. Engl.* **2014**, *53*, 13376–13379. [CrossRef] [PubMed]
62. Casiraghi, M.; Damian, M.; Lescop, E.; Point, E.; Moncoq, K.; Morellet, N.; Levy, D.; Marie, J.; Guittet, E.; Banères, J.-L.; et al. Functional Modulation of a G Protein-Coupled Receptor Conformational Landscape in a Lipid Bilayer. *J. Am. Chem. Soc.* **2016**, *138*, 11170–11175. [CrossRef] [PubMed]
63. Rajagopal, S.; Rajagopal, K.; Lefkowitz, R.J. Teaching old receptors new tricks: Biasing seven-transmembrane receptors. *Nat. Rev. Drug Discov.* **2010**, *9*, 373–386. [CrossRef]
64. Reiter, E.; Ahn, S.; Shukla, A.K.; Lefkowitz, R.J. Molecular mechanism of β-arrestin-biased agonism at seven-transmembrane receptors. *Annu. Rev. Pharm. Toxicol.* **2012**, *52*, 179–197. [CrossRef]
65. Standfuss, J.; Edwards, P.C.; D'Antona, A.; Fransen, M.; Xie, G.; Oprian, D.D.; Schertler, G.F. The structural basis of agonist-induced activation in constitutively active rhodopsin. *Nature* **2011**, *471*, 656–660. [CrossRef]
66. Okude, J.; Ueda, T.; Kofuku, Y.; Sato, M.; Nobuyama, N.; Kondo, K.; Shiraishi, Y.; Mizumura, T.; Onishi, K.; Natsume, M.; et al. Identification of a Conformational Equilibrium That Determines the Efficacy and Functional Selectivity of the μ-Opioid Receptor. *Angew. Chem. Int. Ed.* **2015**, *54*, 15771–15776. [CrossRef]
67. Shiraishi, Y.; Natsume, M.; Kofuku, Y.; Imai, S.; Nakata, K.; Mizukoshi, T.; Ueda, T.; Iwaï, H.; Shimada, I. Phosphorylation-induced conformation of β2-adrenoceptor related to arrestin recruitment revealed by NMR. *Nat. Commun.* **2018**, *9*, 194. [CrossRef]
68. Yang, F.; Yu, X.; Liu, C.; Qu, C.-X.; Gong, Z.; Liu, H.-D.; Li, F.-H.; Wang, H.-M.; He, D.-F.; Yi, F.; et al. Phospho-selective mechanisms of arrestin conformations and functions revealed by unnatural amino acid incorporation and 19F-NMR. *Nat. Commun.* **2015**, *6*, 8202. [CrossRef]
69. Jones, A.J.Y.; Gabriel, F.; Tandale, A.; Nietlispach, D. Structure and Dynamics of GPCRs in Lipid Membranes: Physical Principles and Experimental Approaches. *Molecules* **2020**, *25*, 4729. [CrossRef]
70. Ferré, G.; Eddy, M.T. Structural biology of human GPCR drugs and endogenous ligands—Insights from NMR spectroscopy. *Methods* **2020**, *180*, 79–88. [CrossRef] [PubMed]
71. Vogel, A.; Bosse, M.; Gauglitz, M.; Wistuba, S.; Schmidt, P.; Kaiser, A.; Gurevich, V.V.; Beck-Sickinger, A.G.; Hildebrand, P.W.; Huster, D. The Dynamics of the Neuropeptide Y Receptor Type 1 Investigated by Solid-State NMR and Molecular Dynamics Simulation. *Molecules* **2020**, *25*, 5489. [CrossRef]
72. Schmidt, P.; Thomas, L.; Müller, P.; Scheidt, H.A.; Huster, D. The G-protein-coupled neuropeptide Y receptor type 2 is highly dynamic in lipid membranes as revealed by solid-state NMR spectroscopy. *Chemistry* **2014**, *20*, 4986–4992. [CrossRef]
73. Alam, A.; Kowal, J.; Broude, E.; Roninson, I.; Locher, K.P. Structural insight into substrate and inhibitor discrimination by human P-glycoprotein. *Science* **2019**, *363*, 753. [CrossRef]
74. Aller, S.G.; Yu, J.; Ward, A.; Weng, Y.; Chittaboina, S.; Zhuo, R.; Harrell, P.M.; Trinh, Y.T.; Zhang, Q.; Urbatsch, I.L.; et al. Structure of P-Glycoprotein Reveals a Molecular Basis for Poly-Specific Drug Binding. *Science* **2009**, *323*, 1718. [CrossRef]
75. Jin, M.S.; Oldham, M.L.; Zhang, Q.; Chen, J. Crystal structure of the multidrug transporter P-glycoprotein from Caenorhabditis elegans. *Nature* **2012**, *490*, 566–569. [CrossRef] [PubMed]
76. Kodan, A.; Yamaguchi, T.; Nakatsu, T.; Sakiyama, K.; Hipolito, C.J.; Fujioka, A.; Hirokane, R.; Ikeguchi, K.; Watanabe, B.; Hiratake, J.; et al. Structural basis for gating mechanisms of a eukaryotic P-glycoprotein homolog. *Proc. Natl. Acad. Sci. USA* **2014**, *111*, 4049. [CrossRef]

77. Kodan, A.; Yamaguchi, T.; Nakatsu, T.; Matsuoka, K.; Kimura, Y.; Ueda, K.; Kato, H. Inward- and outward-facing X-ray crystal structures of homodimeric P-glycoprotein CmABCB1. *Nat. Commun.* **2019**, *10*, 88. [CrossRef] [PubMed]
78. Murakami, S.; Nakashima, R.; Yamashita, E.; Matsumoto, T.; Yamaguchi, A. Crystal structures of a multidrug transporter reveal a functionally rotating mechanism. *Nature* **2006**, *443*, 173–179. [CrossRef] [PubMed]
79. Tam, H.K.; Foong, W.E.; Oswald, C.; Herrmann, A.; Zeng, H.; Pos, K.M. Allosteric drug transport mechanism of multidrug transporter AcrB. *Nat. Commun.* **2021**, *12*, 3889. [CrossRef] [PubMed]
80. Du, D.; Wang, Z.; James, N.R.; Voss, J.E.; Klimont, E.; Ohene-Agyei, T.; Venter, H.; Chiu, W.; Luisi, B.F. Structure of the AcrAB–TolC multidrug efflux pump. *Nature* **2014**, *509*, 512–515. [CrossRef] [PubMed]
81. Schuldiner, S. EmrE, a model for studying evolution and mechanism of ion-coupled transporters. *Biochim. Et. Biophys. Acta Proteins Proteom.* **2009**, *1794*, 748–762. [CrossRef] [PubMed]
82. Penders, B.; Horstman, K.; Vos, R. Proper science in moist biology. *EMBO Rep.* **2007**, *8*, 613. [CrossRef] [PubMed]
83. Miller, C. Pretty Structures, But What About the Data? *Science* **2007**, *315*, 459. [CrossRef]
84. Tate, C.G.; Ubarretxena-Belandia, I.; Baldwin, J.M. Conformational changes in the multidrug transporter EmrE associated with substrate binding. *J. Mol. Biol.* **2003**, *332*, 229–242. [CrossRef]
85. Chen, Y.J.; Pornillos, O.; Lieu, S.; Ma, C.; Chen, A.P.; Chang, G. X-ray structure of EmrE supports dual topology model. *Proc. Natl. Acad. Sci. USA* **2007**, *104*, 18999–19004. [CrossRef] [PubMed]
86. Morrison, E.A.; DeKoster, G.T.; Dutta, S.; Vafabakhsh, R.; Clarkson, M.W.; Bahl, A.; Kern, D.; Ha, T.; Henzler-Wildman, K.A. Antiparallel EmrE exports drugs by exchanging between asymmetric structures. *Nature* **2012**, *481*, 45–50. [CrossRef] [PubMed]
87. Morrison, E.A.; Henzler-Wildman, K.A. Transported Substrate Determines Exchange Rate in the Multidrug Resistance Transporter EmrE. *J. Biol. Chem.* **2014**, *289*, 6825–6836. [CrossRef]
88. Dutta, S.; Morrison, E.A.; Henzler-Wildman, K.A. Blocking dynamics of the SMR transporter EmrE impairs efflux activity. *Biophys. J.* **2014**, *107*, 613–620. [CrossRef]
89. Shcherbakov, A.A.; Hisao, G.; Mandala, V.S.; Thomas, N.E.; Soltani, M.; Salter, E.A.; Davis, J.H.; Henzler-Wildman, K.A.; Hong, M. Structure and dynamics of the drug-bound bacterial transporter EmrE in lipid bilayers. *Nat. Commun.* **2021**, *12*, 172. [CrossRef]
90. Tokunaga, Y.; Viennet, T.; Arthanari, H.; Takeuchi, K. Spotlight on the Ballet of Proteins: The Structural Dynamic Properties of Proteins Illuminated by Solution NMR. *Int. J. Mol. Sci.* **2020**, *21*, 1829. [CrossRef]
91. Arthanari, H.; Takeuchi, K.; Dubey, A.; Wagner, G. Emerging solution NMR methods to illuminate the structural and dynamic properties of proteins. *Curr. Opin. Struct. Biol.* **2019**, *58*, 294–304. [CrossRef] [PubMed]
92. Osawa, M.; Takeuchi, K.; Ueda, T.; Nishida, N.; Shimada, I. Functional dynamics of proteins revealed by solution NMR. *Curr. Opin. Struct. Biol.* **2012**, *22*, 660–669. [CrossRef]
93. Takeuchi, K.; Arthanari, H.; Imai, M.; Wagner, G.; Shimada, I. Nitrogen-detected TROSY yields comparable sensitivity to proton-detected TROSY for non-deuterated, large proteins under physiological salt conditions. *J. Biomol. NMR* **2016**, *64*, 143–151. [CrossRef] [PubMed]
94. Boeszoermenyi, A.; Chhabra, S.; Dubey, A.; Radeva, D.L.; Burdzhiev, N.T.; Chanev, C.D.; Petrov, O.I.; Gelev, V.M.; Zhang, M.; Anklin, C.; et al. Aromatic 19F-13C TROSY: A background-free approach to probe biomolecular structure, function, and dynamics. *Nat. Methods* **2019**, *16*, 333–340. [CrossRef] [PubMed]
95. Tokunaga, Y.; Takeuchi, K.; Okude, J.; Ori, K.; Torizawa, T.; Shimada, I. Structural Fingerprints of an Intact Monoclonal Antibody Acquired under Formulated Storage Conditions via 15N Direct Detection Nuclear Magnetic Resonance. *J. Med. Chem.* **2020**, *63*, 5360–5366. [CrossRef] [PubMed]

membranes

MDPI

Article

Small Residues Inhibit Homo-Dimerization of the Human Carbonic Anhydrase XII Transmembrane Domain

Florian Cymer [1,†] and Dirk Schneider [1,2,*]

1 Department of Chemistry, Biochemistry, Johannes Gutenberg University Mainz, 55128 Mainz, Germany; floriancymer@gmx.de
2 Institute of Molecular Physiology, Johannes Gutenberg University Mainz, 55128 Mainz, Germany
* Correspondence: Dirk.Schneider@uni-mainz.de
† Current address: Department PTDE-A, F. Hoffmann-La Roche Ltd., 4070 Basel, Switzerland.

Abstract: Amino acids with small side chains and motifs of small residues in a distance of four are rather abundant in human single-span transmembrane helices. While interaction of such helices appears to be common, the role of the small residues in mediating and/or stabilizing transmembrane helix oligomers remains mostly elusive. Yet, the mere existence of (small)xxx(small) motifs in transmembrane helices is frequently used to model dimeric TM helix structures. The single transmembrane helix of the human carbonic anhydrases XII contains a large number of amino acids with small side chains, and critical involvement of these small amino acids in dimerization of the transmembrane domain has been suggested. Using the GALLEX assay, we show here that the transmembrane domain indeed forms a strong transmembrane helix oligomer within a biological membrane. However, single or multiple mutations of small residue(s) to isoleucine almost always increased, rather than decreased, the interaction propensities. Reduction of helix flexibility and of protein–lipid contacts caused by a reduced lipid accessible surface area likely results in stabilization of helix–helix interactions within the membrane.

Keywords: carbonic anhydrase XII; GxxxG; transmembrane domain; helix–helix interaction; small amino acids; interaction motif; interaction propensity; GALLEX

1. Introduction

While studied to a great extent in recent decades, the principles guiding folding of α-helical membrane proteins and proper assembly on individual α-helices are still only rudimentarily understood. The human GpA TM helix became a paradigm for studying sequence specificity in TM helix oligomerization. The sequence LIxxGVxxGVxxT drives the interaction of the GpA TM helix [1], and especially the motif of two Gly residues in a distance of four (the GxxxG-motif) has been identified to be key for dimerization [1–3]. In an in vivo screen selecting sequences driving strong homo-dimerization of a random library of TM helices, the GxxxG-motif also appeared [4], which indicated a more generalized importance of this motif in mediating TM helix interactions. In support of this, statistical analyses have shown that this motif is highly overrepresented in TM helices [5]. Further studies have indicated that other residues with small side-chains (Ala, Ser, Cys) can also mediate sequence-specific TM helix dimerization (discussed in [6,7]), and thus the motif is better described as "(small)xxx(small)".

Small residues in a distance of four allow two TM helices to come into close contact enabling the formation of stabilizing Van der Waals interactions of neighboring residue side-chains [6–9]. Furthermore, a close distance positions hydrogen-bond donors, i.e., the side chains of Ser or Cys, in such a way that allows the formation of hydrogen bonds with backbone carbonyl oxygens of the interacting helix [10]. In addition, the formation of hydrogen bonds between the Gly Cα hydrogens and a backbone carbonyl oxygen of the interacting helix may contribute to the formation and stability of a TM helix dimer [11,12].

(Small)xxx(small) motifs typically stabilize right-handed helix dimer structures with an average crossing angle of about −40° [13]. This dimerization motif can be extended by other small residues to a (small)xxx(small)xx(small) "glycine zipper" motif [14]. These intriguing observations have prompted many researchers in the past to search for such motifs in their TM helix of interest, and the mere existence of such a motif was often used as a predictor for TM helix interactions. Yet, not solely the motif but also the sequence context is crucial for (small)xxxx(small)-mediated TM helix dimerization, which clearly is important for ensuring the formation of proper TM helix dimers in cellular membranes where multiple TM helices containing such motifs are present at a time. Yet, while two interacting helices obviously have to have complementary surfaces, interaction might not be pinned down to defined motifs consisting of small residues. In fact, the stability of a GxxxG-mediated TM helix dimer is affected by surrounding residues [15–18], and the involvement of other residues and motifs in sequence-specific dimerization of TM helices has been described in past decades [19–26]. Yet, small residues are rather abundant in human TM helices, especially in single-span TM helices, which even contain on average at least one canonical GxxxG motif [7,27–30]. While the interaction of several single-span TM helices has been analyzed to some extent and appears to be common, the role of the (small)xxx(small) motifs and/or of small residues in mediating and/or stabilizing TM helix dimers remains mostly elusive.

In the present study, we have used the human carbonic anhydrase (CA) XII TM helix as a model to systematically study the impact of replacing small residues in a TM domain on the in vivo interaction propensity of the TM segment. First, we show that the CA XII TM domain does indeed have an intrinsic interaction propensity in vivo. Interaction of the wt and mutated CA XII TM segment was analyzed using the GALLEX-system, a bacterial two-hybrid system, allowing the analysis of TM helix–helix interactions within a biological membrane [31,32]. While the CA XII TM domain forms a strong TM helix dimer in a biological membrane, single or multiple mutations of small residue(s) to the β-branched amino acid Ile almost always led to increased interaction propensities. Thus, small residues in the CA XII TM helix block, rather than mediate, TM helix dimerization within a biological membrane. As discussed, this could be due to changes in helix dynamics and/or a decrease of the lipid accessible surface area. Our study highlights that TM helix dimerization and the structure of TM helix dimers should be predicted with caution when mainly based on the mere existence of (small)xxx(small) motifs within the TM domain.

2. Materials and Methods

2.1. Plasmid Construction

Synthetic oligonucleotide cassettes coding for the human CA XII TM domain were ligated into a *Nhe*I and *BamH*I restriction-digested pBLM plasmid [32], resulting in the plasmid pBLMCAXIITM. Correct ligation was confirmed by DNA sequencing. Mutations were introduced via the Quick Change site-directed mutagenesis kit (Agilent, Santa Clara, CA, USA) following the supplier's instructions using the pBLMCAXIITM plasmid as a template. Mutations were confirmed by sequencing.

2.2. GALLEX Assay

The GALLEX system was used to measure CA XII TM helix oligomerization as described in detail elsewhere [31,32]. Briefly, the pBLMCAXIITM construct and variants were used to transform *E. coli* SU101 cells. Addition of 100 μM IPTG resulted in the expression of a chimeric protein consisting of an N-terminal LexA DNA-binding domain fused to the CA XII TM domain and the maltose-binding protein at the C-terminus. As a mock control, we measured the empty pBLM plasmid in the GALLEX system. As a positive control in the GALLEX assay, dimerization of the well-studied and strongly interacting glycophorin A (GpA) TM domain was analyzed. As a negative control, the weakly interacting G83I GpA-variant was assayed.

Proper membrane insertion was tested for all expressed chimeric proteins by extracting soluble and membrane-attached proteins from isolated *E. coli* membranes with 0.1 M NaOH. Only proteins integrated within the membrane remain in the membrane after this treatment. Membrane-integrated chimeric proteins were detected via Western blot analysis using a maltose-binding protein-specific antibody (New England Biolabs, Frankfurt, Germany).

Proper transmembrane orientation of the chimeric proteins was tested by plating the maltose-binding protein-deficient *E. coli* strain NT326 [33], transformed with the pBLMCAXIITM and control plasmids, on M9 minimal media agar plates. The plates contained IPTG to induce recombinant protein expression, 100 μg/ml ampicillin as a selection marker, and maltose as the only carbon source. Bacterial growth can only occur if the maltose-binding protein is located within the periplasm and thus the chimeric protein is inserted in the correct topology.

2.3. *Software*

A model of the human CA XII TM helix dimer was created using the PREDDIMER server [34]. Sequences were aligned using the T-COFFEE algorithm [35] and carbonic anhydrase sequences of different mammalian species were derived from the ExPASy proteomic server [36]. The lipid accessible surface area was calculated with the GETAREA program [37] using the previously generated CA XII TM domain model and energy-minimized models of CA XII TM domain variants. In order to probe the lipid accessible surface area, a probe radius of 1.88 Å was chosen, which corresponds to the radius of a methyl group and has been used for this purpose previously [38].

3. Results and Discussion

The mere existence of a (small)xxx(small) motif is often used to predict a dimeric TM helix structure. For example, the structures of the soluble domains of the human CA IX and XII have been solved in recent years, and the structures of the TM regions have been modeled [39,40]. In tumors, these isoenzymes are overexpressed and contribute to an elevated tumor progression by acidifying the surrounding extracellular space, leading to elevated tissue penetration [41,42]. The human CA XII is a bitopic membrane protein located in the plasma membrane of different tissues. While the crystal structure of the CA XII extracellular domain has been solved [39], the structures of the short intracellular domain-containing putative phosphorylation sites and of the TM domain are still elusive. As the catalytically active domains are dimeric in solution, and the TM domains are rich in small residues and contain (small)xxx(small) motifs, a dimeric TM helix structure was assumed and modeled using the GpA dimer structure as a template [39], however, without experimental proof. Actually, it is currently still unknown whether the CA XII TM domain dimerizes at all in a biological membrane.

More than 40% of the amino acids within the predicted human CA XII TM domain have small side-chains (Figure 1C). The overall sequences of mammalian CA XII TM domains are highly homologous, and small residues are well-conserved. The high abundance of small amino acids potentially enables close packing of TM helices.

To study the CA XII TM helix interaction propensity in a biological membrane in detail, we performed GALLEX measurements. The GALLEX system allows measuring the interaction propensity of TM helices in vivo, within the *E. coli* inner membrane [32]. In the GALLEX system, a TM helix of interest (here, the CA XII TM helix) is genetically fused at its N-terminus to the DNA-binding domain of the *E. coli* LexA protein and at its C-terminus to the *E. coli* MalE protein, which facilitates membrane insertion. Only a homo-dimeric LexA DNA-binding domain can bind to a promoter/operator region located within the *E. coli* genome and thereby repress the activity of the *lacZ* reporter gene (coding for β-galactosidase). As dimerization of the DNA-binding domain is mediated by interactions of the fused TM domains, the determined activity of the reporter directly correlates with a given interaction propensity. Besides studying the interactions of CA XII

TM helices, as controls, we also determined the interaction propensity of the well-studied human GpA TM helix, which forms a highly stable TM helix dimer, as well as a GpA G83I variant, which interacts rather weakly due to the mutation of a Gly residue, which is critically involved in TM helix dimerization [1–3]. Noteworthily, while measured in a bacterial membrane, GALLEX and related genetic systems, such as ToxR, TOXCAT, and POSSYCCAT (or variations of these), have frequently been used in the past to study interactions of human TM proteins, e.g., in [3,10,15–18,25,28,29,43]. Nevertheless, while differences in a given interaction propensity caused by a mutation will likely be observed in a non-native, yet biological, membrane environment, the actual interaction strength might vary depending on the exact membrane lipid composition [6].

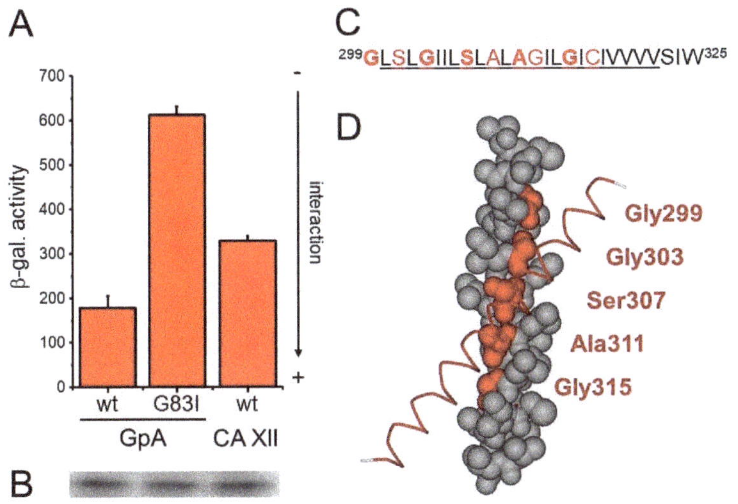

Figure 1. Human CA XII TM helix–helix interaction. (**A**) When measured with the GALLEX assay, the wild-type CA XII TM domain has a significant interaction propensity compared to the strongly interacting TM domain of human GpA and the weakly interacting GpA variant G83I. The interaction propensities are an average of three independent measurements and the standard deviation is shown. (**B**) Identical amounts of isolated total membrane proteins of cells analyzed in (**A**) were separated in an SDS gel, and the expressed amount of the Ca XII TM fusion protein was tested via Western blot analyses using an antibody directed against the MalE protein. All analyzed proteins were expressed at similar levels. (**C**) Sequence of the human CA XII TM region. The TM regions as predicted by TMHMM are underlined. Small residues are highlighted in red. Small residues at a distance of four are highlighted in bold. (**D**) Model of the human CA XII TM domain dimer predicted by the PREDDIMER server. CPK representation of one TM domain with highlighted and labeled small residues Gly299, Gly303, Ser307, Gly311, and Gly315. The interacting helix is shown as a Cα wire.

The interaction propensity of the wt CA XII TM domain is less pronounced than the interaction propensity of the strongly interacting GpA TM helix, yet significantly stronger than the weakly interacting GpA G83I variant (Figure 1A). Thus, these data demonstrate that the human CA XII TM domain has an intrinsic propensity to self-interact within a biological membrane, which most likely contributes to the stability of the dimeric full-length protein.

Putative structures of the human CA XII TM helix dimer have been predicted recently using the program CATM [44]. Three prominent structures have been modeled, which all have a crossing of $-47°$ to $-48°$. In the top-scoring model, the residues Ala311 and Gly315 are at the helix crossing point and thus are predicted to be key for dimerization. While Models 2 and 3 differ slightly, in both models, the crossing point of the TM helices mainly involves the residues Gly299 and Gly303. When the structure of the putative CA XII TM helix dimer was modeled using the program PREDDIMER [34], the top-scoring model also predicted a right-handed helix dimer with a crossing angle of $-45°$. Yet, interaction of the helices is mainly mediated/stabilized by the residues Gly303 and Ser307 (Figure 1D). Further dimeric (left- as well as right-handed) structures were also predicted, where other small residues are involved in helix–helix interactions. However, essentially none of the models agreed with the CATM predictions. With THOIPA (Transmembrane Homodimer Interface Prediction Algorithm), which predicts interfacial residues from evolutionary sequence data [45], the residues Gly299 and Gly303 are predicted to lay at the helix–helix interface, in line with Models 2 and 3 predicted by CATM.

Actually, the small residues Gly299, Gly303, Ser307, Ala311, and Gly315 (highlighted in Figure 1D) form a continuous groove along the CA XII TM helix. Packing of the bulkier residues Leu302, Leu306, Leu310, and Leu314, which form a ridge on the adjacent helix, into this groove might allow close packing of two helices, and diverse small residues might be involved in the formation of different CA XII TM helix dimer structures, which might be involved in regulation of the CA XII activity.

In order to further elucidate the function of the small residues forming the continuous groove in the TM domain, we consecutively mutated the small residues highlighted in Figure 2B to Ile. We anticipated that the replacement of small amino acids by the bulkier Ile residue would disturb close packing interactions and thereby disrupt TM helix/helix interactions, a strategy used several times before (e.g., [1,43,46]. Surprisingly, all mutations of the selected small residues to Ile led to either no changes or to slightly increased interaction propensities (Figure 2A), indicating that the voids or grooves formed by the small residues are eventually not directly involved in TM helix dimerization. To further analyze this, we subsequently mutated multiple small amino acids within the CA XII TM domain in parallel and monitored the interaction tendencies of the variants. As shown in Figure 2C, the interaction tendencies of the double and even triple mutants did not decrease at all, yet a significant increase in the interaction propensities was observed for all mutated TM sequences. Simultaneous mutations of the small amino acids at positions 299/303, 303/307, 307/311, or 311/315 resulted in a stronger interaction compared to the wt or the single mutated TM helices (Figure 2C). Introducing three mutations in parallel at positions 307/311/315 even resulted in an additional increase in the interaction propensity. Solely the triple mutant 299/303/307 displayed a slightly reduced interaction propensity in comparison to the respective double mutants. However, this triple mutant still had a stronger interaction propensity than the wt TM helix. Together, these results clearly demonstrate that, while the CA XII TM helix has an intrinsic propensity to self-interact, the small amino acids appear to destabilize rather than stabilize an oligomeric TM structure. Noteworthily, the expression level was controlled for all constructs, and the concentration of the individual proteins within the *E. coli* inner membrane was similar (Figures 1B and 2B,D). Furthermore, all proteins had the expected TM topology, with the TM helix C-terminus facing the bacterial periplasm (Figure 2E,F).

The unexpected results of our measurements forced us to reevaluate our initial assumption that the replacement of a small residue by Ile typically disrupts helix–helix interactions. The presented observations indicate that the mutated residues are not a critical part of a well-defined interaction interface, as their mutation does not hinder, but rather increases, the interaction propensity of the CA XII TM domain in most cases. GxxxG and GxxxG-like motifs are well-known to frequently mediate helix–helix interactions, but it is also becoming increasingly apparent that there are a host of unknown modes of helix–helix interactions (for a review, view [6,7,47]).

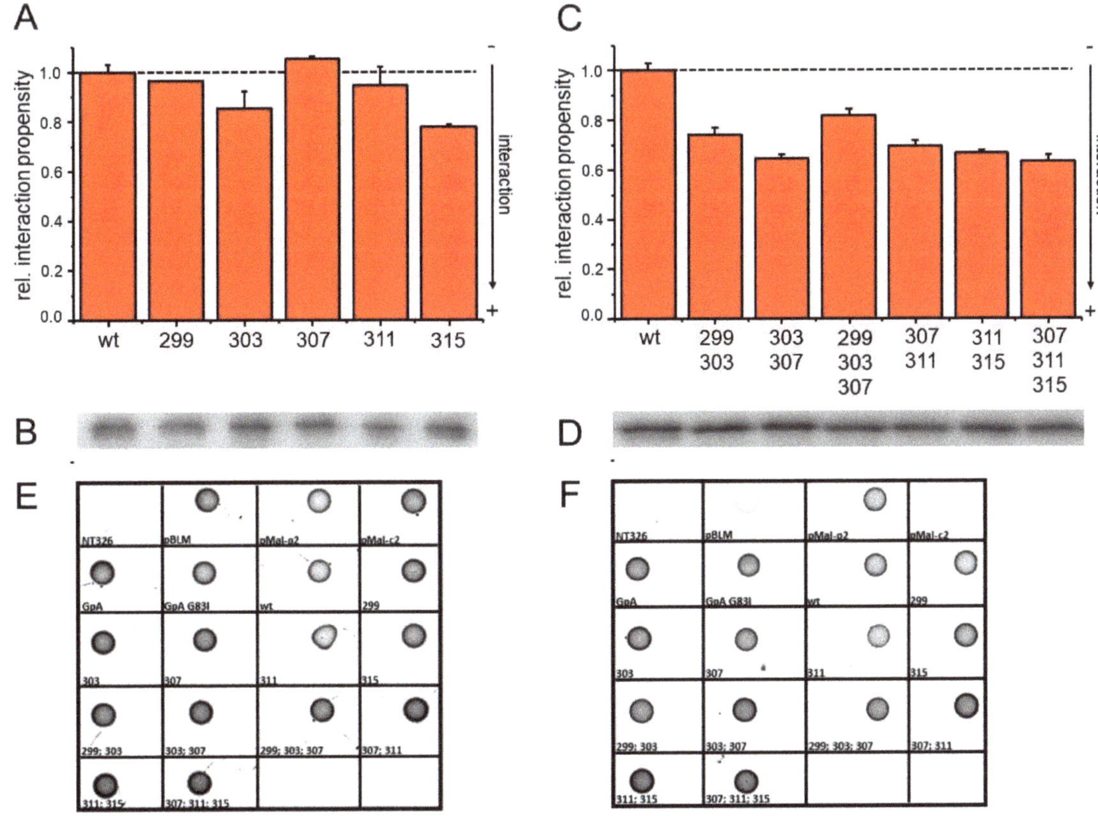

Figure 2. GALLEX measurements of the human CA XII TM domain mutants. (**A**) Interaction tendencies of CA XII TM variants carrying mutations of individual small amino acids to Ile. The position of the mutated amino acid is indicated below each bar. All interaction propensities have been normalized to the wt (set as 1.0). The interaction propensities are an average of three independent measurements, and the standard deviation is shown. (**B**) Identical amounts of isolated total membrane proteins of cells analyzed in (**A**) were separated via SDS-PAGE, and the expressed amount of the CA XII TM fusion protein was analyzed via Western blot analyses using an antibody directed against the MalE protein. All analyzed proteins were expressed at similar levels in membranes. (**C**) GALLEX measurement of CA XII variants carrying mutations of multiple small amino acids to Ile. All interaction propensities have been normalized to the wt (set as 1.0). The interaction propensities are an average of three independent measurements, and the standard deviation is shown. (**D**) Identical amounts of isolated total membrane proteins of cells analyzed in (**C**) were separated in an SDS gel, and the expressed amount of the Ca XII TM fusion protein was analyzed via Western blot analyses using an antibody directed against the MalE protein. All analyzed proteins were expressed at similar levels in membranes. (**E**, **F**) Maltose complementation assay to determine the TM topology of the analyzed chimeric proteins. The correct topology of the analyzed chimeric proteins within the *E. coli* membrane was tested using a maltose complementation assay. MalE-deficient *E. coli* NT326 cells carrying different plasmids are able to grow on M9-agar plates containing glucose as the sole carbon source (**E**), but only grow on maltose plates (**F**) if the construct is correctly oriented in the membrane and the MalE domain is located in the *E. coli* periplasm. The plates contained 100 µg/ml ampicillin and a final concentration of 100 µM IPTG. On both plates, growth of NT326 cells, which have not been transformed with a plasmid, is not seen. Cells containing the empty pBLM plasmid and cells carrying the pMal-c2 plasmid (which leads to a cytoplasmic localization of the MalE domain) are unable to grow on maltose plates. Growth of all used GALLEX constructs on maltose-only plates confirms membrane insertion and a correct TM orientation of the CA XII TM domains.

The TM domain of the human CA XII consists of more than 40% of amino acids with small side chains, leading to the formation of a rather flexible TM helix structure. In fact, as for Gly-containing TM helices, it has been shown that Gly increases the flexibility of a given TM helix [48] and thus weakens the regular "stiff" helix geometry. In TM helices, the helical propensity of an amino acid correlates with its side-chain hydrophobicity [49]. Replacing small residues with a low helical propensity in membranes by Ile, a residue that severely promotes the formation of ideal TM α-helices [49], likely reduces the flexibility and structural dynamics of the CA XII TM helix, shifting the monomer–dimer equilibrium towards the dimer. Furthermore, intra-helical interactions between β-branched amino acids might additionally contribute to the overall stability of a TM helix [5,49]. Since the increase in the interaction propensity caused by the introduction of β-branched residues seems to depend on the distance to other such residues, the increase in interaction propensity of single Ile mutants should also depend on the sequence context with respect to the presence of other β-branched side chains. Indeed, the interaction propensity of the CA XII variant having a single mutation to Ile at position 303 is slightly increased. Interestingly, in the immediate vicinity of residue 303, two Ile are located at positions 304 and 305, and Ile303 might interact with the neighboring side chains. A far more pronounced increase in the interaction propensity was observed when Ile was introduced at position 315, and in the distance of four, a Val residue is located at position 319. Intrahelical interactions of β-branched side chains are strongest in i, i + 4 and weaker in i, i + 3 [50]. The multiple mutations to Ile always led to a positioning of Ile in the pattern i, i + 4. Here, a strong increase in the interaction propensity was always observed, whereas single mutations to Ile at these positions are frequently without effect. Noteworthily, in the present study, exclusively mutations to Ile were analyzed, and it is feasible to assume interactions with other β-branched side chains. Consequently, mutations to other amino acids can clearly have a different effect on the observed interaction propensities.

In the case of the well-studied GpA TM helix dimer, the interaction propensity of the (mutated) GpA TM helix was inversely related to the interfacial lipid accessibility of the GpA helix dimer [38]. Thus, we also analyzed a potential correlation of the apolar lipid accessible surface area of the analyzed CA XII TM segments and the measured interaction propensities. Interactions between CA XII TM helix side-chains and lipids might stabilize the structure of the monomeric TM helix, thereby inhibiting the formation of TM helix dimers. In fact, a correlation between the lipid accessible surface area of the TM helices analyzed in this study and the observed interaction propensity was found (Figure 3). Successive introduction of Ile residues resulted in a less lipid-solvated TM helix, promoting helix–helix interactions.

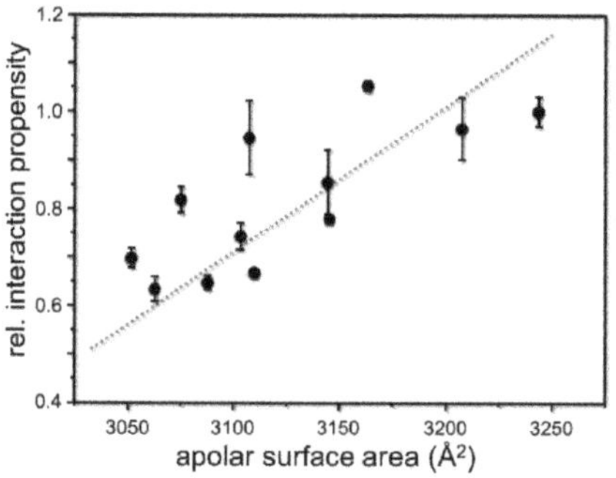

Figure 3. The lipid accessible apolar surface area correlates with the normalized interaction propensity of the wild-type CA XII TM segment and its variants. A linear fit of the correlation is shown. (r^2 = 0.75. The normalized standard deviations of the three independent measurements from the GALLEX experiments are indicated. The lipid accessible apolar surface area was calculated using the GETAREA program [37] and energy-minimized modeled structures of the respective CA XII TM helix variant.

4. Conclusions

Our mutational analyses suggest that the numerous small amino acids are not critically involved in the dimerization of the CA XII TM helix, but rather destabilize the oligomeric structure. Small amino acid side chains might enhance the flexibility of the CA XII TM helix backbone, which has an impact on the lipid shell surrounding the helix. Introduction of Ile residues results in a less flexible TM helix structure, which is less lipid solvated, and thus reduction of protein–lipid contacts and of helix flexibility likely promotes interactions of the CA XII TM helix. Future studies to analyze the impact of small residues and small-xxx-small motives, especially with single-span human TM helices that contain multiple small residues and even contain on average at least one canonical GxxxG motif [7,27–30], will further enhance our knowledge of the role of small residues in TM helix association. Nevertheless, the mere existence and evolutionary conservation of (multiple) small residues and/or of (small)xxx(small) motifs in TM helices is probably not a good predictor for assuming GpA-like TM helix–helix interactions, and structural modeling using such motifs as the key predictor likely falls too short.

Author Contributions: Conceptualization, F.C. and D.S.; methodology, F.C.; validation, F.C. and D.S.; formal analysis, F.C. and D.S.; investigation, F.C.; resources, D.S.; data curation, F.C.; writing—original draft preparation, F.C. and D.S.; writing—review and editing, D.S.; visualization, F.C. and D.S.; supervision, D.S.; project administration, D.S.; funding acquisition, D.S. All authors have read and agreed to the published version of the manuscript.

Funding: This work was supported by grants from the Deutsche Forschungsgemeinschaft (SCHN 690/2-3) and by DynaMem (state of Rhineland-Palatinate).

Data Availability Statement: The data that support the findings of this study are available on re-quest.

Acknowledgments: We thank Hildegard Pearson for critically reading the manuscript.

Conflicts of Interest: The authors declare no conflict of interest.

References

1. Lemmon, M.A.; Flanagan, J.M.; Treutlein, H.R.; Zhang, J.; Engelman, D.M. Sequence specificity in the dimerization of transmembrane alpha-helices. *Biochemistry* **1992**, *31*, 12719–12725. [CrossRef]
2. MacKenzie, K.R.; Prestegard, J.H.; Engelman, D.M. A transmembrane helix dimer: Structure and implications. *Science* **1997**, *276*, 131–133. [CrossRef]
3. Brosig, B.; Langosch, D. The dimerization motif of the glycophorin a transmembrane segment in membranes: Importance of glycine residues. *Protein. Sci.* **1998**, *7*, 1052–1056. [CrossRef] [PubMed]
4. Russ, W.P.; Engelman, D.M. The GxxxG motif: A framework for transmembrane helix-helix association. *J. Mol. Biol.* **2000**, *296*, 911–919. [CrossRef] [PubMed]
5. Senes, A.; Gerstein, M.; Engelman, D.M. Statistical analysis of amino acid patterns in transmembrane helices: The GxxxG motif occurs frequently and in association with beta-branched residues at neighboring positions. *J. Mol. Biol.* **2000**, *296*, 921–936. [CrossRef] [PubMed]
6. Cymer, F.; Veerappan, A.; Schneider, D. Transmembrane helix-helix interactions are modulated by the sequence context and by lipid bilayer properties. *Biochim. Biophys. Acta* **2012**, *1818*, 963–973. [CrossRef]
7. Teese, M.G.; Langosch, D. Role of GxxxG Motifs in Transmembrane Domain Interactions. *Biochemistry* **2015**, *54*, 5125–5135. [CrossRef]
8. Schneider, D. Rendezvous in a membrane: Close packing, hydrogen bonding, and the formation of transmembrane helix oligomers. *FEBS Lett.* **2004**, *577*, 5–8. [CrossRef]
9. Senes, A.; Engel, D.E.; DeGrado, W.F. Folding of helical membrane proteins: The role of polar, GxxxG-like and proline motifs. *Curr. Opin. Struct. Biol.* **2004**, *14*, 465–479. [CrossRef]
10. Schneider, D.; Engelman, D.M. Motifs of Two Small Residues can ASSIST but are not sufficient to Mediate Transmembrane Helix Interactions. *J. Mol. Biol.* **2004**, *343*, 799–804. [CrossRef]
11. Senes, A.; Ubarretxena-Belandia, I.; Engelman, D.M. The Calpha—H...O hydrogen bond: A determinant of stability and specificity in transmembrane helix interactions. *Proc. Natl. Acad. Sci. USA* **2001**, *98*, 9056–9061. [CrossRef] [PubMed]
12. Arbely, E.; Arkin, I.T. Experimental measurement of the strength of a C alpha—H...O bond in a lipid bilayer. *J. Am. Chem. Soc.* **2004**, *126*, 5362–5363. [CrossRef]
13. Walters, R.F.S.; DeGrado, W.F. Helix-packing motifs in membrane proteins. *Proc. Natl. Acad. Sci. USA* **2006**, *103*, 13658–13663. [CrossRef] [PubMed]
14. Kim, S.; Jeon, T.J.; Oberai, A.; Yang, D.; Schmidt, J.J.; Bowie, J.U. Transmembrane glycine zippers: Physiological and pathological roles in membrane proteins. *Proc. Natl. Acad. Sci. USA* **2005**, *102*, 14278–14283. [CrossRef]
15. Herrmann, J.R.; Fuchs, A.; Panitz, J.C.; Eckert, T.; Unterreitmeier, S.; Frishman, D.; Langosch, D. Ionic Interactions Promote Transmembrane Helix-Helix Association Depending on Sequence Context. *J. Mol. Biol.* **2010**, *396*, 452–461. [CrossRef]
16. Herrmann, J.R.; Panitz, J.C.; Unterreitmeier, S.; Fuchs, A.; Frishman, D.; Langosch, D. Complex Patterns of Histidine, Hydroxylated Amino Acids and the GxxxG Motif Mediate High-affinity Transmembrane Domain Interactions. *J. Mol. Biol.* **2008**, *385*, 912–923. [CrossRef] [PubMed]

17. Unterreitmeier, S.; Fuchs, A.; Schaffler, T.; Heym, R.G.; Frishman, D.; Langosch, D. Phenylalanine promotes interaction of transmembrane domains via GxxxG motifs. *J. Mol. Biol.* **2007**, *374*, 705–718. [CrossRef]
18. Melnyk, R.A.; Kim, S.; Curran, A.R.; Engelman, D.M.; Bowie, J.U.; Deber, C.M. The affinity of GXXXG motifs in transmembrane helix-helix interactions is modulated by long-range communication. *J. Biol. Chem.* **2004**, *279*, 16591–16597. [CrossRef]
19. Gurezka, R.; Langosch, D. In vitro selection of membrane-spanning leucine zipper protein-protein interaction motifs using POSSYCCAT. *J. Biol. Chem.* **2001**, *276*, 45580–45587. [CrossRef]
20. Dawson, J.P.; Weinger, J.S.; Engelman, D.M. Motifs of serine and threonine can drive association of transmembrane helices. *J. Mol. Biol.* **2002**, *316*, 799–805. [CrossRef]
21. Zhou, F.X.; Merianos, H.J.; Brunger, A.T.; Engelman, D.M. Polar residues drive association of polyleucine transmembrane helices. *Proc. Natl. Acad. Sci. USA* **2001**, *98*, 2250–2255. [CrossRef] [PubMed]
22. Gratkowski, H.; Lear, J.D.; De Grado, W.F. Polar side chains drive the association of model transmembrane peptides. *Proc. Natl. Acad. Sci. USA* **2001**, *98*, 880–885. [CrossRef] [PubMed]
23. Johnson, R.M.; Hecht, K.; Deber, C.M. Aromatic and cation-pi interactions enhance helix-helix association in a membrane environment. *Biochemistry* **2007**, *46*, 9208–9214. [CrossRef] [PubMed]
24. Sal-Man, N.; Gerber, D.; Bloch, I.; Shai, Y. Specificity in transmembrane helix-helix interactions mediated by aromatic residues. *J. Biol. Chem.* **2007**, *282*, 19753–19761. [CrossRef]
25. Sal-Man, N.; Gerber, D.; Shai, Y. The composition rather than position of polar residues (QxxS) drives aspartate receptor transmembrane domain dimerization in vivo. *Biochemistry* **2004**, *43*, 2309–2313. [CrossRef]
26. Steindorf, D.; Schneider, D. In vivo selection of heterotypically interacting transmembrane helices: Complementary helix surfaces, rather than conserved interaction motifs, drive formation of transmembrane hetero-dimers. *Biochim. Et Biophys. Acta Biomembr.* **2017**, *1859*, 245–256. [CrossRef]
27. Hubert, P.; Sawma, P.; Duneau, J.-P.; Khao, J.; Henin, J.; Bagnard, D.; Sturgis, J. Single-spanning transmembrane domains in cell growth and cell-cell interactions. *Cell Adhes. Migr.* **2010**, *4*, 313–324. [CrossRef]
28. Kirrbach, J.; Krugliak, M.; Ried, C.L.; Pagel, P.; Arkin, I.T.; Langosch, D. Self-interaction of transmembrane helices representing pre-clusters from the human single-span membrane proteins. *Bioinformatics* **2013**, *29*, 1623–1630. [CrossRef]
29. Ried, C.L.; Kube, S.; Kirrbach, J.; Langosch, D. Homotypic interaction and amino acid distribution of unilaterally conserved transmembrane helices. *J. Mol. Biol.* **2012**, *420*, 251–257. [CrossRef]
30. Langosch, D.; Arkin, I.T. Interaction and conformational dynamics of membrane-spanning protein helices. *Protein Sci.* **2009**, *18*, 1343–1358. [CrossRef]
31. Cymer, F.; Sanders, C.R.; Schneider, D. Analyzing oligomerization of individual transmembrane helices and of entire membrane proteins in *E. coli*: A hitchhiker's guide to GALLEX. *Methods Mol. Biol.* **2013**, *932*, 259–276. [PubMed]
32. Schneider, D.; Engelman, D.M. GALLEX, a Measurement of Heterologous Association of Transmembrane Helices in a Biological Membrane. *J. Biol. Chem.* **2003**, *278*, 3105–3111. [CrossRef] [PubMed]
33. Treptow, N.A.; Shuman, H.A. Genetic evidence for substrate and periplasmic-binding-protein recognition by the MalF and MalG proteins, cytoplasmic membrane components of the Escherichia coli maltose transport system. *J. Bacteriol.* **1985**, *163*, 654–660. [CrossRef] [PubMed]
34. Polyansky, A.A.; Chugunov, A.O.; Volynsky, P.E.; Krylov, N.A.; Nolde, D.E.; Efremov, R.G. PREDDIMER: A web server for prediction of transmembrane helical dimers. *Bioinformatics* **2014**, *30*, 889–890. [CrossRef]
35. Notredame, C.; Higgins, D.G.; Heringa, J. T-Coffee: A novel method for fast and accurate multiple sequence alignment. *J. Mol. Biol.* **2000**, *302*, 205–217. [CrossRef]
36. Gasteiger, E.; Gattiker, A.; Hoogland, C.; Ivanyi, I.; Appel, R.D.; Bairoch, A. ExPASy: The proteomics server for in-depth protein knowledge and analysis. *Nucleic Acids Res.* **2003**, *31*, 3784–3788. [CrossRef]
37. Fraczkiewicz, R.; Braun, W. Exact and efficient analytical calculation of the accessible surface areas and their gradients for macromolecules. *J. Comp. Chem.* **1998**, *19*, 319–333. [CrossRef]
38. Cunningham, F.; Poulsen, B.E.; Ip, W.; Deber, C.M. Beta-branched residues adjacent to GG4 motifs promote the efficient association of glycophorin A transmembrane helices. *Biopolymers* **2011**, *96*, 340–347. [CrossRef]
39. Whittington, D.A.; Waheed, A.; Ulmasov, B.; Shah, G.N.; Grubb, J.H.; Sly, W.S.; Christianson, D.W. Crystal structure of the dimeric extracellular domain of human carbonic anhydrase XII, a bitopic membrane protein overexpressed in certain cancer tumor cells. *Proc. Natl. Acad. Sci. USA* **2001**, *98*, 9545–9550. [CrossRef]
40. Alterio, V.; Hilvo, M.; di Fiore, A.; Supuran, C.T.; Pan, P.; Parkkila, S.; Scaloni, A.; Pastorek, J.; Pastorekova, S.; Pedone, C.; et al. Crystal structure of the catalytic domain of the tumor-associated human carbonic anhydrase IX. *Proc. Natl. Acad. Sci. USA* **2009**, *106*, 16233–16238. [CrossRef]
41. Benej, M.; Pastorekova, S.; Pastorek, J. Carbonic anhydrase IX: Regulation and role in cancer. *Subcell Biochem.* **2014**, *75*, 199–219. [PubMed]
42. Waheed, A.; Sly, W.S. Carbonic anhydrase XII functions in health and disease. *Gene* **2017**, *623*, 33–40. [CrossRef]
43. Escher, C.; Cymer, F.; Schneider, D. Two GxxxG-like motifs facilitate promiscuous interactions of the human ErbB transmembrane domains. *J. Mol. Biol.* **2009**, *389*, 10–16. [CrossRef]
44. Anderson, S.M.; Mueller, B.K.; Lange, E.J.; Senes, A. Combination of Calpha-H Hydrogen Bonds and van der Waals Packing Modulates the Stability of GxxxG-Mediated Dimers in Membranes. *J. Am. Chem. Soc.* **2017**, *139*, 15774–15783. [CrossRef]

45. Xiao, Y.; Zeng, B.; Berner, N.; Frishman, D.; Langosch, D.; Teese, M.G. Experimental determination and data-driven prediction of homotypic transmembrane domain interfaces. *Comput. Struct. Biotechnol. J.* **2020**, *18*, 3230–3242. [CrossRef]
46. MacKenzie, K.R.; Engelman, D.M. Structure-based prediction of the stability of transmembrane helix-helix interactions: The sequence dependence of glycophorin a dimerization. *Proc. Natl. Acad. Sci. USA* **1998**, *95*, 3583–3590. [CrossRef] [PubMed]
47. Li, E.; Wimley, W.C.; Hristova, K. Transmembrane helix dimerization: Beyond the search for sequence motifs. *Biochim. Biophys. Acta* **2012**, *1818*, 183–193. [CrossRef]
48. Hogel, P.; Gotz, A.; Kuhne, F.; Ebert, M.; Stelzer, W.; Rand, K.D.; Scharnagl, C.; Langosch, D. Glycine Perturbs Local and Global Conformational Flexibility of a Transmembrane Helix. *Biochemistry* **2018**, *57*, 1326–1337. [CrossRef] [PubMed]
49. Li, S.C.; Deber, C.M. A measure of helical propensity for amino acids in membrane environments. *Nat. Struct. Biol.* **1994**, *1*, 558. [CrossRef]
50. Padmanabhan, S.; Baldwin, R.L. Tests for helix-stabilizing interactions between various nonpolar side chains in alanine-based peptides. *Protein. Sci.* **1994**, *3*, 1992–1997. [CrossRef]

MDPI AG
Grosspeteranlage 5
4052 Basel
Switzerland
Tel.: +41 61 683 77 34

Membranes Editorial Office
E-mail: membranes@mdpi.com
www.mdpi.com/journal/membranes

www.ingramcontent.com/pod-product-compliance
Lightning Source LLC
LaVergne TN
LVHW070450061126
851858LV00011B/203
9783725880096